Pichia Protocols

METHODS IN MOLECULAR BIOLOGY™

John M. Walker, Series Editor

103. ***Pichia* Protocols,** edited by *David R. Higgins and James M. Cregg, 1998*
102. **Bioluminescence Methods and Protocols,** edited by *Robert A. LaRossa, 1998*
101. **Myobacteria Protocols,** edited by *Tanya Parish and Neil G. Stoker, 1998*
100. **Nitric Oxide Protocols,** edited by *Michael A. Titheradge, 1998*
99. **Human Cytokines and Cytokine Receptors**, edited by *Reno Debets, 1998*
98. **Forensic DNA Profiling Protocols**, edited by *Patrick J. Lincoln and James M. Thomson, 1998*
97. **Molecular Embryology:** *Methods and Protocols,* edited by *Paul T. Sharpe, 1998*
96. **Adhesion Proteins Protocols**, edited by *Elisabetta Dejana, 1998*
95. **DNA Topology and DNA Topoisomerases:** *II. Enzymology and Topoisomerase Targetted Drugs,* edited by *Mary-Ann Bjornsti, 1998*
94. **DNA Topology and DNA Topoisomerases:** *I. DNA Topology and Enzyme Purification,* edited by *Mary-Ann Bjornsti, 1998*
93. **Protein Phosphatase Protocols**, edited by *John W. Ludlow, 1997*
92. **PCR in Bioanalysis**, edited by *Stephen Meltzer, 1997*
91. **Flow Cytometry Protocols**, edited by *Mark J. Jaroszeski, 1998*
90. **Drug–DNA Interactions:** *Methods, Case Studies, and Protocols,* edited by *Keith R. Fox, 1997*
89. **Retinoid Protocols**, edited by *Christopher Redfern, 1997*
88. **Protein Targeting Protocols**, edited by *Roger A. Clegg, 1997*
87. **Combinatorial Peptide Library Protocols**, edited by *Shmuel Cabilly, 1998*
86. **RNA Isolation and Characterization Protocols**, edited by *Ralph Rapley, 1997*
85. **Differential Display Methods and Protocols**, edited by *Peng Liang and Arthur B. Pardee, 1997*
84. **Transmembrane Signaling Protocols**, edited by *Dafna Bar-Sagi, 1998*
83. **Receptor Signal Transduction Protocols,** edited by *R. A. J. Challiss, 1997*
82. ***Arabadopsis* Protocols,** edited by *José M Martinez-Zapater and Julio Salinas, 1998*
81. **Plant Virology Protocols,** edited by *Gary D. Foster, 1998*
80. **Immunochemical Protocols,** *second edition,* edited by *John Pound, 1998*
79. **Polyamine Protocols,** edited by *David M. L. Morgan, 1998*
78. **Antibacterial Peptide Protocols,** edited by *William M. Shafer, 1997*
77. **Protein Synthesis:** *Methods and Protocols,* edited by *Robin Martin, 1998*
76. **Glycoanalysis Protocols,** edited by *Elizabeth F. Hounsel, 1998*
75. **Basic Cell Culture Protocols,** edited by *Jeffrey W. Pollard and John M. Walker, 1997*
74. **Ribozyme Protocols,** edited by *Philip C. Turner, 1997*
73. **Neuropeptide Protocols,** edited by *G. Brent Irvine and Carvell H. Williams, 1997*
72. **Neurotransmitter Methods,** edited by *Richard C. Rayne, 1997*
71. **PRINS and *In Situ* PCR Protocols,** edited by *John R. Gosden, 1997*
70. **Sequence Data Analysis Guidebook,** edited by *Simon R. Swindell, 1997*
69. **cDNA Library Protocols,** edited by *Ian G. Cowell and Caroline A. Austin, 1997*
68. **Gene Isolation and Mapping Protocols,** edited by *Jacqueline Boultwood, 1997*
67. **PCR Cloning Protocols:** *From Molecular Cloning to Genetic Engineering,* edited by *Bruce A. White, 1996*
66. **Epitope Mapping Protocols,** edited by *Glenn E. Morris, 1996*
65. **PCR Sequencing Protocols,** edited by *Ralph Rapley, 1996*
64. **Protein Sequencing Protocols,** edited by *Bryan J. Smith, 1996*
63. **Recombinant Proteins:** *Detection and Isolation Protocols,* edited by *Rocky S. Tuan, 1996*
62. **Recombinant Gene Expression Protocols,** edited by *Rocky S. Tuan, 1996*
61. **Protein and Peptide Analysis by Mass Spectrometry,** edited by *John R. Chapman, 1996*
60. **Protein NMR Protocols,** edited by *David G. Reid, 1996*
59. **Protein Purification Protocols,** edited by *Shawn Doonan, 1996*
58. **Basic DNA and RNA Protocols,** edited by *Adrian J. Harwood, 1996*
57. **In Vitro Mutagenesis Protocols,** edited by *Michael K. Trower, 1996*
56. **Crystallographic Methods and Protocols,** edited by *Christopher Jones, Barbara Mulloy, and Mark Sanderson, 1996*
55. **Plant Cell Electroporation and Electrofusion Protocols,** edited by *Jac A. Nickoloff, 1995*
54. **YAC Protocols,** edited by *David Markie, 1995*
53. **Yeast Protocols:** *Methods in Cell and Molecular Biology,* edited by *Ivor H. Evans, 1996*
52. **Capillary Electrophoresis:** *Principles, Instrumentation, and Applications,* edited by *Kevin D. Altria, 1996*
51. **Antibody Engineering Protocols,** edited by *Sudhir Paul, 1995*
50. **Species Diagnostics Protocols:** *PCR and Other Nucleic Acid Methods,* edited by *Justin P. Clapp, 1996*
49. **Plant Gene Transfer and Expression Protocols,** edited by *Heddwyn Jones, 1995*
48. **Animal Cell Electroporation and Electrofusion Protocols,** edited by *Jac A. Nickoloff, 1995*
47. **Electroporation Protocols for Microorganisms,** edited by *Jac A. Nickoloff, 1995*
46. **Diagnostic Bacteriology Protocols,** edited by *Jenny Howard and David M. Whitcombe, 1995*
45. **Monoclonal Antibody Protocols,** edited by *William C. Davis, 1995*
44. ***Agrobacterium* Protocols,** edited by *Kevan M. A. Gartland and Michael R. Davey, 1995*
43. **In Vitro Toxicity Testing Protocols,** edited by *Sheila O'Hare and Chris K. Atterwill, 1995*
42. **ELISA:** *Theory and Practice*, by *John R. Crowther, 1995*
41. **Signal Transduction Protocols,** edited by *David A. Kendall and Stephen J. Hill, 1995*
40. **Protein Stability and Folding:** *Theory and Practice,* edited by *Bret A. Shirley, 1995*
39. **Baculovirus Expression Protocols,** edited by *Christopher D. Richardson, 1995*
38. **Cryopreservation and Freeze-Drying Protocols,** edited by *John G. Day and Mark R. McLellan, 1995*
37. **In Vitro Transcription and Translation Protocols,** edited by *Martin J. Tymms, 1995*
36. **Peptide Analysis Protocols,** edited by *Ben M. Dunn and Michael W. Pennington, 1994*

METHODS IN MOLECULAR BIOLOGY™

Pichia Protocols

Edited by

David R. Higgins

Chiron Technologies/Center for Gene Therapy, San Diego, CA

and

James M. Cregg

Oregon Graduate Institute of Science and Technology, Portland, OR

Humana Press **Totowa, New Jersey**

999 Riverview Drive, Suite 208
Totowa, New Jersey 07512

This publication is printed on acid-free paper. ∞
ANSI Z39.48-1984 (American Standards Institute)
Permanence of Paper for Printed Library Materials.

Cover illustration: Figure 1A,B from Chapter 12, "Secretion of scFv Antibody Fragments," by Hermann Gram, Rita Schmitz, and Rüdiger Ridder.

Cover design by Patricia F. Cleary.

For additional copies, pricing for bulk purchases, and/or information about other Humana titles, contact Humana at the above address or at any of the following numbers: Tel.: 973-256-1699; Fax: 973-256-8341; E-mail: humana@humanapr.com; or visit our Website: http://humanapress.com

Printed in the United States of America. 10 9 8 7 6 5 4 3 2

Library of Congress Cataloging in Publication Data

Main entry under title:

Methods in molecular biology™.

Pichia protocols/edited by David R. Higgins and James M. Cregg.
p. cm.—(Methods in molecular biology; 103)
Includes index.
ISBN 0-89603-421-6 (alk. paper)
1. Pichia pastoris. 2. Yeast fungi—Biotechnology. 3. Genetic vectors. I. Higgins, David R. II. Cregg, James M. III. Series: Methods in molecular biology (Totowa, NJ); 103.
[DNLM: 1. Pichia—genetics. W1 ME9616J v. 103 1998/QW 180.5.A8 P592 1998]
TP248.27.Y43P53 1998
660.6'2—dc21
DNLM/DLC 98-6428
for Library of Congress CIP

Preface

The past five years have seen an exponential growth in the use of the methylotrophic yeast *Pichia pastoris* as a heterologous protein expression system. The existence of yeasts that grow on methanol as a sole source of carbon and energy was unknown just 30 years ago, in contrast to baker's yeast, *Saccharomyces cerevisiae*, which has been exploited by humans for several millennia. However, as a result of intensive efforts during the 1970s by Phillips Petroleum bioengineers to develop *P. pastoris* for the production of single-cell protein and during the 1980s by molecular biologists at the Salk Institute Biotechnology/Industrial Associates (SIBIA), *P. pastoris* leap-frogged into a state-of-the-art molecular and cellular biological system for basic and applied research.

Pichia *Protocols* represents the collective expertise of the international *P. pastoris* scientific community. It is designed as an instruction manual to facilitate research, development, and exploitation of *P. pastoris* by current and future users. The book contains detailed protocols on all aspects of the *P. pastoris* system, from the construction of expression vectors to the production of foreign proteins in high-density fermentor cultures. In addition, the book provides in-depth descriptions of selected protein expression stories. These stories provide step-by-step templates for the inexperienced on how to use the *P. pastoris* system as well as important insights into problems encountered in expression of some foreign genes and their solutions. By following the instructions provided by the authors here, it should be possible for researchers with only minimal backgrounds in molecular biology to construct a *P. pastoris* expression strain and produce a heterologous protein in the yeast.

Although the generation of proprietary biological systems is now commonplace, the *P. pastoris* expression system, born of industrial research, is one of the first and best examples of proprietary technology development. The system is distributed by Invitrogen Corporation of Carlsbad, CA, and is freely available to the academic research community. However, use for commercial purposes requires a license from the patent holder, Research Corporation Technologies (RCT) of Tucson, AZ. Questions regarding license agreements should be directed to RCT.

The editors would like to thank the many people involved in creating Pichia *Protocols*, the first collected set of protocols and knowledge to be assembled on *P. pastoris*. More than 45 authors representing 13 laboratories from five countries participated in the creation of this volume. In the process of putting it together, we have acquired a greater appreciation of this organism as well as an appreciation of the spirit and cooperative nature of the *P. pastoris* research community. We would like to thank the series editor of *Methods in Molecular Biology*, John Walker, for responding to the many requests of researchers and possessing the foresight to see the value in creating such a book. Its sustained value comes from the contributions of all the authors who have contributed chapters. We thank Terrie Hadfield of the Oregon Graduate Institute of Science and Technology for help in editing each of the chapter manuscripts (and their revisions). We acknowledge the assistance of Kal Masri and Nancy Christie in searching the databases for the *P. pastoris* expression table presented in the Appendix. Finally, we would like to thank the editorial staff at Humana Press, led by Tom Lanigan, President, for their advice, expert suggestions, and mostly, their patience.

David R. Higgins
James M. Cregg

Contents

Contributors

RUSSELL A. BRIERLEY • *Cephalon, Inc., Westchester, PA*
RICH BUCKHOLZ • *Glaxo-Wellcome, Research Triangle Park, NC*
KATIE BUSSER • *Research and Development, Invitrogen Corporation, Carlsbad, CA*
VIJAY CHIRUVOLU • *Department of Food Science and Technology, University of Nebraska, Lincoln, NE*
JEFF CLARE • *Medicines Research Centre, Glaxo-Wellcome, Stevenage, UK*
JOHN COMISKEY • *Research and Development, Invitrogen Corporation, Carlsbad, CA*
JAMES M. CREGG • *Department of Biochemistry and Molecular Biology, Oregon Graduate Institute of Science and Technology, Portland, OR*
JOSÉ A. CREMATA • *Center for Genetic Engineering and Biotechnology, Havana, Cuba*
YPE ELGERSMA • *Department of Biology, University of California at San Diego, La Jolla, CA*
KLAUS NICO FABER • *Department of Biology, University of California at San Diego, La Jolla, CA*
ROSSANA GARCÍA • *Center for Genetic Engineering and Biotechnology, Havana, Cuba*
MARTIN A. G. GLEESON • *Research and Development, Invitrogen Corporation, Carlsbad, CA*
HERMANN GRAM • *Department of Drug Metabolism, Pharmacokinetics, and Preclinical Research, Sandoz Pharma, Basel, Switzerland*
WINFRIED HAASE • *Max Planck Institute for Biophysics, Frankfurt, Germany*
JOHN A. HEYMAN • *Department of Biology, University of California at San Diego, La Jolla, CA*
DAVID R. HIGGINS • *Chiron Technologies/Center for Gene Therapy, San Diego, CA*
JAMES P. HOEFFLER • *Research and Development, Invitrogen Corporation, Carlsbad, CA*
MONIQUE JOHNSON • *Department of Biochemistry and Molecular Biology, Oregon Graduate Institute of Science and Technology, Portland, OR*

ANTONIUS KOLLER • *Department of Biology, University of California at San Diego, La Jolla, CA*

ELIZABETH A. KOMIVES • *Department of Chemistry and Biochemistry, University of California at San Diego, La Jolla, CA*

GEORG H. LÜERS • *Department of Biology, University of California at San Diego, La Jolla, CA*

MICHAEL MEAGHER • *Department of Food Science and Technology, University of Nebraska, Lincoln, NE*

DAVID P. MEININGER • *Department of Chemistry and Biochemistry, University of California at San Diego, La Jolla, CA*

DAVID J. MILES • *Research and Development, Invitrogen Corporation, Carlsbad, CA*

RAQUEL MONTESINO • *Center for Genetic Engineering and Biotechnology, Havana, Cuba*

WILLIAM M. NUTTLEY • *Department of Biology, University of California at San Diego, La Jolla, CA*

THOMAS J. PURCELL • *Research and Development, Invitrogen Corporation, Carlsbad, CA*

OMAR QUINTERO • *Center for Genetic Engineering and Biotechnology, Havana, Cuba*

HELMUT REILÄNDER • *Max Planck Institute for Biophysics, Frankfurt, Germany*

RÜDIGER RIDDER • *Department of Drug Metabolism, Pharmacokinetics, and Preclinical Research, Sandoz Pharma, Basel, Switzerland*

MIKE ROMANOS • *Medicines Research Centre, Glaxo-Wellcome, Stevenage, UK*

KIMBERLY A. RUSSELL • *Department of Biochemistry and Molecular Biology, Oregon Graduate Institute of Science and Technology, Portland, OR*

RITA SCHMITZ • *Department of Drug Metabolism, Pharmacokinetics, and Preclinical Research, Sandoz Pharma, Basel, Switzerland*

CAROL SCORER • *Medicines Research Centre, Glaxo-Wellcome, Stevenage, UK*

SHIGANG SHEN • *Department of Biochemistry and Molecular Biology, Oregon Graduate Institute of Science and Technology, Portland, OR*

KOTI SREEKRISHNA • *Procter and Gamble, Ross, OH*

CHRISTINE STALDER • *Research and Development, Invitrogen Corporation, Carlsbad, CA*

JAYNE STRATTON • *Department of Food Science and Technology, University of Nebraska, Lincoln, NE*

SURESH SUBRAMANI • *Department of Biology, University of California at San Diego, La Jolla, CA*

STANLEY R. TERLECKY • *Department of Biology, University of California at San Diego, La Jolla, CA*
HANS R. WATERHAM • *Department of Biochemistry and Molecular Biology, Oregon Graduate Institute of Science and Technology, Portland, OR*
H. MARKUS WEISS • *Max Planck Institute for Biophysics, Frankfurt, Germany*
THIBAUT J. WENZEL • *Department of Biology, University of California at San Diego, La Jolla, CA*
CHRISTOPHER E. WHITE • *Department of Chemistry and Biochemistry, University of California at San Diego, La Jolla, CA*
PETER S. WHITTIER • *Research and Development, Invitrogen Corporation, Carlsbad, CA*

1

Introduction to *Pichia pastoris*

David R. Higgins and James M. Cregg

1. Introduction

Pichia pastoris has become a highly successful system for the expression of heterologous genes. Several factors have contributed to its rapid acceptance, the most important of which include:

1. A promoter derived from the alcohol oxidase I (*AOX1*) gene of *P. pastoris* that is uniquely suited for the controlled expression of foreign genes;
2. The similarity of techniques needed for the molecular genetic manipulation of *P. pastoris* to those of *Saccharomyces cerevisiae*, one of the most well-characterized experimental systems in modern biology;
3. The strong preference of *P. pastoris* for respiratory growth, a key physiological trait that greatly facilitates its culturing at high cell densities relative to fermentative yeasts; and
4. A 1993 decision by Phillips Petroleum Company (continued by Research Corporation Technologies [RTC]) to release the *P. pastoris* expression system to academic research laboratories, the consequence of which has been an explosion in the knowledge base of the system (**Fig. 1**).

As listed in the Appendix to this volume, more than 100 different proteins have been successfully produced in *P. pastoris*.

As a yeast, *P. pastoris* is a single-celled microorganism that is easy to manipulate and culture. However, it is also a eukaryote and capable of many of the posttranslational modifications performed by higher eukaryotic cells, such as proteolytic processing, folding, disulfide bond formation, and glycosylation. Thus, many proteins that end up as inactive inclusion bodies in bacterial systems are produced as biologically active molecules in *P. pastoris*. The *P. pastoris* system is also generally regarded as being faster, easier, and less expensive to use than expression systems derived from higher eukaryotes, such

From: *Methods in Molecular Biology, Vol. 103:* Pichia *Protocols*
Edited by: D. R. Higgins and J. M. Cregg © Humana Press Inc., Totowa, NJ

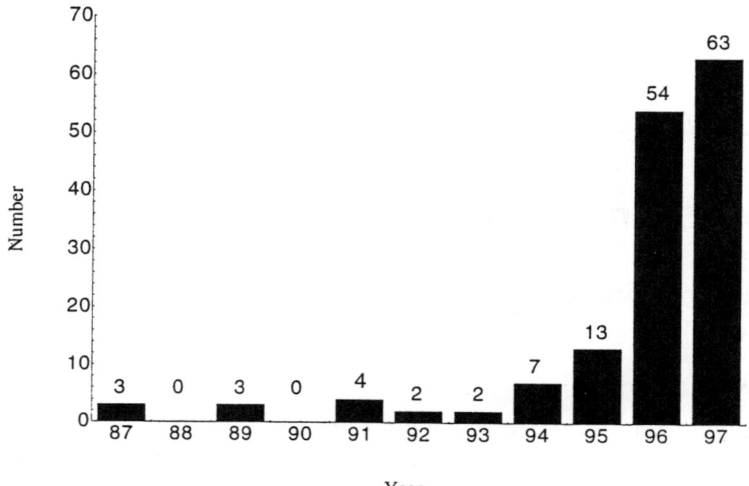

Fig. 1. Graph showing number of publications describing the expression of a foreign protein in *P. pastoris* each year from 1985 to 1997.

as insect and mammalian tissue culture cell systems, and usually gives higher expression levels.

A second role played by *P. pastoris* in research is not directly related to its use as a protein expression system. *P. pastoris* serves as a useful model system to investigate certain areas of modern cell biology, including the molecular mechanisms involved in the import and assembly of peroxisomes (Chapter 10), the selective autophagic degradation of peroxisomes, and the organization and function of the secretory pathway in eukaryotes.

In this chapter, we review basic aspects of the *P. pastoris* expression system and highlight where useful information on the system can be found in this book. Further information on the *P. pastoris* system can be found in the numerous reviews describing the system ***(1–9)*** and the *Pichia* Expression Kit Instruction Manual (Invitrogen Corporation, Carlsbad, CA). The DNA sequence of many *P. pastoris* expression vectors and other useful information can be found on the Invitrogen web site (http://www.invitrogen.com).

2. A Brief History of the *P. pastoris* Expression System

The ability of certain yeast species to utilize methanol as a sole source of carbon and energy was discovered less than 30 years ago by Koichi Ogata ***(10)***. Because methanol could be inexpensively synthesized from natural gas (methane), there was immediate interest in exploiting these organisms for the generation of yeast biomass or single-cell protein (SCP) to be marketed primarily as high protein animal feed. During the 1970s, Phillips Petroleum Company of Bartlesville, OK, developed media and methods for growing *P. pastoris* on methanol in continuous culture at high cell densities (>130 g/L dry cell weight) ***(11)***. However, during this same period, the cost of methane increased dramatically because of the oil crisis, and the cost of soy beans—the major alternative source of animal feed protein—decreased. As a result, the SCP process was never economically competitive.

In the early 1980s, Phillips Petroleum Company contracted with the Salk Institute Biotechnology/Industrial Associates, Inc. (SIBIA), a biotechnology company located in La Jolla, CA, to develop *P. pastoris* as a heterologous gene expression system. Researchers at SIBIA isolated the *AOX1* gene (and its promoter) and developed vectors, strains, and methods for molecular genetic manipulation of *P. pastoris* ***(12–17,*** Chapters 2, 3, and 6***)***. The combination of strong regulated expression under control of the *AOX1* promoter, along with the fermentation media and methods developed for the SCP process, resulted in strikingly high levels of foreign proteins in *P. pastoris*. In 1993, Phillips Petroleum sold its patent position with the *P. pastoris* expression system to RCT, the current patent holder. In addition, Phillips Petroleum licensed Invitrogen to sell components of the system to researchers worldwide, an arrangement that continues under RCT.

3. *P. pastoris* as a Methylotrophic Yeast

P. pastoris is one of approximately a dozen yeast species representing four different genera capable of metabolizing methanol ***(18)***. The other genera include *Candida*, *Hansenula*, and *Torulopsis*. The methanol metabolic pathway appears to be the same in all yeasts and involves a unique set of pathway enzymes ***(19)***. The first step in the metabolism of methanol is the oxidation of methanol to formaldehyde, generating hydrogen peroxide in the process, by the enzyme alcohol oxidase (AOX). To avoid hydrogen peroxide toxicity, this first step in methanol metabolism takes place within a specialized organelle, called the peroxisome, that sequesters toxic hydrogen peroxide away from the rest of the cell. AOX is a homo-octomer with each subunit containing one noncovalently bound FAD (flavin adenine dinucleotide) cofactor. Alcohol oxidase has a poor affinity for O_2, and methylotrophic yeasts appear to compensate for this deficiency by synthesizing large amounts of the enzyme.

There are two genes in *P. pastoris* that code for AOX—*AOX1* and *AOX2*—but the *AOX1* gene is responsible for the vast majority of alcohol oxidase activity in the cell ***(16)***. Expression of the *AOX1* gene is tightly regulated and induced by methanol to high levels. In methanol-grown shake-flask cultures, this level is typically ~5% of total soluble protein but can be ≥30% in cells fed methanol at growth limiting rates in fermentor cultures ***(20)***. Expression of the *AOX1* gene is controlled at the level of transcription ***(12,14,16)***. In methanol-grown cells, ~5% of polyA$^+$ RNA is from the *AOX1* gene, whereas in cells grown on other carbon sources, the AOX1 message is undetectable. The regulation of the *AOX1* gene is similar to the regulation of the *GAL1* gene of *S. cerevisiae* in that control appears to involve two mechanisms: a repression/derepression mechanism plus an induction mechanism. However, unlike *GAL1* regulation, derepressing conditions (e.g., the absence of a repressing carbon source, such as glucose in the medium) do not result in substantial transcription of the *AOX1* gene. The presence of methanol appears to be essential to induce high levels of transcription ***(14)***.

4. Secretion of Heterologous Proteins

With *P. pastoris*, heterologous proteins can either be expressed intracellularly or secreted into the medium. Because *P. pastoris* secretes only low levels of endogenous proteins and because its culture medium contains no added proteins, a secreted heterologous protein comprises the vast majority of the total protein in the medium ***(21,22)***. Thus, secretion serves as a major first step in purification, separating the foreign protein from the bulk of cellular proteins. However, the option of secretion is usually limited to foreign proteins that are normally secreted by their native hosts. Secretion requires the presence of a signal sequence on the foreign protein to target it to the secretory pathway. Although several different secretion signal sequences have been used successfully, including the native secretion signal present on some heterologous proteins, success has been variable. The secretion signal sequence from the *S. cerevisiae* α-factor prepro peptide has been used with the most success.

5. Common Expression Strains

All *P. pastoris* expression strains are derivatives of NRRL-Y 11430 (Northern Regional Research Laboratories, Peoria, IL) (**Table 1**). Most have a mutation in the histidinol dehydrogenase gene (*HIS4*) to allow for selection of expression vectors containing *HIS4* upon transformation ***(13)***. Other biosynthetic gene/auxotrophic mutant host marker combinations are also available, but used less frequently (Chapter 2). All of these strains grow on complex media but require supplementation with histidine (or other appropriate nutrient) for growth on minimal media.

Three types of host strains are available that vary with regard to their ability to utilize methanol resulting from deletions in one or both *AOX* genes. Strains with deleted *AOX* genes sometimes are better producers of a foreign protein than wild-type strains ***(21,23,24)***. These strains also require much less methanol to induce expression, which can be useful in large fermentor cultures where a large amount of methanol is sometimes considered a significant fire hazard.

Table 1
***P. pastoris* Expression Host Strains**

Strain name	Genotype	Phenotype	Reference
Y-11430	Wild-type	NRRL[a]	
GS115	*his4*	Mut^+ His^-	***13***
KM71	*aox1Δ::SARG4 his4 arg4*	Mut^s His^-	***14***
MC100-3	*aox1Δ::SARG4 aox2Δ::Phis4 his4 arg4*	Mut^- His^-	***16***
SMD1168	*pep4Δ his4*	Mut^+ His^- Protease-deficient	Chapter 7
SMD1165	*prb1 his4*	Mut^+ His^- Protease-deficient	Chapter 7
SMD1163	*pep4 prb1 his4*	Mut^+ His^- Protease-deficient	Chapter 7

[a]Northern Regional Research Laboratories, Peoria, IL.

However, the most commonly used expression host is GS115 (*his4*), which is wild-type with regard to the *AOX1* and *AOX2* genes and grows on methanol at the wild-type rate (methanol utilization plus [Mut^+] phenotype). KM71 (*his4 arg4 aox1Δ::ARG4*) is a strain in which the chromosomal *AOX1* gene is largely deleted and replaced with the *S. cerevisiae ARG4* gene ***(15)***. As a result, this strain must rely on the much weaker *AOX2* gene for AOX and grows on methanol at a slow rate [methanol utilization slow (Mut^s) phenotype]. With many *P. pastoris* expression vectors, it is possible to insert an expression cassette and simultaneously delete the *AOX1* gene of a Mut^+ strain *(**23**,* Chapters 5 and 13*)*. The third host MC100-3 (*his4 arg4 aox1Δ::SARG4 aox2Δ::Phis4*) is deleted for both *AOX* genes and is totally unable to grow on methanol [methanol utilization minus (Mut^-) phenotype] *(**16,24**,* Chapter 9*)*.

Some secreted foreign proteins are unstable in the *P. pastoris* culture medium in which they are rapidly degraded by proteases. Major vacuolar proteases appear to be a significant factor in degradation, particularly in fermentor cultures, because of the high cell density environment in combination with the lysis of a small percentage of cells. The use of host strains that are defective in these proteases has proven to help reduce degradation in several instances (Chapters 7, 11, and 14). SMD1163 (*his4 pep4 prb1*), SMD1165 (*his4 prb1*), and SMD1168 (*his4 pep4*) are protease-deficient strains that may provide a more suitable environment for expression of certain heterologous proteins. The

PEP4 gene encodes proteinase A, a vacuolar aspartyl protease required for the activation of other vacuolar proteases, such as carboxypeptidase Y and proteinase B. Proteinase B, prior to processing and activation by proteinase A, has about half the activity of the processed enzyme. The *PRB1* gene codes for proteinase B. Therefore, *pep4* mutants display a substantial decrease or elimination in proteinase A and carboxypeptidase Y activities, and partial reduction in proteinase B activity. In the *prb1* mutant, only proteinase B activity is eliminated, whereas *pep4 prb1* double mutants show a substantial reduction or elimination in all three of these protease activities.

6. Expression Vectors

Plasmid vectors designed for heterologous protein expression in *P. pastoris* have several common features (**Table 2**). The foreign gene expression cassette is one of those and is composed of DNA sequences containing the *P. pastoris AOX1* promoter, followed by one or more unique restriction sites for insertion of the foreign gene, followed by the transcriptional termination sequence from the *P. pastoris AOX1* gene that directs efficient 3' processing and polyadenylation of the mRNAs. Many of these vectors also include the *P. pastoris HIS4* gene as a selectable marker for transformation into *his4* mutant hosts of *P. pastoris*, as well as sequences required for plasmid replication and maintenance in bacteria (i.e., ColE1 replication origin and ampicillin-resistance gene). Some vectors also contain *AOX1* 3' flanking sequences that are derived from a region of the *P. pastoris* genome that lies immediately 3' of the *AOX1* gene and can be used to direct fragments containing a foreign gene expression cassette to integration at the *AOX1* locus by gene replacement (or gene insertion 3' to *AOX1* gene). This is discussed in more detail in **Subheading 7.** and Chapter 13.

Additional features that are present in certain *P. pastoris* expression vectors serve as tools for specialized functions. For secretion of foreign proteins, vectors have been constructed that contain a DNA sequence immediately following the *AOX1* promoter that encodes a secretion signal. The most frequently used of these is the *S. cerevisiae* α-factor prepro signal sequence ***(25,26,*** Chapter 5***)***. However, vectors containing the signal sequence derived from the *P. pastoris* acid phosphatase gene (*PHO1*) are also available.

Vectors with dominant drug-resistance markers that allow for enrichment of strains that receive multiple copies of foreign gene expression cassettes during transformations have been developed. One set of vectors (pPIC3K and pPIC9K) contains the bacterial kanamycin-resistance gene and confers resistance to high levels of G418 on strains that contain multiple copies of these vectors (***26,*** Chapter 5). Another set of vectors (the pPICZ series) contains the *Sh ble* gene from *Streptoalloteichus hindustanus* (Chapter 4). This gene is small (375 bp) and confers resistance to the drug Zeocin in *Escherichia coli*, yeasts (including

Table 2
Common *P. pastoris* Expression Vectors

Vector name	Selectable markers	Features	References
Intracellular			
pHIL-D2	*HIS4*	*Not*I sites for *AOX1* gene replacement	Sreekrishna, personal communication
pAO815	*HIS4*	Expression cassette bounded by *Bam*HI and *Bgl*II sites for generation of multicopy expression vector	*(2)*
pPIC3K	*HIS4* and kan^r	Multiple cloning sites for insertion of foreign genes; G418 selection for multicopy strains	*(33)*
pPICZ	ble^r	Multiple cloning sites for insertion of foreign genes; Zeocin selection for multicopy strains; potential for fusion of foreign protein to His_6 and *myc* epitope tags	Chapter 5
pHWO10	*HIS4*	Expression controlled by constitutive *GAPp*	*(27)*
pGAPZ	ble^r	Expression controlled by constitutive *GAPp*; multiple cloning site or insertion of foreign genes; Zeocin selection for multicopy strains; potential for fusion of foreign protein to His_6 and *myc* epitope tags	Invitrogen
Secretion			
pHIL-S1	*HIS4*	*AOX1p* fused to *PHO1* secretion signal; *Xho*I, *Eco*RI, and *Bam*HI sites available for insertion of foreign genes	Sreekrishna, personal communication; Invitrogen
pPIC9K	*HIS4* and kan^r	*AOX1p* fused to α-MF prepro signal sequence; *Xho*I (not unique), *Eco*RI, *Not*I, *Sna*BI and *Avr*II sites available for insertion of foreign genes; G418 selection for multicopy strains	*(33)*
pPICZα	ble^r	*AOX1p* fused to α-MF prepro signal sequence; multiple cloning site for insertion of foreign genes; Zeocin selection for multicopy strains; potential for fusion of foreign protein to His_6 and *myc* epitope tags	Chapter 5
pGAPZα	ble^r	Expression controlled by constitutive *GAPp*; *GAPp* fused to α-MF prepro signal sequence; multiple cloning site for insertion of foreign genes; Zeocin selection for multicopy strains; potential for fusion of foreign protein to His_6 and *myc* epitope tags	Invitrogen

P. pastoris), and other eukaryotes. Because the *ble* gene serves as the selectable marker for both *E. coli* and *P. pastoris*, the Zeo^R vectors are much smaller (~3 kb) and easier to manipulate than other *P. pastoris* expression vectors. These vectors also contain a multiple cloning site (MCS) with several unique restriction sites for convenience of foreign gene insertion and sequences encoding the His_6 and *myc* epitopes so that foreign proteins can be easily epitope-tagged at their carboxyl termini, if desired.

Another feature present on certain vectors (e.g., pAO815 and the pPICZ vector series) is designed to facilitate the construction of expression vectors with multiple expression cassette copies (Chapter 11). Multiple copies of an expression cassette are introduced in these vectors by inserting an expression cassette bounded by a *Bam*HI and a *Bgl*II site into the *Bam*HI site of a vector already containing a single expression cassette copy. The resulting *Bam*HI/*Bgl*II junction between the two cassettes can no longer be cleaved by either enzyme allowing for the insertion of another *Bam*HI–*Bgl*II-bounded cassette into the same vector to generate a vector with three cassette copies. The process of addition is repeated until 6–8 copies of a cassette are present in a single final vector that is then transformed into the *P. pastoris* host strain.

Finally, vectors containing a constitutive *P. pastoris* promoter derived from the *P. pastoris* glyceraldehyde-3-phosphate dehydrogenase gene (*GAP*) have recently become available ***(27)***. The *GAP* promoter is a convenient alternative to the *AOX1* promoter for expression of genes whose products are not toxic to *P. pastoris*. In addition, its use does not involve the use of methanol, which may be problematic in some instances.

7. Integration of Vectors into the *P. pastoris* Genome

As in *S. cerevisiae*, linear vector DNAs can generate stable transformants of *P. pastoris* via homologous recombination between sequences shared by the vector and host genome *(**13,23,*** Chapters 5 and 13*)*. Such integrants show strong stability in the absence of selective pressure even when present as multiple copies. All *P. pastoris* expression vectors carry at least one *P. pastoris* DNA segment (the *AOX1* or *GAP* promoter fragment) with unique restriction sites that can be cleaved and used to direct the vector to integrate into the host genome by a single crossover type insertion event (**Fig. 2A**). Vectors containing the *P. pastoris HIS4* gene can also be directed to integrate into the *P. pastoris* genomic *his4* locus.

Expression vectors that contain *3'AOX1* sequences can be integrated into the *P. pastoris* genome by a single crossover event at either *AOX1* or *HIS4* loci or by a gene replacement (Ω insertion) event at *AOX1* (**Fig. 2B**). The latter event arises from crossovers at both the *AOX1* promoter and *3'AOX1* regions of the vector and genome, and results in the deletion of the *AOX1* coding region (i.e.,

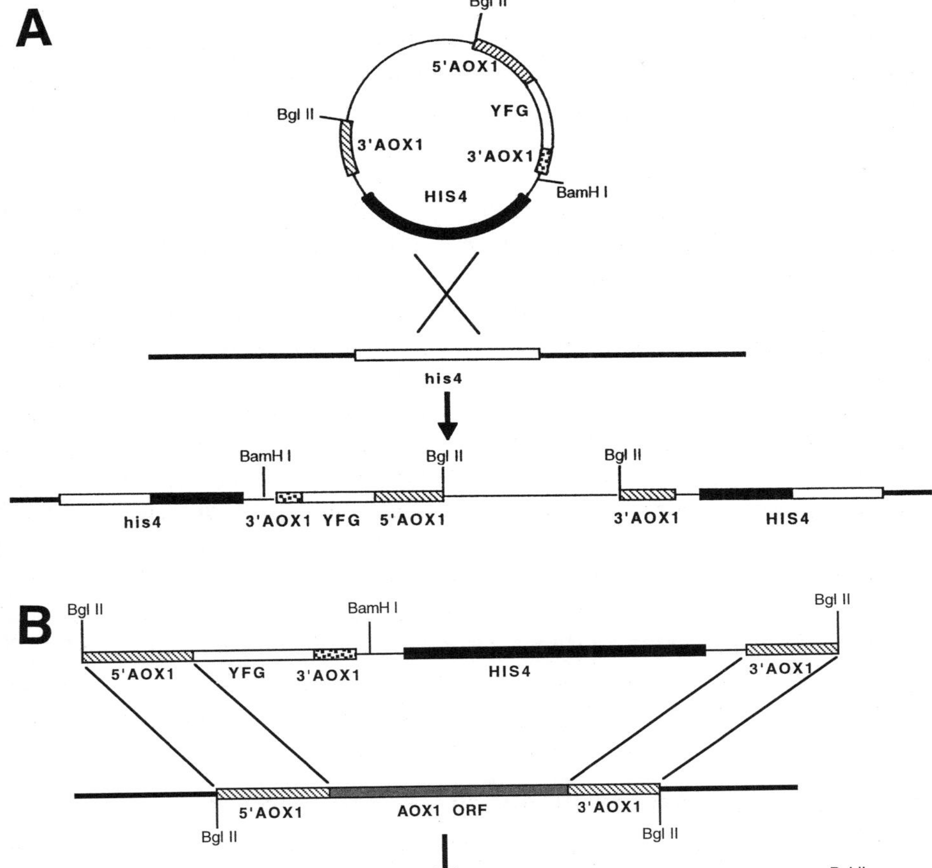

Fig. 2. Integration of expression vectors into the *P. pastoris* genome. **(A)** Single crossover integration into the *his4* locus. **(B)** Integration of vector fragment by replacement of *AOX1* gene.

gene replacement). Transformants resulting from such an *AOX1* replacement event are phenotypically His$^+$ and Mut^s. As described in **Subheading 6.**, such Mut^s strains sometimes express higher levels of foreign protein. In addition, a Mut^s phenotype serves as a convenient indicator to confirm the presence of an integrated expression cassette in the *P. pastoris* genome.

With either single crossover or gene replacement integration strategies and selection for His$^+$ transformants, a significant percentage of transformants will not contain the expression vector. This appears to be the result of gene conversion events between the *HIS4* gene on the vector and the *P. pastoris his4* locus such that the wild-type *HIS4* gene recombines into the genome without any additional vector sequences. These events account for 10–50% of His$^+$ transformant colonies and appear to occur at highest frequency when using electroporation to introduce vector DNAs.

Multiple gene insertion events at a single locus occur spontaneously at a low but detectable frequency—between 1 and 10% of His$^+$ transformants ***(28,*** Chapters 4, 5, 13, and 14*)*. Multicopy events can occur as gene insertions either at the *AOX1* or *his4* loci and can be detected by DNA analysis methods (e.g., PCR, Southern/dot blotting, or differential hybridization) ***(29,30)*** or by methods that directly examine levels of the foreign protein (e.g., activity assay, sodium dodecyl sulfate polyacrylamide gel electrophoresis [SDS-PAGE], or colony immunoblotting) ***(28,31)***. As mentioned in **Subheading 6.**, it is possible to enrich transformant populations for ones that have multiple copies of an expression vector by use of either a G418^R or Zeo^R gene-containing vector and selecting for hyper-resistance to the appropriate drug (***26,*** Chapters 4 and 5). It is important to note that, with the G418^R vectors, it is essential to first select for His$^+$ transformants and to then screen for ones that are resistant to G418^R. With Zeo^R vectors, it is possible to directly select for hyper-Zeo-resistant transformants. Most drug-resistant strains resulting from either the G418^R or Zeo^R selection methods contain between one and five copies of the expression vector. To find strains with 20 or more copies, it is usually necessary to screen at least 50–100 drug-resistant strains.

8. Posttranslational Modifications

P. pastoris has the potential to perform many of the posttranslational modifications typically associated with higher eukaryotes. These include processing of signal sequences (both pre- and prepro-type), folding, disulfide bridge formation (Chapter 7), and O- and N-linked glycosylation.

Glycosylation of secreted foreign (higher) eukaryotic proteins by *P. pastoris* and other fungi can be problematic. In mammals, *O*-linked oligosaccharides are composed of a variety of sugars, including *N*-acetylgalactosamine, galactose, and sialic acid. In contrast, lower eukaryotes, including *P. pastoris*, add *O*-oligosaccharides solely composed of mannose (Man) residues (Chapter 11). The number of Man residues per chain, their manner of linkage, and the frequency and specificity of *O*-glycosylation in *P. pastoris* have yet to be determined. One should not assume that, because a protein is not *O*-glycosylated by its native host, *P. pastoris* will not glycosylate it. *P. pastoris* added O-linked mannose to ~15% of human IGF-1 protein, although this protein is not glycosylated at all in humans. Furthermore, one should not assume that the specific Ser and Thr residue(s) selected for *O*-glycosylation by *P. pastoris* will be the same as the native host.

N-glycosylation in *P. pastoris* and other fungi is also different than in higher eukaryotes (*4*, Chapters 8 and 14). In all eukaryotes, it begins in the endoplasmic reticulum with the transfer of a lipid-linked oligosaccharide unit, $Glc_3Man_9GlcNAc_2$ (Glc = glucose; GlcNAc = *N*-acetylglucosamine), to asparagine at the recognition sequence Asn-X-Ser/Thr. This oligosaccharide core unit is subsequently trimmed to $Man_8GlcNAc_2$. It is at this point that lower and higher eukaryotic glycosylation patterns begin to differ. The mammalian Golgi apparatus performs a series of trimming and addition reactions that generates oligosaccharides composed of either $Man_{5-6}GlcNAc_2$ (high-mannose type), a mixture of several different sugars (complex type), or a combination of both (hybrid type). Two distinct patterns of *N*-glycosylation have been observed on foreign proteins secreted by *P. pastoris*. Some proteins, such as *S. cerevisiae* invertase, are secreted with carbohydrate structures similar in size and structure to the core unit ($Man_{8-11}GlcNAc_2$) *(21,32)*.

Others foreign proteins secreted from *P. pastoris* receive much more carbohydrate and appear by SDS-PAGE and western blotting to be hyperglycosylated (Chapter 14). Interestingly, *P. pastoris* does not appear to be capable of adding α1,3-terminal mannose to oligosaccharides (R. Trimble, personal communication). This contrasts with *S. cerevisiae* oligosaccharides, in which α1,3-linked terminal mannose is common. Aside from the probable absence of α1,3-linked mannose, little is known regarding the structure of *P. pastoris* outer-chain oligosaccharides. Furthermore, it is also not clear why outer chains are added to some *P. pastoris*-secreted proteins and not others nor how outer chain addition may be prevented.

N-linked high-mannose oligosaccharides added to proteins by yeasts represent a significant problem in the use of foreign-secreted proteins by the pharmaceutical industry. They can be exceedingly antigenic when introduced intravenously into mammals and are rapidly cleared from the blood by the liver. An additional problem caused by the differences between yeast and mammalian N-linked glycosylation patterns is that the long outer chains can potentially interfere with the folding or function of a foreign protein.

9. Expression in Fermentor Cultures

Although a few foreign proteins have expressed well in *P. pastoris* shake-flask cultures, expression levels in shake-flasks are typically low relative to what is obtainable in fermenter cultures. One reason fermenter culturing is necessary is that only in the controlled environment of a fermenter is it possible to grow the organism to high cell densities (>100 g/L dry cell weight or 500 OD_{600} U/mL). Especially for secreted proteins, the concentration of product in the medium is roughly proportional to the concentration of cells in culture. A second reason is that the level of transcription initiated from the *AOX1* promoter can be 3–5 times greater in *P. pastoris* cells fed methanol at growth-limiting rates in fermenter culture relative to cells grown in excess methanol. Thus, even for intracellularly expressed proteins, yields of product from a given strain as a percentage of total cellular proteins are significantly higher from fermenter cultured cells. A third reason is that methanol metabolism utilizes oxygen at a high rate, and expression of foreign genes is negatively affected by oxygen limitation. Only in the controlled environment of a fermenter is it feasible to accurately monitor and adjust oxygen levels in the culture medium. Thus, most users of the *P. pastoris* expression system should expect to produce their foreign protein in fermenters.

A hallmark of the *P. pastoris* system is the ease by which expression strains scale up from shake-flask to high-density fermenter cultures. Considerable effort has gone into the optimization of high cell density fermentation techniques for expression strains, and, as a result, a variety of fed-batch and continuous culture schemes are available (Chapters 7, 9, 11, and 13). All schemes involve the initial growth of strains in a defined medium on glycerol. During this period, growth is rapid but heterologous gene expression is fully repressed. Upon depletion of glycerol, a transition phase is initiated in which additional glycerol is fed to cultures at a growth-limiting rate. Finally, methanol, or a mixture of glycerol and methanol, is fed to cultures to induce expression. The time of harvest, typically the peak concentration of a foreign protein, is determined empirically for each protein.

High-density fermentation of *P. pastoris* expression strains is especially attractive for the production of secreted proteins, because their concentration in the culture medium should increase with cell density. Unfortunately, the concentrations of other cellular materials, particularly proteases, increase as well. Three strategies have proven effective in minimizing the proteolytic instability of foreign proteins secreted into the *P. pastoris* culture medium. One is the addition of amino acid-rich supplements, such as peptone or casamino acids, to the culture medium that appear to reduce product degradation by acting as excess substrates for one or more problem proteases ***(26)***. A second is changing the culture medium pH ***(26)***. *P. pastoris* is capable of growing across a relatively broad pH range from 3.0 to 7.0, which allows considerable leeway in adjusting the pH to one that is not optimal for a problem protease. A third is the use of a protease-deficient *P. pastoris* host strain (**Subheading 5.** of this chapter and Chapters 7 and 11).

10. Conclusion

Based on available data, there is an ~50–75% probability of expressing your protein of interest in *P. pastoris* at a reasonable level. The biggest hurdle seems to be generating initial success—that is, expressing your protein at *any* level. After success at this stage, there are well-defined parameters that can be manipulated to optimize expression, and it is often at this stage that attractive levels of expression are achieved. Although there are relatively few examples of expression of ≥10 g/L, there are many examples of expression in the ≥1 g/L range, ranking the *P. pastoris* expression system as one of the most productive eukaryotic expression systems available. Likewise, there are examples of proteins that have been successfully expressed in *P. pastoris* that were completely unsuccessful in baculovirus or *S. cerevisiae* expression systems, making the *P. pastoris* system an important alternative to have available in the protein expression "toolbox".

References

1. Romanos, M. A., Scorer, C. A., and Clare, J. J. (1992) Foreign gene expression in yeast: a review. *Yeast* **8,** 423–488.
2. Cregg, J. M., Vedvick, T. S., and Raschke, W. C. (1993) Recent advances in the expression of foreign genes in *Pichia pastoris. Bio/Technology* **11,** 905–910.
3. Romanos, M. (1995) Advances in the use of *Pichia pastoris pastoris* for high-level expression. *Curr. Opin. Biotechnol.* **6,** 527–533.
4. Cregg, J. M. (1998) Expression in the methylotrophic yeast *Pichia pastoris*, in *Nature: The Palette for the Art of Expression* (Fernandez, J. and Hoeffler, J., eds.), Academic, San Diego, in press.
5. Cregg, J. M. and Higgins, D. R. (1995) Production of foreign proteins in the yeast *Pichia pastoris. Can. J. Bot.* **73(Suppl. 1),** S981–S987.
6. Sreekrishna, K., Brankamp, R. G., Kropp, K. E., Blankenship, D. T., Tsay, J. T., Smith, P. L., Wierschke, J. D., Subramaniam, A., and Birkenberger, L. A. (1997) Strategies for optimal synthesis and secretion of heterologous proteins in the methylotrophic yeast *Pichia pastoris. Gene* **190,** 55–62.
7. Gellissen, G. and Hollenberg, C. P. (1997) Application of yeasts in gene expression studies: a comparison of *Saccharomyces cerevisiae, Hansenula polymorpha* and *Kluyveromyces lactis*—a review. *Gene* **190,** 87–97.
8. Higgins, D. R. (1995) Overview of protein expression in *Pichia pastoris*, in *Current Protocols in Protein Science, Supplement 2* (Wingfield, P. T., ed.), Wiley, New York, pp. 5.7.1–5.7.16.
9. Sreekrishna, K. (1993) Strategies for optimizing protein expression and secretion in the methylotrophic yeast *Pichia pastoris*, in *Industrial Microorganisms: Basic and Applied Molecular Genetics* (Baltz, R. H., Hegeman, G. D., and Skatrud, P. L., eds.), American Society for Microbiology, Washington, DC, pp. 119–126.
10. Ogata, K., Nishikawa, H., and Ohsugi, M. (1969) A yeast capable of utilizing methanol. *Agric. Biol. Chem.* **33,** 1519,1520.
11. Wegner, G. (1990) Emerging applications of methylotrophic yeasts. *FEMS Microbiol. Rev.* **87,** 279–284.

12. Ellis, S. B., Brust, P. F., Koutz, P. J., Waters, A. F., Harpold, M. M., and Gingeras, T. R. (1985) Isolation of alcohol oxidase and two other methanol regulatable genes from the yeast *Pichia pastoris*. *Mol. Cell. Biol.* **5,** 1111–1121.
13. Cregg, J. M., Barringer, K. J., Hessler, A. Y., and Madden, K. R. (1985) *Pichia pastoris* as a host system for transformations. *Mol. Cell. Biol.* **5,** 3376–3385.
14. Tschopp, J. F., Brust, P. F., Cregg, J. M., Stillman, C. A., and Gingeras, T. R. (1987) Expression of the *lacZ* gene from two methanol-regulated promoters in *Pichia pastoris*. *Nucleic Acids Res.* **15,** 3859–3876.
15. Cregg, J. M. and Madden, K. R. (1987) Development of yeast transformation systems and construction of methanol-utilization-defective mutants of *Pichia pastoris* gene disruption, in *Biological Research on Yeasts,* vol. II (Stewart, G. G., Russell, I., Klein, R. D., and Hiebsch, R. R., eds.), CRC, Boca Raton, FL, pp. 1–18.
16. Cregg, J. M., Madden, K. R., Barringer, K. J., Thill, G. P., and Stillman, C. A. (1989) Functional characterization of the two alcohol oxidase genes from the yeast *Pichia pastoris*. *Mol. Cell. Biol.* **9,** 1316–1323.
17. Koutz, P. J., Davis, G. R., Stillman, C., Barringer, K., Cregg, J. M., and Thill, G. (1989) Structural comparison of the *Pichia pastoris* alcohol oxidase genes. *Yeast* **5,** 167–177.
18. Lee, J.-D. and Komagata, K. (1980) Taxonomic study of methanol-assimilating yeasts. *J. Gen. Appl. Microbiol.* **26,** 133–158.
19. Veenhuis, M., van Dijken, J. P., and Harder, W. (1983) The significance of peroxisomes in the metabolism of one-carbon compounds in yeasts. *Adv. Microb. Physiol.* **24,** 1–82.
20. Couderc, R. and Baratti, J. (1980) Oxidation of methanol by the yeast *Pichia pastoris*: purification and properties of alcohol oxidase. *Agric. Biol. Chem.* **44,** 2279–2289.
21. Tschopp, J. F., Sverlow, G., Kosson, R., Craig, W., and Grinna, L. (1987) High level secretion of glycosylated invertase in the methylotrophic yeast *Pichia pastoris*. *Bio/Technology* **5,** 1305–1308.
22. Barr, K. A., Hopkins, S. A., and Sreekrishna, K. (1992) Protocol for efficient secretion of HSA developed from *Pichia pastoris*. *Pharm. Eng.* **12,** 48–51.
23. Cregg, J. M., Tschopp, J. F., Stillman, C., Siegel, R., Akong, M., Craig, W. S., Buckholz, R. G., Madden, K. R., Kellaris, P. A., Davis, G. R., Smiley, B. L., Cruze, J., Torregrossa, R., Velicelebi, G., and Thill, G. P. (1987) High-level expression and efficient assembly of hepatitis B surface antigen in the methylotrophic yeast *Pichia pastoris*. *Bio/Technology* **5,** 479–485.
24. Chirulova, V., Cregg, J. M., and Meagher, M. M. (1997) Recombinant protein production in an alcohol oxidase-defective strain of *Pichia pastoris* in fed batch fermentations. *Enzyme Microb. Technol.* **21,** 277–283.
25. Larouche, Y., Storme, V., De Muetter, J., Messens, J., and Lauwereys, M. (1994) High-level secretion and very efficient isotopic labeling of tick anticoagulant peptide (TAP) expressed in the methylotrophic yeast *Pichia pastoris*. *Bio/Technology* **12,** 1119–1124.
26. Clare, J. J., Romanos, M. A., Rayment, F. B., Rowedder, J. E., Smith, M. A., Payne, M. M., Sreekrishna, K., and Henwood, C. A. (1991) Production of mouse epidermal growth factor in yeast: high-level secretion using *Pichia pastoris* strains containing multiple gene copies. *Gene* **105,** 205–212.

27. Waterham, H. R., Digan, M. E., Koutz, P. J., Lair, S. L., and Cregg, J. M. (1997) Isolation of the *Pichia pastoris* glyceraldehyde-3-phosphate dehydrogenase gene and regulation and use of its promoter. *Gene* **186,** 37–44.
28. Sreekrishna, K., Nelles, L., Potenz, R., Cruze, J., Mazzaferro, P., Fish, W., Fuke, M., Holden, K., Phelps, D., Wood, P., and Parker, K. (1989) High-level expression, purification, and characterization of recombinant human tumor necrosis factor synthesized in the methylotrophic yeast *Pichia pastoris. Biochemistry* **28,** 4117–4125.
29. Clare, J. J., Rayment, F. B., Ballantine, S. P., Sreekrishna, K., and Romanos, M. A. (1991) High-level expression of tetanus toxin fragment C in *Pichia pastoris* strains containing multiple tandem integrations of the gene. *Bio/Technology* **9,** 455–460.
30. Romanos, M. A., Clare, J. J., Beesley, K. M., Rayment, F. B., Ballantine, S. P., Makoff, A. J., Dougan, G., Fairweather, N. F., and Charles, I. G. (1991) Recombinant *Bordetella pertussis* pertactin (P69) from the yeast *Pichia pastoris*: high-level production and immunological properties. *Vaccine* **9,** 901–906.
31. Wung, J. L., and Gascoigne, N. R. (1996) Antibody screening for secreted proteins expressed in *Pichia pastoris. Bio/Techniques* **21,** 808, 810, 812.
32. Trimble, R. B., Atkinson, P. H., Tschopp, J. F., Townsend, R. R., and Maley, F. (1991) Structure of oligosaccharides on *Saccharomyces SUC2* invertase secreted by the methylotrophic yeast *Pichia pastoris. J. Biol. Chem.* **266,** 22,807–22,817.
33. Scorer, C. A., Clare, J. J., McCombie, W. R., Romanos, M. A., and Sreekrishna, K. (1994) Rapid selection using G418 of high copy number transformants of *Pichia pastoris* for high-level foreign gene expression. *Bio/Technology* **12,** 181–184.

2

Classical Genetic Manipulation

James M. Cregg, Shigang Shen, Monique Johnson, and Hans R. Waterham

1. Introduction

A significant advantage of *Pichia pastoris* as an experimental system is the ability to bring to bear readily both classical and molecular genetic approaches to a research problem. Although the recent advent of yeast molecular genetics has introduced new and exciting capabilities, classical genetics remains the approach of choice in many instances. These include: the generation of mutations in previously unidentified genes (mutagenesis); the removal of unwanted secondary mutations (backcrossing); the assignment of mutations to specific genes (complementation analysis); and the construction of strains with new combinations of mutant alleles. In this chapter, these and other methods for genetic manipulation of *P. pastoris* are described.

To comprehend the genetic strategies employed with *P. pastoris*, it is first necessary to understand basic features of the life cycle of this yeast ***(1,2)***. *P. pastoris* is an ascomycetous budding yeast that most commonly exists in a vegetative haploid state (**Fig. 1**). On nitrogen limitation, mating occurs and diploid cells are formed. Since cells of the same strain can readily mate with each other, *P. pastoris* is by definition homothallic. (However, it is probable that *P. pastoris* has more than one mating type that switches at high frequency and that mating occurs only between haploid cells of the opposite mating type. In the related yeast *Pichia methanolica* [a.k.a. *Pichia pinus*], the existence of two mating types has been demonstrated by the isolation of mating type interconversion mutants, which are heterothallic ***[3,4]***.) After mating, the resulting diploid products can be maintained in that state by shifting them to a standard vegetative growth medium. Alternatively, they can be made to proceed through meiosis and to the production of asci containing four haploid spores.

From: *Methods in Molecular Biology, Vol. 103:* Pichia *Protocols*
Edited by: D. R. Higgins and J. M. Cregg © Humana Press Inc., Totowa, NJ

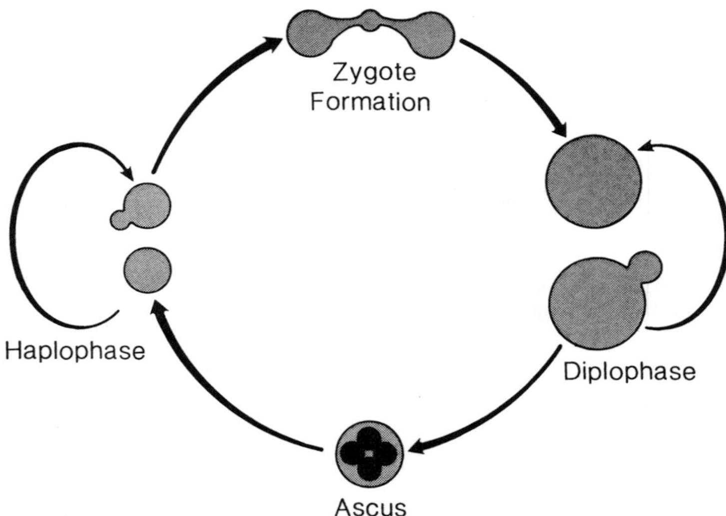

Fig. 1. Diagram of the *P. pastoris* life cycle.

The key feature of the *P. pastoris* life cycle that permits genetic manipulation is its physiological regulation of mating. *P. pastoris* is most stable in its vegetative haploid state, a great advantage in the isolation and phenotypic characterization of mutants. (In wild-type homothallic strains of *Saccharomyces cerevisiae*, the reverse is true: haploid cells are unstable and rapidly mate to form diploids *[5]*.) To cross *P. pastoris*, selected pairs of complementarily marked parental strains are mixed and subjected to nitrogen limitation for a time period sufficient to initiate mating. The strains are then shifted to a nonlimiting medium supplemented with a combination of nutrients that select for growth of hybrid diploid strains, and against the growth of the haploid parental strains and self-mated diploid strains. To initiate meiosis and sporulation, diploid strains are simply returned to a nitrogen-limited medium.

2. Materials

2.1. Strains and Media

All *P. pastoris* strains are derivatives of the wild-type strain NRRL-11430 (Northern Regional Research Laboratories, Peoria, IL). Auxotrophically marked strains are convenient for selection of diploid strains, and a collection of such strains is listed in **Table 1**. The identity of the biosynthetic genes affected in these strains is known for only three of the mutant groups: *his4*, histidinol dehydrogenase; *arg4*, argininosuccinate lyase; and *ura3*, orotidine-5'-phosphate decarboxylase. The *ura5* group strains are resistant to 5-fluoroorotic acid and,

Table 1
Auxotrophic Mutants of *P. pastoris*

Representative strain	Geneotype	Representative strain	Geneotype
JC233	*his1*	JC239	*met1*
JC234	*his2*	JC240	*met2*[a]
GS115	*his4*	JC241	*met3*[a]
		JC242	*met4*[a]
JC247	*arg1*		
JC248	*arg2*	JC220	*ade1*
GS190	*arg4*	JC221	*ade2*
		JC222	*ade3*
JC235	*lys1 his4*	JC223	*ade4*
JC236	*lys2 his4*	JC224	*ade5*
JC237	*lys3*	JC225	*ade6*
		JC226	*ade7 his4*
JC251	*pro1*		
JC252	*pro2*	JC254	*ura3*
		JC255	*ura5*

[a]Mutants in these groups will grow when supplemented with either methionine or cysteine.

therefore, are thought to be defective in the homolog of the *S. cerevisiae* orotidine-5'-phosphate pyrophosphorylase gene (*URA5*), mutants of which are also resistant to the drug ***(6)***. *P. pastoris ade1* strains are pink and, therefore, may be defective in the homolog of the *S. cerevisiae ADE1* (PR-aminoimidazolesuccinocarboxamide synthase) or *ADE2* (PR-aminoimidazole carboxylase) genes ***(7)***.

P. pastoris strains are grown at 30°C in either YPD medium (1% yeast extract, 2% peptone, 2% glucose) or YNB medium (0.67% yeast nitrogen base without amino acids) supplemented with either 0.4% glucose or 0.5% methanol. Amino acids and nucleotides are added to 50 μg/mL as required. Mating (sporulation) medium contains 0.5% sodium acetate, 1% potassium chloride, and 1% glucose. Uracil-requiring mutants are selected on 5-FOA medium, which is composed of YNB glucose medium supplemented with 50 μg/mL uracil and 750 μg/mL 5-fluoroorotic acid (PCR, Inc., Gainesville, FL). For solid media, agar is added to 2%.

2.2. Reagents for Mutagenesis

1. 1 mL of a 10 mg/mL solution of *N*-methyl-*N'*-nitro-*N*-nitrosoguanidine (NTG) (Sigma Chemical, St. Louis, MO) in acetone, stored frozen at –20°C (*see* **Note 1**).
2. 1 L mutagenesis buffer: 50 m*M* potassium phosphate buffer, pH 7.0.
3. 2 L 10% Na thiosulfate.

3. Methods

3.1. Long-Term Strain Storage

1. Viable *P. pastoris* strains are readily stored frozen for long periods (>10 yr). For each strain to be stored, pick a single fresh colony from a plate containing a selective medium, and inoculate the colony into a sterile tube containing 2 mL of YPD liquid medium.
2. After overnight incubation with shaking, transfer 1.2 mL of the culture to a sterile 2.0-mL cryovial containing 0.6 mL of glycerol. Mix the culture and glycerol thoroughly, and freeze at –70°C.
3. To resurrect a stored strain, remove the cryovial from the freezer, immediatley plunge a hot sterile inoculation loop into the frozen culture, transfer a few microliters of the culture from the loop to an agar plate containing an appropriate medium, and immediately return the culture to the freezer (*see* **Note 2**).

3.2. NTG Mutagenesis

1. Inoculate the strain to be mutagenized into a 10-mL preculture ofYPD, and incubate with shaking overnight (*see* **Note 3**).
2. On the next morning, dilute the preculture with fresh medium, and maintain it in logarithmic growth phase (OD_{600} < 1.0) throughout the day.
3. In late afternoon, use a portion of the preculture to inoculate a 500-mL culture of YPD medium in a baffled Fernbach culture flask (or alternatively two 250-mL cultures in 1-L baffled culture flasks) to an OD_{600} of approx 0.005 and incubate with shaking overnight (*see* **Note 4**).
4. On the following morning, harvest the culture at an OD_{600} of approx 1.0 by centrifugation at 5000*g* and 4°C for 5 min, and suspend the culture in 100 mL of cold sterile mutagenesis buffer.
5. Determine the density of the culture and transfer 100 OD_{600} U to each of four 250-mL sterile plastic centrifuge bottles.
6. Adjust the volume in each bottle to 100 mL with the same buffer, and wash the cells once more by centrifugation and resuspension in 50 mL of the buffer.
7. Add aliquots of 100, 200, and 400 μL of the NTG solution to each of three cultures, and hold the fourth as an untreated control. Incubate the cultures for 1 h at 30°C, and stop the mutagenesis by adding 50 mL of a 10% Na thiosulfate solution to each culture.
8. Wash each mutagenized culture by centrifugation once with 100 mL of mutagenesis buffer and once with 100 mL ofYPD medium, and resuspend each in 150 mL of YPD medium.
9. Remove 100-μL samples of each culture, prepare 100- and 10,000-fold serial dilutions of each, and spread 100-μL aliquots of the dilutions on YPD plates to determine the percentage of cells that have survived mutagenesis in each culture. Optimal survival rates are between 2 and 20% of the untreated control culture.
10. Transfer each mutagenized culture to a 500-mL shake flask, and allow cells to recover for 4 h at 30°C with shaking.

11. Concentrate the final cultures by centrifugation, and resuspended in 15 mL of YPD medium in 30% glycerol.
12. Place aliquots of 0.5 mL of NTG-treated samples into sterile microcentrifuge tubes or cryovials, and store frozen at –70°C for future use.
13. In preparation for screening mutagenized cultures, thaw a tube of each, serially dilute in sterile water, spread on a nonselective medium, such as YPD, YNB glucose, or other suitable medium, and determine the concentration of culture required to produce 500–1000 colonies/plate.
14. Screen for mutants by replica plating onto sets of plates containing appropriate diagnostic media.

3.3. Selection for Uracil Auxotrophs Using 5-Fluoroorotic Acid

5-Fluoroorotic acid (5-FOA), a uracil biosynthetic pathway analog, is metabolized to yield a toxic compound by certain enzymes in the pathway ***(8)***. As a result, organisms that are prototrophic for uracil synthesis (Ura^+) are sensitive to 5-FOA, whereas certain Ura^- auxotrophs cannot metabolize the drug and, thus, are resistant to it. Selection for 5-FOA-resistant strains of *P. pastoris* is a highly effective means of isolating Ura^- mutants affected in either of two Ura pathway genes. One of these genes, *URA3*, encodes orotidine-5'-phosphate decarboxylase. The other is likely to be the homolog of the *S. cerevisiae* orotidine-5'-phosphate pyrophosphorylase gene (*URA5*), since *ura5* mutants represent the other complementation group selected by 5-FOA in this yeast.

1. To select for Ura^- strains of *P. pastoris*, spread approx 2 OD_{600} units (~5×10^7 cells) on a 5-FOA plate. Resistant colonies will appear after approx 1 wk at 30°C.
2. Test the 5-FOA-resistant colonies for Ura phenotype by streaking them onto each of two YNB glucose plates, one with and one without uracil. The highest frequency of Ura^- mutants is found in mutagenized cultures like those described above. However, Ura^- stains often exist at a low, but significant frequency within unmutagenized cell populations as well, and can be readily selected by simply suspending cells from a YPD plate in sterile water and spreading them onto a 5-FOA plate.
3. If it is necessary to determine which *URA* gene is defective in new Ura^- strains, the strains can either be subjected to complementation testing against the known *ura* mutants (**Table 1**) or transformed with a vector containing the *P. pastoris* *URA3* gene (*see* Chapters 3 and 7).

3.4. Mating and Selection of Diploids

The mating and selection of diploid strains constitute the core of complementation analysis, and are the first step in strain construction and backcrossing. Because *P. pastoris* is functionally homothallic, the mating type of strains is not a consideration in planning a genetic cross. However, since cells of the same strain will also mate, it is essential that strains to be crossed contain complementary markers that allow for the selective growth of crossed diploids, and against the growth of self-mated diploids or parental strains. Auxotrophic markers are generally most convenient for this purpose, but mutations in any gene that affect the growth phenotype of *P. pastoris*, such as genes required for utilization of a specific carbon source (e.g., methanol or ethanol) or nitrogen source (e.g., methylamine), can be used as well.

1. To begin a mating experiment, select a fresh colony (no more than 1 wk old) of each strain to be mated from a YPD plate using an inoculation loop, and streak across the length of each of two YPD plates (**Fig. 2A**).
2. After overnight incubation, transfer the cell streaks from both plates onto a sterile replica plate velvet such that the streaks from one plate are perpendicular to those on the other.
3. Transfer the cross-streaks from the velvet to a mating medium plate, and incubate overnight to initiate mating.
4. On the next day, replica plate to an appropriate agar medium for the selection of complementing diploid cells. Diploid colonies will arise at the junctions of the streaks after approx 3 d of incubation (**Fig. 2B**). Diploid cells of *P. pastoris* are approximately twice as large as haploid cells and are easily distinguished by examination under a light microscope.
5. Colony-purify diploid strains by streaking at least once for single colonies on diploid selection medium (*see* **Note 5**).

3.5. Sporulation and Spore Analysis

Diploid *P. pastoris* strains efficiently undergo meiosis and sporulation in response to nitrogen limitation.

1. To initiate this phase of the life cycle, transfer freshly grown diploid colonies from a YPD plus glucose plate to a mating (sporulation) plate either by replica plating or with an inoculation loop, and incubate the plate for 3–4 d. After incubation, sporulated samples will have a distinctive tan color relative to the normal white color of vegetative *P. pastoris* colonies. In addition, spores and spore-containing asci are readily visible by phase-contrast light microscopy.
2. To analyze spore products by the random spore method, transfer an inoculation loop full of sporulated material to a 1.5-mL microcentrifuge tube containing 0.7 mL of sterile water, and vortex the mixture.
3. In a fume hood, add 0.7 mL of diethyl ether to the spore preparation, mix thoroughly, and leave standing in the hood for approx 30 min at room temperature. The ether treatment selectively kills vegetative cells remaining in spore preparations.
4. Serially dilute samples of the spore preparation (the bottom aqueous phase) to approx 10^{-4}, and spread 100-µL samples of each dilution onto a nonselective medium (e.g., YPD).
5. After incubation, replica plate colonies from plates that contain in the range of 100–1000 colonies onto a series of plates containing suitable diagnostic media. For example, to analyze the spore products resulting from a cross of GS115 (*his4*)

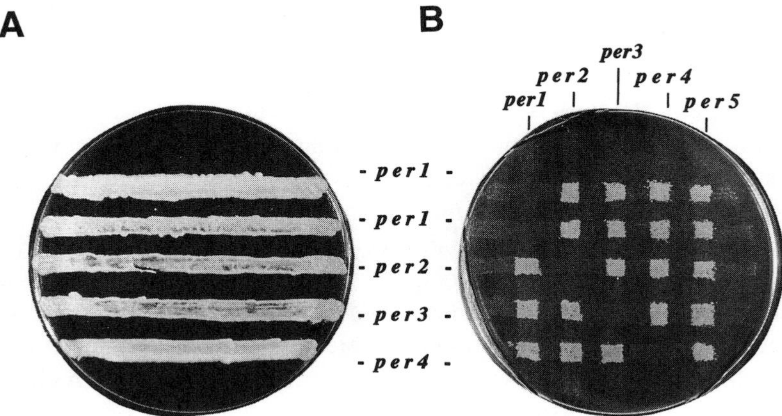

Fig. 2. Complementation analysis plates. **(A)** A YPD medium plate in which five methanol-utilization-defective strains have been streaked in preparation for complementation testing. **(B)** A YNB methanol medium plate on which complementing diploid strains have selectively grown.

and GS190 (*arg4*), appropriate diagnostic media would be YNB plus glucose supplemented with:
   a. No amino acids;
   b. Arginine;
   c. Histidine; and
   d. Arginine and histidine.
6. Compare or score the phenotype of individual colonies on each of the diagnostic plates, and identify ones with the desired phenotype(s).
7. For backcrossing or strain construction, select several colonies that appear to have the appropriate phenotype, streak for single colonies onto a nonselective medium plate, and retest a single colony from each streak on the same set of diagnostic medium-containing plates. This step is important since *P. pastoris* spores adhere tightly to one another, and colonies resulting from spore germination frequently contain cells derived from more than one spore. Another consequence of spore clumping is that markers appear not to segregate 2:2, but to be biased toward the dominant or wild-type phenotype. For example, in the GS115 (*his4*) × GS190 (*arg4*) cross described above, more His$^+$Arg$^+$ spore products will be apparent than the 25% expected in the population, and His$^-$Arg$^-$ spore products will appear to be underrepresented.

3.6. Regeneration of Selectable Markers by Ectopic Gene Conversion

In *P. pastoris*, the number of genetic manipulations (e.g., gene replacements or gene knockouts) that can be performed on a single strain is constrained by the limited number of selectable marker genes that are available. Since each new marker requires considerable effort to develop, a convenient means of regenerating previously used markers is sometimes useful. One general method takes advantage of the high frequency of homologous recombination events in diploid strains of *P. pastoris* undergoing meiosis ***(9)***. In addition to expected recombination events between genes and their homologs at the normal loci on the homologous chromosome, recombination events also occur between genes and homologous copies located at other (ectopic) sites in the genome. Thus, a wild-type *P. pastoris* marker gene inserted into the *P. pastoris* genome at an ectopic location as part of a gene knockout construction can be meiotically stimulated to recombine with its mutant allele located at the normal locus. A frequent result of such events is an ectopic gene conversion in which the wild-type allele at the knockout site is converted to its mutant allele. Spore products that harbor a mutant allele-containing knockout construction are once again auxotrophic for the selectable marker gene, and can be identified by a combination of random spore and Southern blot analyses.

As an example of this strategy, we constructed a *P. pastoris* strain in which the alcohol oxidase genes *AOX1* and *AOX2* had been disrupted with DNA fragments containing the *S. cerevisiae ARG4* (*SARG4*) and *P. pastoris HIS4* (*PHIS4*) genes, respectively (**Fig. 3**, lane 3) ***(9)***. This strain, KM7121 (*arg4 his4 aox1*Δ::*SARG4 aox2*Δ::*PHIS4*) cannot grow on methanol, but is prototrophic and therefore not easily transformed. Ectopic gene conversion events between the wild-type *HIS4* allele at *AOX2* and mutant *his4* alleles at their normal genomic loci were induced by crossing KM7121 with PPF1 (*arg4 his4 AOX1 AOX2*) (**Fig. 3**, lane 1), selecting for diploid strains on YNB methanol medium (**Fig. 3**, lanes 4–7) and sporulating the diploids. Approximately 5% of the resulting spores were, like the parent strain KM7121, unable to grow on methanol, but unlike KM7121, were auxotrophic for histidine. A Southern blot of the *AOX2* locus in these strains confirmed that in each the locus was still disrupted, but with a mutant *Phis4* allele (**Fig. 3**, lanes 13 and 14).

4. Notes

1. NTG is a powerful mutagen and a potent carcinogen. Therefore, great care should be exercised in handling this hazardous compound. Gloves, eye protection, and lab coat should be worn while working with NTG. The compound is most dangerous as a dry powder, and therefore, a particle filter mask should be worn when weighing out the powder. All materials that come in contact with NTG should be soaked overnight in a 10% solution of Na thiosulfate prior to disposal or washing.
2. Freezing kills approx 90% of the cells. However, once frozen, the remaining cells maintain their viability. Thus, it is critical not to allow the frozen culture to thaw, since approx 90% of the remaining viable cells will be killed with each round of freezing.

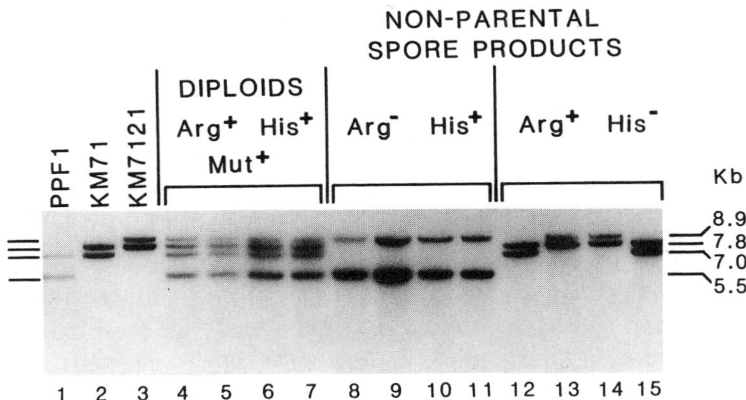

Fig. 3. Southern blot of selected strains resulting from the cross of KM7121 and PPF1. *Eco*RI-digested genomic DNA samples from the following strains are shown: lane 1, PPF1 (*arg4 his4 AOX1 AOX2*); land 2, KM71 (*arg4 his4 aox1Δ::SARG4 AOX2*); lane 3, KM7121 (*arg4 his4 aox1Δ::SARG4 aox2Δ::PHIS4*); lanes 4–7, diploid stains; lanes 8–11, Arg$^-$ His$^+$ Mut$^+$ nonparental spore products (*arg4 his4 AOX1 aox2Δ::PHIS4*); lanes 12 and 15, Arg$^+$ His$^-$ Mut^s nonparental spore products (*arg4 his4 aox1Δ::SARG4 AOX2*); lanes 13 and 14, Arg$^+$ His$^-$ Mut$^-$ ectopic gene conversion spore products (*arg4 his4 aox1Δ::SARG4 aox2Δ::Phis4*).

3. This procedure is a modified version of that described by Gleeson and Sudbury ***(10)*** and Liu et al. ***(2)***. Alternative mutagenesis methods, such as with ethylmethane sulfonate or UV light, may also be effective with *P. pastoris* and are described in Rose et al. ***(11)***, Spencer and Spencer ***(12)***, and Sherman ***(13)***.
4. The starting density of this culture can be adjusted to compensate for changes in the length of the incubation period. Adjust the density assuming that *P. pastoris* has a generation time of between 90 and 120 min at 30°C in YPD medium.
5. *P. pastoris* diploid strains are unstable relative to haploid strains and will sporulate if subjected to the slightest stress (e.g., 1 or 2 wk on a YPD plate at room temperature). Thus, to maintain diploid strains, either transfer frequently to fresh plates or store frozen at –70°C. When working with these strains, check under the microscope frequently to be sure that strains are still diploid.

References

1. Cregg, J. M. (1987) *Genetics* of methylotrophic yeasts, in *Proceedings of the Fifth International Symposium on Microbial Growth on C1 Compounds* (Duine, J. A. and Verseveld, H. W., eds.), Martinus Nijhoff, Dordrecht, The Netherlands, pp. 158–167.

2. Liu, H., Tan, X., Veenhuis, M., McCollum, D., and Cregg, J. M. (1992) An efficient screen for peroxisome-deficient mutants of *Pichia pastoris*. *J. Bacteriol.* **174,** 4943–4951.
3. Tolstorukov, I. I. and Benevolenskii, S. V. (1978) Study of the mechanism of mating and self-diploidization in haploid yeasts *Pichia pinus*: I. Bipolarity of mating. *Genetika* **14,** 519–526.
4. Tolstorukov, I. I. and Benevolenskii, S. V. (1980) Study of the mechanism of mating and self-diploidization in haploid yeasts *Pichia pinus*: II. Mutations in the mating type locus. *Genetika* **16,** 1335–1341.
5. Hicks, J. B. and Herskowitz, I. (1976) Interconversion of yeast mating types. I. Direct observation of the action of the homothalism (*HO*) gene. *Genetics* **83,** 245–258.
6. Boeke, J. D., Trueheart, J., Natsoulis, G., and Fink, G. R. (1987) 5-Fluoroorotic acid as a selective agent in yeast molecular genetics. *Methods Enzymol.* **154,** 164–175.
7. Jones, E. W. and Fink, G. R. (1982) Regulation of amino acid and nucleotide biosynthesis in yeast, in *The Molecular Biology of the Yeast Saccharomyces: Metabolism and Gene Expression* (Strathern, J. N., Jones, E. W. and Broach, J. R., eds.), Cold Spring Harbor Laboratory Press, Cold Spring Harbor, NY, pp. 181–299.
8. Krooth, R. S., Hsiao, W.-L., and Potvin, B. W. (1979) Resistance to 5-fluoroorotic acid and pyrimidine auxotrophy: a new bidirectional selective system for mammalian cells. *Som. Cell Genet.* **5,** 551–569.
9. Cregg, J. M., Madden, K. R., Barringer, K. J., Thill, G. P., and Stillman, C. A. (1989) Functional characterization of the two alcohol oxidase genes from the yeast *Pichia pastoris*. *Mol. Cell. Biol.* **9,** 1316–1323.
10. Gleeson, M. A. and Sudbery, P. E. (1988) Genetic analysis in the methylotrophic yeast *Hansenula polymorpha*. *Yeast* **4,** 293–303.
11. Rose, M. D., Winston, F. and Hieter, P. (1988) *Methods in Yeast Genetics Laboratory Manual.* Cold Spring Harbor Laboratory Press, Cold Spring Harbor, NY.
12. Spencer, J. F. T. and Spencer, D. M. (1988) Yeast genetics, in *Yeast: A Practical Approach* (Campbell, I. and Dufus, J. H., eds.), IRL, Oxford, pp. 65–106.
13. Sherman, F. (1991) Getting started with yeast. *Methods Enzymol.* **194,** 3–21.

3

Transformation

James M. Cregg and Kimberly A. Russell

1. Introduction

1.1. General Characteristics of the Methods

The key to the molecular genetic manipulation of any organism is the ability to introduce and maintain DNA sequences of interest. For *Pichia pastoris*, the fate of introduced DNAs is generally similar to those described for *Saccharomyces cerevisiae*. Vectors can be maintained as autonomously replicating elements or integrated into the *P. pastoris* genome ***(1–4)***. As in *S. cerevisiae*, integration events occur primarily by homologous recombination between sequences shared by the transforming vector and *P. pastoris* genome ***(1)***. Thus, the controlled integration of vector sequences at preselected positions in the genome via yeast gene targeting and gene replacement (gene knockout) strategies are readily performed in *P. pastoris* ***(2)***.

Four methods for introducing DNAs into *P. pastoris* have been described, and vary with regard to convenience, transformation frequencies, and other characteristics (**Table 1**). With any of the four transformation procedures, it is possible to introduce vectors as autonomous elements or to integrate them into the *P. pastoris* genome. The spheroplast generation-polyethylene glycol-$CaCl_2$ (spheroplast) method is the best characterized of the techniques and yields a high frequency of transformants (~10^5/µg), but is laborious and results in transformed colonies that must be recovered from agar embedding ***(1)***. The other three methods utilize intact or whole cells (i.e., they do not involve spheroplasting), and therefore, are more convenient and result in transformants on the surface of agar plates, which are easily picked or replica plated for further analysis. Of the whole-cell methods, electroporation yields transformants at frequencies comparable to those from spheroplasting and is the method of choice for most researchers. However, for laboratories that do not have access

From: *Methods in Molecular Biology, Vol. 103:* Pichia *Protocols*
Edited by: D. R. Higgins and J. M. Cregg © Humana Press Inc., Totowa, NJ

Table 1
General Characteristics of *P. pastoris* Transformation Methods

Method	Transformation frequency, per μg	Convenience factor	Multicopy integration?
Spheroplast	10^5	Low	Yes
Electroporation	10^5	High	Yes
PEG_{1000}[a]	10^3	High	No
LiCl[a]	10^2	High	No

[a]See **Note 12**.

to an electroporation instrument, either of the other whole-cell procedures based on polyethylene glycol or alkali cations generates adequate numbers of transformants for most types of experiments and without the labor of spheroplasting. In this chapter, procedures for each of these four methods are described. However, to aid in understanding these methods, we first describe general features of *P. pastoris* vectors and host strains.

1.2. P. pastoris *Vectors*

All *P. pastoris* vectors are of the shuttle type, i.e., composed of sequences necessary to grow and maintain them selectively in either *Escherichia coli* or *P. pastoris* hosts. *P. pastoris* transformations most commonly involve an auxotrophic mutant host and vectors containing a complementary biosynthetic gene. Selectable markers include:

1. The *P. pastoris* or *S. cerevisiae* histidinol dehydrogenase genes (*PHIS4* and *SHIS4*) ***(1)***; and
2. The argininosuccinate lyase genes from these yeasts (*PARG4* and *SARG4*) ***(2)***; and
3. The *P. pastoris* orotidine-5'-phosphate decarboxylase gene (*URA3*) (*see* **Note 1**).

Recently, a dominant selectable marker based on the Zeocin™ resistance gene has been developed (*see* Chapter 4). Advantages of the Zeocin™ system relative to biosynthetic gene/auxotrophic host systems are that transformations are not limited to specific mutant host strains and that the Zeocin™ resistance gene is also the bacterial selectable marker gene, thus substantially reducing the size of shuttle vectors (*see* Chapter 4).

Autonomous replication of plasmids in *P. pastoris* requires the inclusion of a *P. pastoris*-specific autonomous replication sequence (PARS) ***(1)***. PARSs will maintain vectors as circular elements at an average copy number of approx 10 per cell. However, relative to *S. cerevisiae*, *P. pastoris* appears to be particularly recombinogenic, and even with a PARS, any vector that also contains

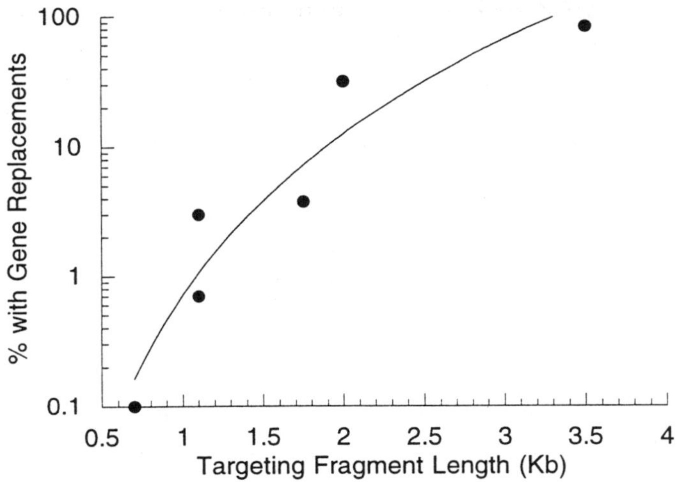

Fig. 1. Graph showing the percentage of total transformants that have undergone gene replacement as a function of the total length of targeting fragments at the termini of gene replacement vectors.

more than approx 0.5 kb of *P. pastoris* DNA will integrate into the *P. pastoris* genome at some point during the first ~100 generations after transformation. Thus, when cloning *P. pastoris* genes by functional complementation, it is important to recover the complementing vector from the yeast as soon as possible. Replication elements analogous to the *S. cerevisiae* 2-μ circle plasmid or centromers have not been described for *P. pastoris*.

1.3. Gene Replacement

In *P. pastoris*, the frequency of gene replacement events is highly dependent on the length of terminal fragments responsible for proper targeting replacement vectors. As shown in **Fig. 1**, the frequency of gene replacement events can be >50% of the total transformant population when the targeting fragments are each >1 kb, but drops precipitously to <0.1% when their total length is <0.5 kb. Gene replacements can also be performed with linear vectors composed of only one terminal-targeting fragment with the other targeting fragment located internally, although the frequency of replacement events which such vectors is substantially reduced.

Gene replacements have been performed with linear vectors that lack a selectable marker gene via cotransformation ***(2)***. This is an especially useful technique in situations where the number of genome manipulations to be performed is greater than the number of selectable markers available for the host strain. For cotransformations, *P. pastoris* cells are simply transformed with a mixture of two DNA vectors: an autonomously replicating vector that contains a selectable marker and an approx 10-fold excess of a linear gene replacement vector. Transformants are selected for the presence of the marker gene phenotype and then screened for ones that also receive the nonselected gene replacement vector. Typically, <1% of the transformants will have undergone a gene replacement event with the nonselected vector, a frequency sufficient to identify cotransformants by the phenotype conferred by the replacement event. After identifying a proper cotransformant, the autonomous vector is cured from the strain by growing it in a nonselective medium.

2. Materials

2.1. Culture Medium

Prior to all transformation procedures, *P. pastoris* strains are cultured in YPD medium (1% yeast extract, 2% peptone, 2% dextrose). Growth of *P. pastoris* in liquid media and on agar media plates is at 30°C, unless otherwise specified.

2.2. Spheroplast Procedure

2.2.1. Stock Solutions

Prepare and autoclave the following:

1. H_2O (1 L).
2. 2 *M* sorbitol (1 L).
3. 250 m*M* EDTA, pH 8.0 (100 mL).
4. 1 *M* Tris-HCl, pH 7.5 (100 mL).
5. 100 m*M* $CaCl_2$ (100 mL).
6. 10X yeast nitrogen base without amino acids (YNB): 6.7 g/100 mL.
7. 20% Glucose (100 mL).
8. Regeneration agar: 55 g sorbitol, 6 g agar, 240 mL H_2O (*see* **Notes 2** and **3**).

Prepare and filter-sterilize the following solutions:

1. 100 m*M* Na citrate, pH 5.8 (100 mL).
2. 40% Polyethylene glycol, 3350 average mol wt (PEG_{3350}) (100 mL).
3. 0.5 *M* dithiothreitol (DTT) (10 mL).
4. Prepare 1 mL of a 4 mg/mL solution of Zymolyase T100 (ICN, Costa Mesa, CA), and dispense in 100-µL aliquots into minicentrifuge tubes.

All of the above solutions can be stored at room temperature, except for DTT and Zymolyase, which must be stored at –20°C (*see* **Note 4**). All are stable for at least 1 yr.

2.2.2. Sterile Working Solutions

1. 1 *M* sorbitol (200 mL).
2. SCE: 1 *M* sorbitol, 10 m*M* Na citrate (pH 5.8), 10 m*M* EDTA (100 mL).

3. CaS: 1 *M* sorbitol, 10 m*M* $CaCl_2$ (100 mL).
4. PEG-CaT: 20% PEG, 10 m*M* $CaCl_2$, 10 m*M* Tris-HCl (pH 7.5) (100 mL).
5. SOS: 1 *M* sorbitol, 0.3X YPD medium, 10 m*M* $CaCl_2$ (20 mL).
6. SED: 1 *M* sorbitol, 25 m*M* EDTA, 50 m*M* DTT (10 mL). Prepare fresh and hold on ice until use.
7. Regeneration medium agar: 240 mL of regeneration agar (from stock remelted via autoclave, microwave, or boiling water bath), 30 mL 10X YNB, 30 mL 20% glucose held in a 45°C water bath until use. Prepare fresh the day of transformation.

2.3. Electroporation

Prepare the following:

1. YPD medium (100 mL) with 20 mL 1 *M* HEPES buffer, pH 8.0.
2. 1 *M* DTT (2.5 mL).
3. H_2O (1 L).
4. 1 *M* sorbitol (100 mL).
5. YND agar plates: 0.67% YNB, 2% glucose, 2% agar (0.5 L/~25 plates).
6. Electroporation instrument: BTX Electro Cell Manipulator 600 (BTX, San Diego, CA); Bio-Rad *Gene* Pulser (Bio-Rad, Hercules, CA); Electroporator II (Invitrogen, Carlsbad, CA); BRL Cell-Porator (Life Technologies, Gaithersburg, MD).
7. Sterile electroporation cuvets.

All solutions should be autoclaved except the DTT and HEPES solutions, which should be filter sterilized.

2.4. Polyethylene Glycol (PEG) Procedure

1. Buffer A: 1 *M* sorbitol, 10 m*M* bicine (pH 8.35), 3% ethylene glycol (100 mL). Filter-sterilize, and store at –20°C until use.
2. Buffer B: 40% PEG 1000 average mol wt (PEG_{1000}; *see* **Note 5**), 0.2 *M* Bicine (pH 8.35) (50 mL). Filter-sterilize, and store at –20°C until use.
3. Buffer C: 0.15 *M* NaCl, 10 m*M* Bicine (pH 8.35) (50 mL). Filter-sterilize and store at –20°C until use.
4. Dimethyl sulfoxide (DMSO) (*see* **Note 6**). Store at –70°C.
5. YND agar plates: 0.67% YNB, 2% glucose, 2% agar (0.5 L/~25 plates).

2.5. Alkali Cation Procedure

Prepare and autoclave the following solutions:

1. H_2O (1 L).
2. TE buffer: 10 m*M* Tris-HCl, pH 7.4, 1 m*M* EDTA (100 mL).
3. LiCl buffer: 0.1 *M* LiCl, 10 m*M* Tris-HCl, pH 7.4, 1 m*M* EDTA (100 mL).
4. PEG + LiCl buffer: 40% PEG_{3350}, 0.1 *M* LiCl, 10 m*M* Tris-HCl, pH 7.4, 1 m*M* EDTA (100 mL).
5. YND agar plates: 0.67% YNB, 2% glucose, 2% agar (0.5 L/~25 plates).

3. Methods

3.1. DNA Preparation

For highest transformation frequencies, *E. coli-P. pastoris* shuttle vector DNAs should be pure and dissolved in water or TE buffer. Most standard procedures for purification of plasmid DNAs, such as those involving CsCl ethidium bromide centrifugation or a commercial plasmid preparation kit (e.g., Qiagen, Hilden, Germany), work well. However, plasmids prepared by "miniprep" methods, such as the alkaline lysis procedure, can be transformed into *P. pastoris* at frequencies adequate for most purposes.

For gene targeting and gene replacement constructions, vectors should be digested with a restriction enzyme(s) that cuts within *P. pastoris* DNA sequences in the vector such that a minimum of 200 bp of *P. pastoris* DNA are available at each terminus to direct targeted integration. Prior to transformation, linear vectors should be extracted with phenol-chloroform-isoamyl alcohol (PCA, 25:24:1), alcohol-precipitated, and dissolved in water or TE buffer. It is typically not necessary to separate a gene replacement fragment from the remaining nontransforming fragment. However, vector fragments purified from agarose gels via standard electroelution procedures or commercially available kits (e.g., QIAEX gel extraction kit, Qiagen) can be transformed into *P. pastoris*.

3.2. Spheroplast Procedure

This procedure is a modified version of that described by Hinnen et al. ***(5)*** and Cregg et al. ***(1)***. The general characteristics of the method are summarized in **Table 1**.

3.2.1. Preparation of Spheroplasts

1. Inoculate a 10-mL YPD culture with a single *P. pastoris* colony of the strain to be transformed from a fresh agar plate, and grow overnight with shaking. This culture can be stored at 4°C for several days.
2. Inoculate three 200-mL YPD cultures into 500-mL baffled culture flasks with 5, 10, and 20 µL of the above culture, and incubate overnight with shaking.
3. In the morning, select the culture that has an OD_{600} of between 0.2 and 0.3 (*see* **Note 7**).
4. Wash the culture by centrifugation at 2000*g* in a 50-mL conical tube at room temperature once with 10 mL of water, once with 10 mL of freshly prepared SED, once with 10 mL of 1 *M* sorbitol, and once with 10 mL of SCE.
5. Add between 1 and 10 µL of Zymolyase T100, and incubate at 30°C without shaking. To monitor spheroplasting, remove 100-µL aliquots of the cells before adding the Zymolyase and at times of 5, 10, 15, 20, 30, and 45 min after addition of the enzyme. Dispense them into a series of glass tubes containing 900 µL of 1% SDS. After addition of each sample to the SDS solution, mix and visually

examine the solution for cell lysis as judged by decreased turbidity and increased viscosity. The optimal time for spheroplasting is between 15 and 30 min. Adjust the amount of Zymolyase in future transformations accordingly (*see* **Note 8**). Proceed with the next step in the procedure as soon as spheroplasts appear ready.

3.2.2. Transformation

1. Wash the spheroplast preparation by centrifugation at 1500*g* for 10 min once in 10 mL of 1 *M* sorbitol and once in 10 mL of CaS. Spheroplasts are very fragile. Therefore, decant supernatants carefully, and resuspend gently by tapping the side of the tube or by gentle pipeting.
2. Centrifuge the preparation, and resuspend spheroplasts in 1.0 mL of CaS.
3. Dispense 100-µL aliquots of spheroplasts into 5-mL polypropylene snap-top Falcon (or similar) tubes. Add DNAs to each tube, and incubate at room temperature for 20 min.
4. Add 1 mL of PEG-CaT to each tube, and incubate for an additional 15 min at room temperature.
5. Centrifuge samples at 1500*g* for 10 min, carefully decant PEG-CaT, and gently resuspend spheroplasts in 200 µL of SOS. Incubate samples at room temperature for 30 min, and then add 800 µL of 1 *M* sorbitol.

3.2.3. Plating

1. Prepare plates containing a 10-mL bottom layer of regeneration medium agar (one for each transformation sample plus at least two additional bottom agar plates for viability testing).
2. Dispense a 10-mL aliquot of molten 45°C regeneration medium agar into a 50-mL polypropylene tube for each transformation sample. Gently mix transformation samples of between 10 µL and 0.5 mL with the regeneration agar, and pour over the bottom agar layer. After the agar solidifies (~10 min), incubate the plates for 4–7 d.
3. Monitor the quality of the spheroplast preparation and regeneration conditions as follows. Remove a 10-µL aliquot of one of the transformation samples, and add to a tube containing 990 µL of 1 *M* sorbitol (10^{-2} dilution). Mix, remove a 10-µL sample of the 10^{-2} dilution, and dilute again by addition to a second tube containing 990 µL of 1 *M* sorbitol (10^{-4} dilution). Spread 100 µL of each dilution on a YPD plate to determine the concentration of remaining whole (unspheroplasted) cells. To determine the concentration of spheroplasts with the potential to regenerate into viable cells, add a second 100-µL aliquot of each dilution to a tube containing 10 mL of molten regeneration medium agar supplemented with 50 µg/mL of the missing nutrient (e.g., histidine). Gently mix; pour over a bottom agar plate. Incubate these control plates with the transformation plates as described in **step 2**. Good spheroplast preparations and regeneration reagents will produce $>1 \times 10^7$ colonies/mL (i.e., >100 colonies on the 10^{-4} spheroplast regeneration plate) and $<1 \times 10^4$ colonies/mL unspheroplasted whole cells (i.e., <10 colonies on the 10^{-2} remaining whole-cell plate).

3.2.4. Recovery of Transformants

Most transformant colonies will be embedded within the top agar. To recover individual embedded transformants, dig them out of the agar with an inoculation loop, and streak for single colonies on the surface of an agar plate containing the selective medium. To recover embedded cells from a large proportion of the embedded colonies in a transformation plate for further analysis (e.g., to screen for ones that have undergone a gene replacement event), the following procedure can be used:

1. Scrape the colony-containing top agar layer into a sterile 50-mL centrifuge tube using a spatula or forceps (sterilized with 70% ethanol). Add 20 mL of sterile water, and mix vigorously to break up the agar and release embedded cells.
2. Filter the suspension through fourfolds of sterile cheesecloth. Rinse cells from the agar by addition of approx 20 mL of water to the agar. Centrifuge the filtrate at 2000*g* for 5 min and decant.
3. Suspend the cells (and a small amount of remaining agar) in 5 mL of sterile water, and vortex vigorously to disperse the cells.
4. Spread dilutions of the cells on selective medium agar plates at a concentration that will generate 100–500 colonies/plate, and incubate for 2–3 d (*see* **Note 7**).
5. Replica plate onto an agar medium appropriate for identifying transformants that have undergone a gene replacement event. For example, when looking for transformants with a gene replacement at *AOX1* and, as a result, you have acquired a methanol-utilization-slow (Mut^s) phenotype, replica plate colonies from YND plates to two sets of plates, one containing YNB plus methanol and the other containing YND. After 1 or 2 d incubation, compare the two sets of plates, and select colonies from the YND plate that are not growing well on the YNB plus methanol plate.

3.3. Electroporation

This procedure is a modified version of that described by Becker and Guarente *(6)*. The characteristics of the method are listed in **Table 1**. Parameters for electroporation with four different instruments are shown in **Table 2** (*see* **Note 9**).

3.3.1. Preparation of Competent Cells

1. Inoculate a 10-mL YPD culture with a single fresh *P. pastoris* colony of the strain to be transformed from an agar plate, and grow overnight with shaking.
2. In the morning, use the overnight culture to inoculate a 500-mL YPD culture in a 2.8-L Fernback culture flask to a starting OD_{600} of 0.1, and grow to an OD_{600} of 1.3–1.5 (*see* **Note 7**).
3. Harvest the culture by centrifugation at 2000*g* at 4°C, and suspend the cells in 100 mL of YPD medium plus HEPES.
4. Add 2.5 mL of 1 *M* DTT and gently mix.
5. Incubate at 30°C for 15 min.

Table 2
Parameters for Electroporation Using Selected Instruments

Instrument	Cuvette gap, mm	Sample volume, μL	Charging voltage, V	Capacitance, μF	Resistance, Ω	Field strength, kV/cm	Pulse length, ~ms	Reference
ECM600[a] (BTX)	2	40	1500	Out	129	7500	5	***6***
Electroporator II (Invitrogen)	2	80	1500	50	200	7500	10	*7*
Gene-Pulser (Bio-Rad)	2	40	1500	25	200	7500	5	***8***
Cell-Porator (BRL)	1.5	20	480	10	Low	2670	NS	***9,10***

[a]For ECM600, select 2.5 kV/RESISTANCE high voltage.
NS = not specified.

6. Bring to 500 mL with cold water. Wash by centrifugation at 4°C once in 250 mL of cold water, once in 20 mL of cold 1 *M* sorbitol, and resuspend in 0.5 mL of cold 1 *M* sorbitol. (Final volume including cells will be 1.0–1.5 mL.)
7. For highest frequencies, transform the cells directly without freezing as described below.
8. To freeze competent cells, distribute in 40-μL aliquots to sterile 1.5-mL minicentrifuge tubes, and place the tubes in a –70°C freezer until use (*see* **Note 10**).

3.3.2. Electroporation

1. Add up to 0.1 μg of DNA sample in no more than 5 μL total volume of water or TE buffer to a tube containing 40 μL of frozen or fresh competent cells, and transfer to a 2-mm gap electroporation cuvet held on ice (*see* **Notes 9** and **10**).
2. Pulse cells according to the parameters suggested for yeast by the manufacturer of the specific electroporation instrument being used (**Table 2**).
3. Immediately add 1 mL of cold 1 *M* sorbitol, and transfer the cuvet contents to a sterile 1.5-mL minicentrifuge tube.
4. Spread selected aliquots onto agar plates containing YND or other selective medium, and incubate for 2–4 d (*see* **Note 11**).

3.4. PEG Procedure

This procedure is a modified version of that described in Klebe et al. ***(11)***. The characteristics of the method are listed in **Table 1** (*see* **Note 12**).

3.4.1. Preparation of Competent Cells

1. Inoculate a 10-mL YPD culture with a single fresh *P. pastoris* colony of the strain to be transformed from an agar plate, and grow overnight with shaking.
2. In the morning, use the overnight culture to inoculate a 100-mL YPD culture to a starting OD_{600} of 0.1, and grow to an OD_{600} of 0.5–0.8 (*see* **Note 7**).
3. Harvest the culture by centrifugation at 2000*g* at room temperature, and wash the cells once in 50 mL of buffer A.
4. Resuspend cells in 4 mL of buffer A, and distribute in 0.2-mL aliquots to sterile 1.5-mL minicentrifuge tubes. Add 11 μL of DMSO to each tube, mix, and quickly freeze cells in a liquid nitrogen bath.
5. Store frozen competent cells at –70°C.

3.4.2. Transformation

1. Add up to 50 μg of DNA sample (in no more than 20 μL total volume) directly to a still-frozen tube of competent cells (*see* **Note 13**). Carrier DNA (40 μg of denatured and sonified salmon sperm DNA) should be included with sub-microgram DNA samples for maximum transformation frequencies.
2. Incubate samples in a 37°C water bath for 5 min. Gently mix samples once or twice during this incubation period.
3. Remove tubes from bath, and add 1.5 mL of buffer B to each. Mix contents thoroughly.

4. Incubate tubes in a 30°C water bath for 1 h.
5. Centrifuge sample tubes at 2000*g* for 10 min at room temperature, and gently resuspend cells in 1.5 mL of buffer C.
6. Centrifuge samples a second time, and resuspend cells gently in 0.2 mL of buffer C.
7. Spread contents of each tube on an agar plate containing selective growth medium, and incubate plates for 3–4 d (*see* **Note 11**).

3.5. Alkali Cation Procedure

This procedure is essentially the same as described by Ito et al. ***(11)***, except that LiCl is used instead of Li acetate. The characteristics of the method are listed in **Table 1** (*see* **Note 12**). Note that unlike the other transformation methods, transformation frequencies are very low for both PARS- and nonPARS-containing circular vectors.

3.5.1. Preparation of Competent Cells

1. Inoculate a 10-mL YPD culture with a single fresh *P. pastoris* colony of the strain to be transformed from an agar plate, and grow overnight with shaking.
2. In the morning, use the overnight culture to inoculate a 50-mL YPD culture to a starting OD_{600} of 0.1, and grow to an OD_{600} of 0.5–0.8 (*see* **Note 7**).
3. Harvest the culture, and wash the cells once with 10 mL of H_2O by centrifugation at 2000*g* at room temperature, once with 10 mL of TE buffer, and once with 20 mL of LiCl buffer.
4. Incubate the cells for 1 h at 30°C. For highest transformation frequencies, transform the cells directly without freezing as described below.
5. To freeze competent cells, distribute in 0.2-mL aliquots to sterile 1.5-mL minicentrifuge tubes. Add 11 µL of DMSO to each tube, mix, and quickly freeze cells in a liquid nitrogen bath. Store frozen competent cells at –70°C.

3.5.2. Transformation

1. For each transformation sample, add the following to a sterile 1.5-mL centrifuge tube: 0.1 mL of competent cells, 0.1–20 µg of vector DNA in no more than a 20-µL vol. For maximum transformation frequencies with submicrogram amounts of DNA, add 10 µg of a carrier DNA (e.g., denatured and sonified salmon sperm DNA).
2. Incubate samples at 30°C for 30 min.
3. Add 0.7 mL of PEG + LiCl solution, and briefly vortex to mix.
4. Heat-shock at 37°C for 5 min.
5. Centrifuge samples at 2000*g* and resuspend in 0.1 mL of H_2O.
6. Spread samples onto selective medium plates, and incubate for 3 d (*see* **Note 11**).

4. Notes

1. *ura3 P. pastoris* strains grow more slowly that the wild type, even on YPD medium supplemented with a large excess of uracil and/or uridine. However, on transformation with the *URA3* gene, they resume growth at the wild-type rate. Because it

is possible to select for or against either Ura^+ or Ura^- phenotypes, the *ura3* host/*URA3* vector marker system is convenient for marker regeneration strategies, such as described in Chapter 7.

2. Spheroplast transformation reagents can be purchased from Invitrogen (Carlsbad, CA).
3. Some lots of sorbitol are much better than others for regeneration of *P. pastoris* spheroplasts. As an alternative to sorbitol, 0.6 *M* KCl can be substituted (13.4 g KCl, 6 g agar, 240 mL H_2O).
4. Zymolyase lots vary with regard to suitability for use in *P. pastoris* transformations. Since the activity in the dry powder is stable for years at –20°C, it is recommended that good lots be set aside exclusively for transformations. As an alternative spheroplasting agent, the snail gut preparation Glusulase (DuPont, Wilmington, DE) can be used. Glusulase is stable for periods of at least a year stored at 4°C. To spheroplast with Glusulase, follow the same procedure, and add approx 20 μL of Glusulase/40–60 OD_{600} units of cells in SCE.
5. The purity of the PEG_{1000} is critical for transformation. Known sources of high-quality PEG_{1000} are Sigma Chemical Company (St. Louis, MO) and Carl Roth GmbH (Karlsruhe, Germany; United States distributor is Atomergic Chemetals Corp., New York, NY).
6. DMSO must be from a fresh unopened bottle or from a stock of DMSO prepared from a new unopened bottle, and stored at –70°C until use.
7. One OD_{600} unit of *P. pastoris* culture equals approx 5×10^7 cells.
8. For highest transformation frequencies, it is necessary to establish the optimal spheroplasting conditions (i.e., the optimal amount of Zymolyase or Glusulase and incubation time), which vary with different lots of enzyme and other reagents and equipment used for cell growth and spheroplasting. Another method of establishing these conditions is to divide the washed cells into equal portions before spheroplasting, and add enzyme to only one tube while holding the other on ice. Spheroplasting is monitored by incubating the tube with enzyme at 30°C and removing samples at selected times to SDS solution. Samples in SDS are then examined spectrophotometrically at OD_{800} and the time required for 70% of the cells to become sensitive to lysis with SDS determined quantitatively. Subsequently, enzyme is added to the remaining tube, which is then incubated at 30°C for the length of time determined for optimal (70%) spheroplasting.
9. In general, procedures described for *S. cerevisiae* also work well for *P. pastoris*. Thus, if your instrument is not listed, use the protocol recommended for electroporation of *S. cerevisiae* with your instrument. *See* technical literature provided by manufacturers for information on the use of specific electroporation devices.
10. If using Invitrogen electroporation cuvets, the volume of cells should be 80 μL. If using BRL 1.5-mm gap chambers, the volume of cells should be 20 μL.
11. *P. pastoris* cells are flocculent (i.e., tend to grow in multicell clumps). As a result, transformant colonies are frequently composed of more than one transformed strain. To avoid the problem of mixed colonies, pick and restreak them for single colonies on selective medium at least once before proceeding with further analysis.

When looking for gene replacement transformants by replica plate screening for a recessive phenotype (e.g., *AOX1* gene replacements with a recessive methanol-utilization slow or Mut^s phenotype), colonies should be recovered from the transformation plates, suspended in water, and replated on selective medium before screening.

12. Multicopy vector integration events are much less frequent with the PEG_{1000} or alkali cation methods than with either the spheroplast or electroporation methods (Koti Sreekrishna, personal communication).
13. Cell competence decreases very rapidly after cells thaw, even if the cells are held on ice. Therefore, adding DNA to frozen samples is critical. To transform large numbers of samples, it is convenient to process them in groups of about six at a time.

References

1. Cregg, J. M., Barringer, K. J., Hessler, A. Y., and Madden K. R. (1985) *Pichia pastoris* as a host system for transformations. *Mol. Cell. Biol.* **5,** 3376–3385.
2. Cregg, J. M., Madden, K. R., Barringer, K. J., Thill, G. P., and Stillman, C. A. (1989) Functional characterization of the two alcohol oxidase genes from the yeast *Pichia pastoris*. *Mol. Cell. Biol.* **9,** 1316–1323.
3. Liu, H., Tan, X., Wilson, K., Veenhuis, M., and Cregg, J. M. (1995) *PER3*, a gene required for peroxisome biogenesis in *Pichia pastoris*, encodes a peroxisomal membrane protein involved in protein import. *J. Biol. Chem.* **270,** 10,940–10,951.
4. Waterham, H. R., de Vries, Y., Russell, K. A., Xie, W., Veenhuis, M., and Cregg, J. M. (1996) The *Pichia pastoris PER6* gene product is a peroxisomal integral membrane protein essential for peroxisome biogenesis and has sequence similarity to PAF-1. *Mol. Cell. Biol.* **16,** 2527–2536.
5. Hinnen, A., Hicks, J. B., and Fink, G. R. (1978) Transformation of yeast. *Proc. Natl. Acad. Sci. USA* **75,** 1929–1934.
6. Becker, D. M. and Guarente, L. (1991) High-efficiency transformation of yeast by electroporation. *Methods Enzymol.* **194,** 182–187.
7. Pichia Expression Kit Instruction Manual, Version E, Invitrogen, San Diego, CA, p. 63.
8. Grey, M. and Brendel, M. (1995) Ten-minute electrotransformation of *Saccharomyces cerevisiae*, in *Methods in Molecular Biology, vol. 47: Electroporation Protocols for Microorganisms* (Nickoloff, J. A., ed.), Humana, Totowa, NJ, pp. 269–272.
9. Stowers, L., Gautsch, J., Dana, R., and Hoekstra, M. F. (1995) *Yeast* transformation and preparation of frozen spheroplasts for electroporation, in *Methods in Molecular Biology, vol. 47: Electroporation Protocols for Microorganisms* (Nickoloff, J. A., ed.), Humana, Totowa, NJ, pp. 261–267.
10. Lorow-Murray, D. and Jesse, J. (1991) High efficiency transformation of *Saccharomyces cerevisiae* by electroporation. *Focus* **13,** 65–68.
11. Klebe, R. J., Harriss, J. V., Sharp, Z. D., and Douglas, M. G. (1983) A general method for polyethylene-glycol-induced genetic transformation of bacteria and yeast. *Gene* **25,** 333–341.
12. Ito, H., Fukuda, Y., Murata, K., and Kimura, A. (1983) Transformation of intact yeast cells treated with alkali cations. *J. Bacteriol.* **153,** 163–168.

4

Small Vectors for Expression Based on Dominant Drug Resistance with Direct Multicopy Selection

David R. Higgins, Katie Busser, John Comiskey, Peter S. Whittier, Thomas J. Purcell, and James P. Hoeffler

1. Introduction

The most commonly used vectors for heterologous protein expression in *Pichia pastoris* carry its wild-type *HIS4* gene and the bacterial ampicillin resistance gene as selectable markers ***(1–6)***. The *HIS4* gene is relatively large (3 kb), with ill-defined functional boundaries, and its use limits the vectors to selection in *his4* auxotrophic strains. *HIS4*-based vectors are large, typically between 7 and 10 kb in size, creating difficulties for routine cloning manipulations, especially when generating constructions with large or multiple insertions. The *Escherichia coli* kanamycin gene (from Tn 903) present on some *P. pastoris* expression vectors functions as a second marker on some *HIS4*-based vectors to identify multicopy (i.e., G418 hyperresistant) recombinant strains. However, it cannot be used efficiently as a direct selectable marker for transformation of *P. pastoris* to G418 resistance, and its presence leads to a further increase in vector size. The combination of the *HIS4*, ampicillin, and kanamycin genes adds up to 5.5 kb of "space" in a typical expression vector.

Generation of significantly smaller vectors has been achieved by using a single dominant selectable marker that functions both in *E. coli* and *P. pastoris* replacing the *HIS4*, ampicillin, and kanamycin genes ***(7)***. This marker is the *Sh ble* gene from *Streptoalloteichus hindustanus*, which is small with an open reading frame of 375 bp encoding a 13,665-Dalton protein. The *Sh ble* gene product confers resistance to the drug Zeocin in *E. coli*, yeast (including *P. pastoris*), and other eukaryotes ***(8–14)***. Zeocin is a bleomycin-like compound that kills cells by introducing lethal double-strand breaks in

From: *Methods in Molecular Biology, Vol. 103:* Pichia *Protocols*
Edited by: D. R. Higgins and J. M. Cregg © Humana Press Inc., Totowa, NJ

chromosomal DNA *(15)*. The Zeocin-resistance protein (the product of the *Sh ble* gene) confers resistance stoichiometrically, not enzymatically, by binding to and inactivating the drug. Natural sensitivity of *P. pastoris* to Zeocin, and the resistance conferred by expression of the *Sh ble* gene has proven successful and provides a dominant selection system that is strain-, genotype-, and ploidy-independent *(7)*. Because this small Zeocin resistance gene substitutes for the *HIS4*, ampicillin, and kanamycin genes, a net reduction in vector size of approx 3.5 kb is achieved. Furthermore, selection of Zeocin-resistant transformants at high concentrations of Zeocin (i.e., Zeocin hyperresistant transformants) generates an enrichment in recombinant strains with multiple copies of the integrated vector. Direct selection of Zeocin hyperresistant transformants can be used routinely to generate a population of multicopy clones that may ultimately result in an increase in the level of heterologous protein production *(7)*.

The generation of recombinant strains with multiple copies of the expression plasmid integrated into the genome has been shown to result in an increase in heterologous protein production via a gene dosage effect for a number of different heterologous genes *(5,16–21)*. In addition to screening for relatively rare spontaneous events, there are several methods to generate intentionally recombinant strains containing multiple plasmid copies, including in vitro multimerization (Chapter 13) and G418 hyperresistance screening of His^+ transformants (Chapter 5).

The bacterial kanamycin gene acts as a marker that can be used to delineate further His^+ transformants into strains that carry multiple copies of the heterologous gene. Each expression cassette introduces a copy of the bacterial kanamycin gene. This gene confers resistance to G418, multiple copies of which make recombinant strains hyperresistant to the G418. Hyperresistance to G418 allows for phenotypic identification of strains carrying multiple copies of the kanamycin gene and, by linkage, multiple copies of the expression cassette *(21)*.

Presumably, the same molecular biology occurs for Zeocin resistance: Zeocin hyperresistant clones arise from multiple copies of the Zeocin resistance gene, which in turn corresponds to multiple plasmid copies and thus multiple copies of the heterologous gene being expressed. If expression levels of the heterologous gene product respond to a gene dosage effect, higher levels of heterologous protein can result. The use of the Zeocin-resistant maker has the added benefit that Zeocin hyperresistant transformants can be selected directly, whereas G418 hyperresistant clones must be screened secondarily after initial His^+ prototrophic transformation selection. Also, the product of the Zeocin resistance gene functions stoichiometrically, not enzymatically, as does the kanamycin gene product, which may lead to a more efficient copy number correlation with a Zeocin hyperresistant phenotype *(7)*.

2. Materials

2.1. *E. coli* and *P. pastoris* Strains

E. coli (plasmid maintenance): TOP10 (F^- *mcr*A Δ[*mrr-hsd*RMS-*mcr*BC] φ80*lac*ZΔM15 Δ*lac*X74 *deo*R *rec*A1 *ara*D139 Δ[*ara-leu*]7697 *gal*U *gal*K *rps*L [Str^R] *end*A1 *nup*G) (*see* **Note 1**). *P. pastoris*: GS115 (Mut^+ *his4*); X-33 (Mut^+ His^+ derivative of GS115 generated by transformation of GS115 with the 3-kb *Nae*I-*Bam*H1 *his4* from plasmid pAO815); KM71 (Mut^s; *his4 arg4 aox1*Δ::*ARG4*), SMD1168 (Mut^+; *his4 pep4*) (*see* **Note 2**).

2.2. Zeocin Resistance-Based Plasmids

Vectors pPICZ A, B, and C, and pPICZα A, B, and C represent six *P. pastoris* expression vectors available from Invitrogen Corporation (San Diego, CA) (**Fig. 1**) *(7)*. Selection for these vectors in both *P. pastoris* and *E. coli* is based on a single small dominant selectable marker that confers resistance to the drug Zeocin. To confer Zeocin resistance in *P. pastoris*, a 1267-bp *Bam*HI–*Hin*dIII fragment from vector pUT352 was used as a selectable marker in expression vector constructions. This fragment contains the She ble open reading frame (ORF) preceded by the *Saccharomyces cerevisiae TEF1* gene promoter and a synthetic *E. coli* promoter, EM7. The *Sh ble* ORF is followed by the *S. cerevisiae CYC1* gene 3' transcription processing signal sequence *(14,22)*. This fragment has been shown to efficiently confer Zeocin resistance in both *E. coli* (driven by the EM7 promoter) and *P. pastoris* (by the *TEF1* promoter) *(7,23)*.

2.3. Media

1. ½X salt LB: 0.5% NaCl, 1% tryptone, 0.5% yeast extract (for solid media, 1.5% agar) supplemented with Zeocin to a final concentration of 50 μg/mL (*see* **Note 3**).
2. YPD: 1% yeast extract, 2% peptone, 2% dextrose (for solid media, 2% agar).
3. YPDS: 1% yeast extract, 2% peptone, 2% dextrose, 1 *M* sorbitol (for solid media, 2% agar) supplemented with Zeocin to a final concentration of 100 μg/mL (*see* **Note 4**).
4. YPM: 1% yeast extract, 2% peptone, 0.5% methanol (for solid media, 2% agar).
5. BMGY: 1% yeast extract, 2% peptone, 1.34% yeast nitrogen base with ammonium sulfate without amino acids, 1% glycerol, 100 m*M* sodium phosphate buffer, pH 6.0.
6. BMMY: 1% yeast extract, 2% peptone, 1.34% yeast nitrogen base with ammonium sulfate without amino acids, 0.5% methanol, 100 m*M* sodium phosphate buffer, pH 6.0.
7. Zeocin: Used as a stock solution of 100 mg/mL in sterile water; should be added to media solutions that are ≤60°C. Zeocin stock solution should be stored at –20°C.

2.4. Zeocin

Zeocin, a member of the bleomycin family of drugs, has cytotoxic effects on most aerobically growing cells *(15)*. Its mode of action is by intercalation into

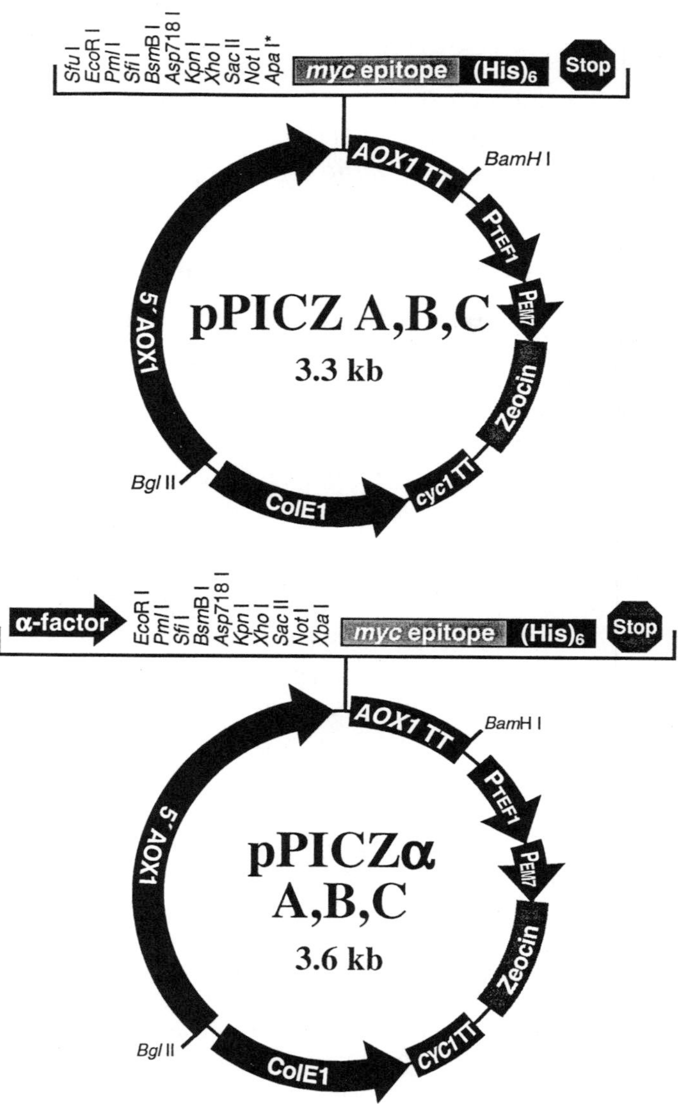

Fig. 1. Graphic maps of vectors pPICZ A, B, C and pPICZα A, B, C. A series of six *P. pastoris* expression vectors have been constructed that are between 3.3 and 3.5 kb in size, yet still contain all of the elements for protein expression and detection in *P. pastoris*. Selection for these vectors in both *P. pastoris* and *E. coli* is based on a single small dominant selectable marker that confers resistance to the drug Zeocin. Expression of the Zeocin resistance gene is driven by the *S. cerevisiae* promoter *TEF1* and is followed by the *S. cerevisiae CYC1* transcription processing and polyadenylation sequence. *(cont.)*

DNA inducing double-strand breaks, which accumulate to lethal levels. Zeocin is available from Invitrogen Corporation (*see* **Note 5**).

2.5. Yeast Transformation

1. 1 *M* sorbitol, filter-sterilized, ice-cold.
2. Ultra-pure H_2O, sterile, ice-cold.
3. 5–10 µg linearized plasmid DNA, after restriction digestion: phenol extracted, ethanol-precipitated, resuspended in sterile H_2O to ~1 µg/µL (*see* **Note 6**).
4. Electroporation device (*see* **Note 7**).
5. Sterile electroporation cuvets, 0.2 cm (*see* **Note 7**).

2.6. Quantitative Slot Blot Analysis

1. 5–10 µg genomic DNA in 50 µL H_2O (*see* **Note 8**).
2. Slot-blot apparatus (e.g., Bio-Rad, Hercules, CA).
3. Appropriate nylon membrane (e.g., Zeta probe from Bio-Rad).
4. 0.4 *N* NaOH.
5. DNA denaturing buffer: 0.4 *N* NaOH, 10 m*M* EDTA.
6. 2X SSC made from 20X SSC stock solution diluted 1/10 in ultra-pure water (20X SSC: 175.3 g NaCl, 88.2 g Na citrate, 1 L final volume, pH 7.0; adjust pH using solutions of 10 *N* NaOH).
7. Prehybridization/hybridization buffer: 20 m*M* Na_2HPO_4, 7% SDS, 1X Denhardt's, 10 µg/mL denatured sheared salmon sperm DNA.
8. Suitable labeled probe (*see* **Note 9**).
9. Low-stringency wash solution: 2X SSPE, 0.1% SDS, room temperature.
10. Medium-stringency wash solution: 1X SSPE, 0.1% SDS, 65°C.
11. High-stringency wash solution: 0.1X SSPE, 0.1% SDS, 65°C.
12. 100X Denhardt's solution: 2% Ficoll 400 (w/v), 2% polyvinylpyrrolidone (w/v), 20 mg/mL BSA (w/v), Pentax Fraction V.
13. 20X SSPE: 175.3 g NaCl, 27.6 g NaH_2PO_4, 7.4 g EDTA, in 1 L, adjusted to pH 7.4, with NaOH.
14. XAR X-ray film (Eastman Kodak, Rochester, NY) or equivalent.

Fig. 1 *(cont.)*. Versions with and without the *S. cerevisiae* α-factor prepro sequence for targeting expressed protein for secretion exist as well as an expanded multiple cloning site with 10 unique restriction sites. Optional C-terminal *myc* epitope and polyhistidine tags are included that can be used for recombinant protein detection and purification, respectively. These vectors contain a unique *Bgl*II site 5' to the *AOX1* promoter sequence and a unique *Bam*HI site 3' to the *P. pastoris* "transcription termination" sequence to facilitate the generation of in vitro multimers. Three different versions exist with reading frame shifts between the MCS and C-terminal tag (pPICZ A, B, C) or reading frame shifts between the α-factor prepro sequence and the MCS (pPICZα A, B, C).

3. Methods

3.1. *E. coli* and *P. pastoris* Transformation via Electroporation and Direct Multicopy Selection Using Zeocin

Plasmid DNA was transformed into chemically competent or electro-competent *E. coli* TOP10 cells (Invitrogen; *see* **Note 1**). Selection of Zeocin-resistant *E. coli* transformants is on ½X salt LB + 50 μg/mL Zeocin. *P. pastoris* is transformed by electroporation using the method of Becker and Guarente ***(24)*** (*see* **Note 10**).

Electroporation protocol:

1. Grow 500 mL YPD culture of *P. pastoris* to an OD_{600} = 1.3–1.5.
2. Collect cells by centrifugation at 5000*g*, 4°C, 5 min (centrifugations can be done in 1 × 500 mL centrifuge bottle or split into 2 × 250 mL bottles; *see* **Note 11**).
3. Resuspend cell pellet in 500 mL of ice-cold sterile water.
4. Collect cells by centrifugation at 5000*g*, 4°C, 5 min, as before.
5. Resuspend cell pellet in 250 mL of ice-cold sterile water.
6. Collect cells by centrifugation at 5000*g*, 4°C, 5 min, as before.
7. Resuspend cell pellet in 25 mL of sterile ice-cold 1 *M* sorbitol (transfer to a smaller prechilled centrifuge tube).
8. Collect cells by centrifugation at 5000*g*, 4°C, 5 min.
9. Resuspend cell pellet in 1 mL of sterile ice-cold 1 *M* sorbitol (keep on ice).
10. In a sterile 1.7-mL microfuge tube, combine 80 μL of cells with 10 μL of linearized transforming DNA (1–10 μg total).
11. Let cell/DNA mixture set on ice for 5–10 min.
12. Transfer cell/DNA mixture to an ice-cold 0.2-cm electroporation cuvet.
13. Deliver electric pulse using an Electroporator II device (Invitrogen) set at 200 Ω, 50 μF, and 1500 V (*see* **Note 7**).
14. Immediately after pulse, add 1 mL of sterile ice-cold 1 *M* sorbitol to the cuvet.
15. Transfer mixture to a room-temperature 15-mL Falcon 2059 tube (or equivalent).
16. Incubate tube at 30°C, without shaking, for 1 h.
17. Spread different volumes of cells (10, 25, 50, 100, 200 μL) onto YPDS + 100 μg/mL Zeocin plates for selection of Zeocin-resistant transformants (*see* **Note 12**).
18. Incubate plates at 30°C.
19. Colonies will appear in 2–3 d.
20. Pick well-isolated colonies and subclone (reisolate new independent colonies) on YPDS + 100 μg/mL Zeocin plates (*see* **Note 13**).

Because the *Sh ble* gene product that confers resistance to Zeocin functions stoichiometrically instead of enzymatically, the use of Zeocin hyperresistance selection has proven an efficient means to generate recombinant strains containing multiple plasmid copies. Zeocin hyperresistance often corresponds to multiple plasmid copies, meaning multiple copies of the heterologous gene and the possibility of higher levels of heterologous protein expression. Direct selection of Zeocin-resistant transformants at increasing Zeocin concentration followed by an analysis of plasmid copy number and protein expression levels from clones representing different resistance levels can be used to identify clones that produce higher levels of heterologous protein on a per-cell or per-culture volume.

Zeocin hyperresistant transformants can be selected directly after transformation by electroporation using the protocol just described, followed by plating on media containing higher concentrations of Zeocin. The number of Zeocin-resistant colonies that are generated by electroporation transformation with Zeocin resistance-based plasmids decreases as the concentration of Zeocin increases. For selection of Zeocin hyperresistant transformants, we have plated the transformation mix on YPDS + Zeocin at 100, 500, 1000, and 2000 μg/mL. Representative colonies that arise on each of these different Zeocin concentrations can be screened for plasmid copy number.

The number of colonies generated at each Zeocin concentration can best be modulated by the volume of transformation mix plated, and a range of volumes from 25, 50, 100, to 200 μL is typically sufficient to generate plates that have many well-separated Zeocin-resistant colonies. In general, the number of colonies appearing on plates containing Zeocin at 2000 μg/mL is only ~1% of that appearing on plates containing Zeocin at 100 μg/mL). This suggests that a subpopulation of total Zeocin-resistant transformants is hyperresistant to Zeocin and is likely to represent multicopy events. These clones can be further screened by slot-blot and Southern blot analyses (*see* **Subheading 3.2.**).

Analysis of the Zeocin-resistant phenotype (i.e., assaying for hyperresistance) by patching colonies to plates containing Zeocin at different concentrations does not work, because the relatively thick layer of cells placed on the media surface results in the growth of false-positive (i.e., false hyperresistant) colonies. The Zeocin-resistant phenotype interpretation is sensitive to the number and density of cells being assayed, because dead cells titrate out the available Zeocin from the media, and allow the remaining cells to grow and appear resistant and/or hyperresistant.

Generally, the addition of Zeocin to liquid cultures for growth in subsequent analysis and protein expression experiments has not been necessary for plasmid maintenance. However, an exhaustive analysis of the long-term stability of plasmid multicopy arrays with these vectors, either in the presence or absence of Zeocin in the growth media, has not been carried out.

His$^-$ auxotrophic (*his4*) *P. pastoris* strains transformed to Zeocin resistance with the pPICZ series of plasmids remain His$^-$ auxotrophs. It has been our experience that His$^-$ auxotrophic strains do not do as well in fermentation in minimal media; presumably histidine becomes severely limiting. To address this, an His$^+$ prototrophic strain isogenic to GS115 was generated, called X-33, that is ideally suited for expression in fermentation without supplemental histidine.

3.2. Copy Number Estimation by Quantitative Slot-Blot Analysis

Differential hybridization, i.e., different intensity of hybridization signal based simply on more plasmid copies hybridizing to more labeled probe, can be used as a first screen to identify multicopy strains. An example of this technique is shown in **Fig. 2** and has been used successfully in a number of other cases *(7,19,20)*. Genomic DNA from *P. pastoris* strains to be tested is generated as described in Chapter 6 (*see* also **refs.** *23* and *25*). An appropriate nylon membrane should be chosen to allow immobilization of the DNA with alkaline fixation per manufacturer's instructions. Aliquots of 25 µL of denatured genomic DNA are applied to each slot of a vacuum slot-blot apparatus per manufacturer's instructions. Duplicate filters should be made loading the same amount of DNA in corresponding slots.

1. Combine 50 µL (0.1–1 µg) of genomic DNA with 1.5 mL of DNA denaturing buffer.
2. Place at 100°C for 5 min; fast-cool in ice.
3. Set up slot-blot apparatus per manufacturer's instructions.
4. Apply half of the denatured genomic DNA solution (~0.75 mL) to one slot, repeating for each sample. Apply vacuum to pull DNA onto membrane.
5. After the liquid has been pulled through each slot, and without prolonged drying, add 0.5 mL 0.4 *N* NaOH.
6. After this liquid has been pulled through each slot, remove the filter and place on clean filter paper.
7. Repeat **steps 6–9**, creating a duplicate filter.
8. Air-dry filters for ~20–30 min.
9. Wash filters in a bath of 2X SSC at room temperature for 1 min.
10. Air dry filters for ~30 min. Filters can be stored in sealed bags for several weeks or can be used immediately for hybridization.
11. Prepared slot-blots are prehybridized in separate bags in 25 mL of prehybridization/hybridization buffer/bag, at 60–65°C for 2–4 h.
12. Two probes should be generated, one specific to the *AOX1* gene and the other specific to the *HIS4* gene (*see* **Notes 9** and **14**). Immediately before use, denature probe by heating at 100°C for 5 min.
13. After prehybridization, remove fluid and replace with 5 mL of fresh prehybridization/hybridization buffer that has been prewarmed to 60–65°C.
14. To one bag containing one blot, add *AOX1* probe; to other bag, add *HIS4* probe.
15. Hybridize overnight (16–20 h) at 60–65°C with constant slow shaking.
16. After overnight hybridization, wash blots. A series of increasing stringency washes is recommended. The exact wash stringencies used will vary with the probe used and its melting temperature. For relatively large (200–900 bp) double-strand DNA fragments or for the oligonucleotide probes listed in **Note 14**, the following conditions work well: first wash, 2X SSPE, 0.1% SDS, room temperature, 10 min; second wash, 1X SSPE, 0.1% SDS, 65°C, 30 min; third and fourth washes, 0.1% SSPE, 01.% SDS, 65°C, 30 min each.

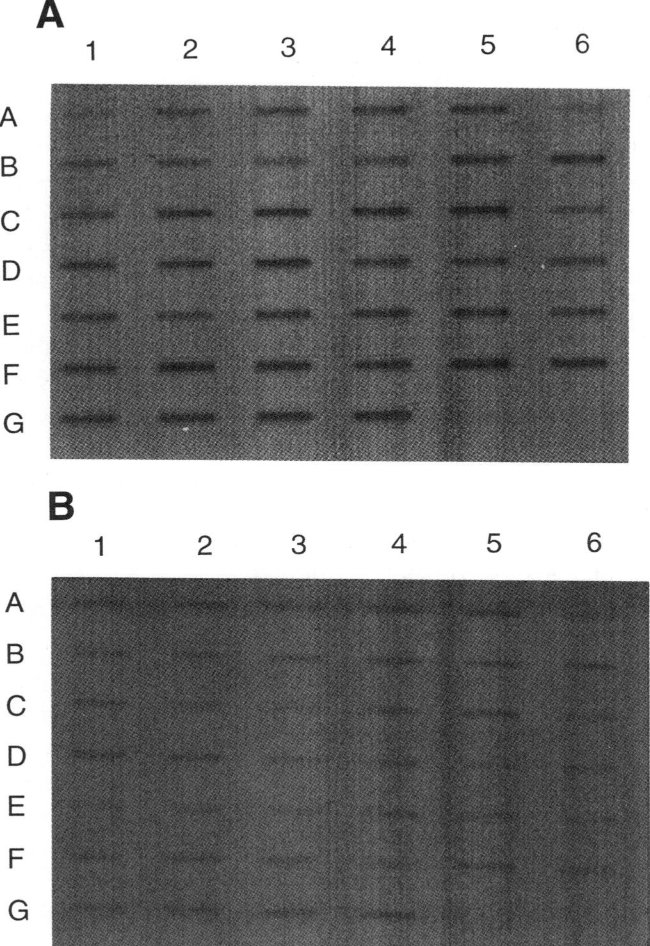

Fig. 2. Plasmid copy number estimation by quantitative slot-blot analysis. Equal volume aliquots of a genomic DNA preparation from different cultures grown to approximately the same density were applied to slots of a slot-blot apparatus. Duplicate filters were made. One filter was probed with a ^{32}P-end-labeled *AOX1*-specific oligonucleotide (**A**), the other with a ^{32}P-end-labeled *HIS4*-specific oligonucleotide (**B**). The intensity of each slot in the *HIS4*-probed blot was uniform (B), indicating equal loading of genomic DNA in each slot since *HIS4* is a known single-copy gene. The pattern of the *AOX1*-probed blot differs from slot to slot. Since the *HIS4* blot indicated that slots were equally loaded with genomic DNA, increased hybridization signal should correspond to increase *AOX1* target sequence, and thereby increase the heterologous gene copy number. Slot 1A: untransformed strain. Slots 2A, 3A, 4A, 5A, 6A, 1B, 2B, 3B, 4B: transformants selected on 100 µg/mL Zeocin. Slots 5B, 6B, 1C, 2C, 3C, 4C, 5C, 6C, 1D, 2D: transformants selected on 500 µg/mL Zeocin. Slots 3D, 4D, 5D, 6D, 1E, 2E, 3E, 4E, 5E, 6E: transformants selected on 1000 µg/mL Zeocin. Slots 1F, 2F, 3F, 4F, 5F, 6F, 1G, 2G, 3G, 4G: transformants selected on 2000 µg/mL Zeocin. Slots 5G, 6G: blank.

17. Washed membranes should be wrapped in plastic wrap and placed against suitable X-ray film. A variety of exposures from 30 min to overnight should be used.

Putative multicopy strains identified using this technique can be further analyzed either by directly comparing levels of heterologous protein produced or by determining the precise plasmid copy number by Southern blot analysis ***(26,27)***. For example, in a typical analysis to determine plasmid copy number via Southern blotting, genomic DNA isolated from strains initially identified as multicopy via slot-blot can be digested with *Eco*RI, separated on a 0.8% agarose gel, Southern blotted, and hybridized with an *AOX1* promoter-specific probe. *Eco*RI-digested genomic DNA from untransformed strain GS115 probed with an *AOX1*-specific probe lights up a 5.5-kb fragment. The banding pattern on a Southern blot of multicopy strains will differ from an untransformed strain in a predictable and representative way. If the genomic DNA is digested with a restriction enzyme that does not cut within the plasmid (or insert), the size of the band on the Southern blot will increase in molecular weight by the unit size of the plasmid (plus insert) for each additional copy present in the genome. If the genomic DNA is digested with a restriction enzyme that does cut within the plasmid (or insert), the banding pattern will change (from one to two or three bands), and the intensity of one of the bands will increase with an increase in plasmid copy number. These differences can be used to distinguish 0, 1, 2, 3, and so forth plasmid copy numbers (*see* **ref. *15*** for specific examples of this analysis).

4. Notes

1. Any commercially available or user-made *E. coli*-competent cells should work if the strain does not contain a bleomycin resistance gene (e.g., Tn10-containing strains).
2. The wild-type phenotype of *P. pastoris* is Zeocin-sensitive.
3. Use of >0.5% NaCl will inhibit the activity of Zeocin.
4. Omission of 1 *M* sorbitol from plates used for selection of Zeocin-resistant transformants can result in a decrease of the number of colonies generated by as much as 10-fold.
5. Phleomycin may function as an alternative to Zeocin, although it has not been tested for transformation or use in plasmid multicopy selection in *P. pastoris*. Phleomycin is an alternate preparation of the same compound and is available from Sigma (St. Louis, MO), but Phleomycin is reported to be five times more potent than Zeocin.
6. Digestion of pPICZ plasmid DNA prior to transformation: Digest for several hours with desired enzyme in optimal buffer; check small aliquot of digestion on agarose gel to confirm that digestion is ≥80% complete; heat-inactivate reaction (if enzyme is subject to heat inactivation); extract with equal volume of phenol:chloroform:isoamyl alcohol (25:24:1); ethanol-precipitate digested plas-

mid DNA; pellet and wash pellet in 80% ethanol; dry pellet; resuspend in sterile water to ~1 μg/μL DNA concentration.

7. The electroporation device used in developing this protocol was the Electroporator II from Invitrogen Corporation (San Diego, CA). Equivalent electroporator and cuvet devices can substitute, such as those available from Bio-Rad.
8. Genomic DNA used for slot-blot analysis does not have to be restriction enzyme digested, but care should be taken to generate a completely dissolved and uniform solution of genomic DNA.
9. Any suitable labeled nucleic acid probe can be used: radioactive (^{32}P) or nonradioactive, oligonucleotide, double-strand DNA fragment, or riboprobe. For the work presented here, ^{32}P-end-labeled oligonucleotides were used as probes for the slot-blots, and random-primed ^{32}P double-strand DNA fragment probes were used for Southern blot analysis. Follow manufacturer's recommendations for probe preparation and use.
10. Chemical means of transformation of *P. pastoris* (e.g., LiCl) are compatible with Zeocin selection. However, spheroplast transformation is not compatible with Zeocin selection. Protocols for transformation of *P. pastoris* by electroporation using other devices are presented in Chapter 3 of this volume.
11. Any combination of suitable centrifuge, rotor, and bottles can be used, e.g., 500-mL bottles in Sorvall GS3 rotor at 5500 rpm = 5000*g*; 250-mL bottles in Sorvall GSA rotor at 5500 rpm = 5000*g*; 35-mL tubes in Sorvall SS-34 rotor at 6500 rpm = 5000*g*.
12. Cells plated at too high a density will reduce the effectiveness of the Zeocin selection, since dead cells will absorb the drug, thereby allowing growth of non-sensitive cells. Plating a range of volumes of the final transformation solutions will yield some plates at optimal densities.
13. Choose well-isolated colonies from regions of plates that do not have a background film of cells. Whenever possible, choose colonies from the lowest density (i.e., lowest volume) plated.
14. Sequence of oligonucleotides used as probes for slot-blots:
    5'-GACTGGTTCCAATTGACAAGC, hybridizes to *AOX1* 5'-UTR.
    5'-CTTGAGAAATTCTGAAGCCG, hybridizes to *HIS4* 5'-UTR.

References

1. Cregg, J. M., Barringer, K. J., and Hessler, A. Y. (1985) *Pichia pastoris* as a host system for transformations. *Mol. Cell. Biol.* **5,** 3376–3385.
2. Cregg, J. M. and Higgins, D. R. (1995) Production of foreign proteins in the yeast *Pichia pastoris*. *Can. J. Botany* **73(Suppl.),** 5981–5987.
3. Cregg, J. M., Vedvick, T. S., and Raschke, W. C. (1993) Recent advances in the expression of foreign genes in *Pichia pastoris*. *Bio/Technology* **11,** 905–910.
4. Higgins, D. R. (1995) Heterologous protein expression in the methylotrophic yeast *Pichia pastoris*, in *Current Protocols in Protein Science*, Suppl. 2 (Wingfield, P. T., ed.), Wiley, New York, pp. 5.7.1–5.7.16.
5. Romanos, M. (1995) Advances in the use of *Pichia pastoris* pastoris for high-level expression. *Curr. Opinion Biotechnol.* **6,** 527–533.

6. Romanos, M. A., Scorer, C. A., and Clare, J. J. (1992) Foreign gene expression in yeast: a review. *Yeast* **8,** 423–488.
7. Higgins, D. R., Busser, K., Comiskey, J., Whittier, P. S., Purcell, T. J., and Hoeffler, J. P. (1998) Small vectors for protein expression in *Pichia pastoris* based on dominant drug resistance with direct multicopy selection and tags for heterologous protein detection and purification. Submitted.
8. Baron, M., Reynes, J. P., Stassi, D., and Tiraby, G. (1992) A selectable bifunctional β-galactosidase::phleomycin-resistance fusion protein as a potential marker for eukaryotic cells. *Gene* **114,** 239–243.
9. Calmels, T., Parriche, M., Burand, H., and Tiraby, G. (1991) High efficiency transformation of *Tolypocladium* geodes conidiospores to phleomycin resistance. *Curr. Genet.* **20,** 309–314.
10. Drocourt, D., Calmels, T. P. G., Reynes, J. P., Baron, M., and Tiraby, G. (1990) Cassettes of the *Streptoalloteichus hindustanus ble* gene for transformation of lower and higher eukaryotes to phleomycin resistance. *Nucleic Acids Res.* **18,** 4009.
11. Gatignol, A., Durand, H., and Tiraby, G. (1988) Bleomycin resistance conferred by a drug-binding protein. *FEBS Lett.* **230,** 171–175.
12. Mulsant, P., Tiraby, G., Kallerhoff, J., and Perret, J. (1989) Phleomycin resistance as a dominant selectable marker in CHO cells. *Somat. Cell Mol. Genet.* **14,** 243–252.
13. Perez, P., Tiraby, G., Kallerhoff, J., and Perret, J. (1989) Phleomycin resistance as a dominant selectable marker for plant cell transformation. *Plant Mol. Biol.* **13,** 365–373.
14. Wenzel, T. J., Migliazza, A., Steensma, H. Y., and VanDenBerg, J. A. (1992) Efficient selection of phleomycin-resistant *Saccharomyces cerevisiae* transformants. *Yeast* **8,** 667,668.
15. Berdy, J. (1980) Bleomycin-type antibiotics, in *Amino Acid and Peptide Antibiotics* (Berdy, J., ed.), CRC, Boca Raton, FL, pp. 459–497.
16. Brierley, R. A., Davis, G. R., and Holtz, G. C. (1994) Production of insulin-like growth factor-1 in methylotrophic yeast cells. US Patent 5,324,639.
17. Clare, J. J., Rayment, F. B., Ballantine, S. P., Sreekrishna, K., and Romanos, M. A. (1991) High-level expression of tetanus toxin fragment c in *Pichia pastoris* strains containing multiple tandem integrations of the gene. *Bio/Technology* **9,** 455–460.
18. Clare, J. J., Romanos, M. A., Rayment, F. B., Rowedder, J. E., Smith, M. A., Payne, M. M., Sreekrishna, K., and Henwood, C. A. (1991) Production of epidermal growth factor in yeast: high-level secretion using *Pichia pastoris* strains containing multiple gene copies. *Gene* **105,** 205–212.
19. Romanos, M. A., Clare, J. J., Beesley, K. M., Rayment, F. B., Ballantine, S. P., Makoff, A. J., Dougan, G., Fairweather, N. F., and Charles, I. G. (1991) Recombinant *Bordetella pertussis* peractin (P69) from the yeast *Pichia pastoris*: high-level production and immunological properties. *Vaccine* **9,** 901–906.
20. Scorer, C. A., Buckholz, R. G., Clare, J. J., and Romanos, M. A. (1993) The intracellular production and secretion of HIV-1 envelope protein in the methylotrophic yeast *Pichia pastoris*. *Gene* **136,** 111–119.

21. Scorer, C. A., Clare, J. J., McCombie, W. R., Romanos, M. A., and Sreekrishna, K. (1994) Rapid selection using G418 of high copy number transformants of *Pichia pastoris* for high-level foreign gene expression. *Bio/Technology* **12,** 181–184.
22. Henikoff, S. and Cohen, E. H. (1984) Sequences responsible for transcription termination on a gene segment in *Saccharomyces cerevisiae. Mol. Cell. Biol.* **4,** 1515–1520.
23. Strathern, J. N. and Higgins, D. R. (1991) Recovery of plasmids from yeast into *Escherichia coli*: shuttle vectors. *Methods Enzymol.* **194,** 319–329.
24. Becker, D. M. and Guarente, L. (1991) High-efficiency transformation of yeast by electroporation. *Methods Enzymol.* **194,** 182–187.
25. Holm, C., Meeks-Wagner, D. W., Fangman, W. L., and Botstein, D. (1986) A rapid, efficient method for isolating DNA from yeast. *Gene* **42,** 169–173.
26. Sambrook, J., Fritsch, E. F., and Maniatis, T. (1989) *Molecular Cloning: A Laboratory Manual*, 2nd ed., Cold Spring Harbor Laboratory, Cold Spring Harbor, NY.
27. Southern, E. M. (1975) Detection of specific sequences among DNA fragments separated by gel electrophoresis. *J. Mol. Biol.* **98,** 503–517.

5

The Generation of Multicopy Recombinant Strains

Mike Romanos, Carol Scorer, Koti Sreekrishna, and Jeff Clare

1. Introduction

The extremely high levels of alcohol oxidase produced from the native *AOX1* gene in *Pichia pastoris* (5–30% of cell protein on induction) suggested that single-copy *AOX1*-promoter expression vectors would be sufficient for efficient foreign gene expression. Therefore, the first strategy adopted for generating recombinant strains was to replace the *AOX1* gene with a single copy of the foreign gene expression cassette (transplacement), since this type of transformant is the most stable. Some of the earliest studies supported this strategy, e.g., expression of β-galactosidase or hepatitis B surface antigen was efficient and was not improved by increasing vector copy number ***(1,2)***.

However, in many subsequent cases, expression levels using single-copy transformants were disappointingly low, and numerous examples have now accumulated (**Table 1**) in which multicopy transformants have been used to increase yields dramatically ***(3–7)***. Indeed, for intracellular expression, vector copy number is usually the most important factor affecting product yield. This was demonstrated in a detailed optimization study of tetanus toxin fragment C expression ***(5)***, which showed that protein levels increased with increasing copy number (**Fig. 1**), although site of integration and Mut phenotype had at most a minor effect. A more recent study of foreign gene mRNA levels suggested that this effect is general ***(8)***. Transformants expressing HIV-1 ENV with increasing vector copy number (from 1 to 12) had foreign gene mRNA levels that increased progressively, exceeding *AOX1* mRNA levels at three copies and greatly exceeding them at 12 copies; a similar result was also shown for fragment C ***(8)***. Transcript level may frequently limit foreign gene expression; therefore, it is logical to maximize copy number routinely. Even if mRNA level is not limiting, a higher transcript level may partially overcome other barriers,

From: *Methods in Molecular Biology, Vol. 103:* Pichia *Protocols*
Edited by: D. R. Higgins and J. M. Cregg © Humana Press Inc., Totowa, NJ

Table 1
Some Examples of High-Level Expression Resulting from Multicopy Integration

Protein expressed	Intracellular or secreted	Copy number	Yield	Increase in yield[a]	Ref.
Tumor necrosis factor	I	>20	25%	200×	*4*
Tetanus toxin fragment C	I	14	27%	6×	*5*
Pertactin	I	21	10%	n/a	*6*
Aprotinin	S	5	0.90 g/L	7×	*11*
Murine EGF	S	19	0.45 g/L	13×	*7*
IGF-1	S	6	0.50 g/L	5×	*10*

[a]Increase in yield over single-copy clones in comparable fermenter inductions. The increases are usually more pronounced in shake-flask inductions.

such as suboptimal 5'-untranslated sequence, mRNA secondary structure, or protein instability. Obviously, these factors still have a significant effect, since final yields of different proteins vary greatly even with multicopy clones. Difference in protein stability is a major factor, and the most likely explanation for the unusually high level of accumulation of alcohol oxidase from a single copy of the gene.

With secreted proteins, the effects of gene dosage are not as simple. There are many examples where product yield has been improved using multiple vector copies (e.g., **refs.** ***9*** and ***10***), and indeed in several cases, the maximal copy number tested was optimal, e.g., with murine EGF ***(7)***. However, it has usually been observed that a too-high copy number reduces yield, i.e., an optimal rather than maximal copy number is required ***(11,12)***. Parallel results have been seen with *Saccharomyces cerevisiae*, where increasing promoter strength can adversely affect the yield of many secreted proteins, and it would appear that less efficiently secreted proteins block the secretory pathway at higher expression levels ***(13)***.

1.1. Types of Multicopy Transformants

Since different types of multicopy transformants can be obtained, we will briefly describe the options available for generating them. Most *P. pastoris* expression vectors are of a similar design (**Fig. 2**) and can integrate into the genome in one of two general ways (depending on where the DNA is cut prior to transformation): by insertion of the entire plasmid at the *HIS4* or *AOX1* locus, or by replacement of *AOX1* by the expression cassette (transplacement). The different types of transformant that can be generated and their properties are summarized in **Table 2**.

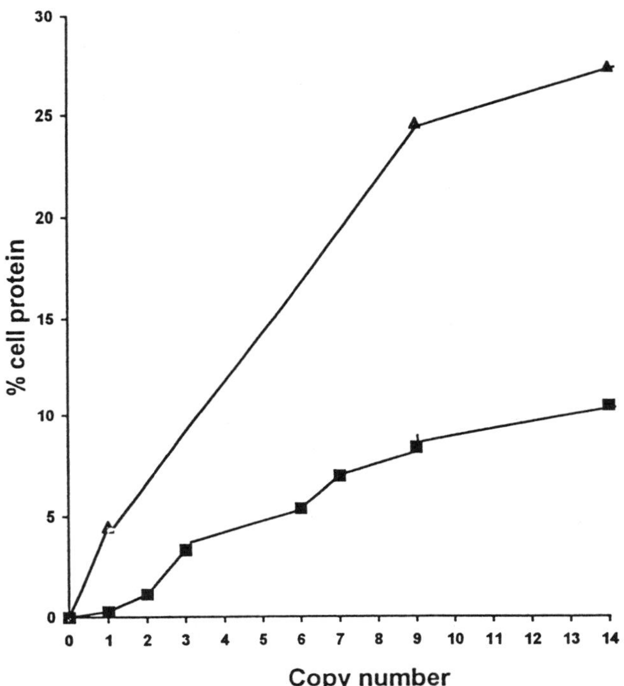

Fig. 1. Correlation of gene dosage with expression level for tetanus toxin fragment C.

1.1.1. Integration at AOX1 or HIS4

Linearization of the vector, by cutting either 5' to the *AOX1* promoter (e.g., at the *Sac*I site) or within the *HIS4* marker, directs integration of the plasmid to the homologous sites in the genome; multicopy transformants (up to 10) can arise from repeated recombination events. Integration at *AOX1*, using *Sac*I digestion, is an efficient straightforward way to generate recombinant clones for expression. We would recommend not using *HIS4* integrants, since they can eliminate the foreign gene by recombination while retaining the His$^+$ phenotype.

1.1.2. Transplacement at AOX1

Digestion to give a DNA fragment with both ends homologous to *AOX1* leads to replacement of genomic *AOX1* (transplacement), generating an *aox1* strain (Mut^s phenotype: Methanol utilization-slow vs the Mut$^+$ phenotype of the wild-type strain). Transplacement theoretically yields only single-copy

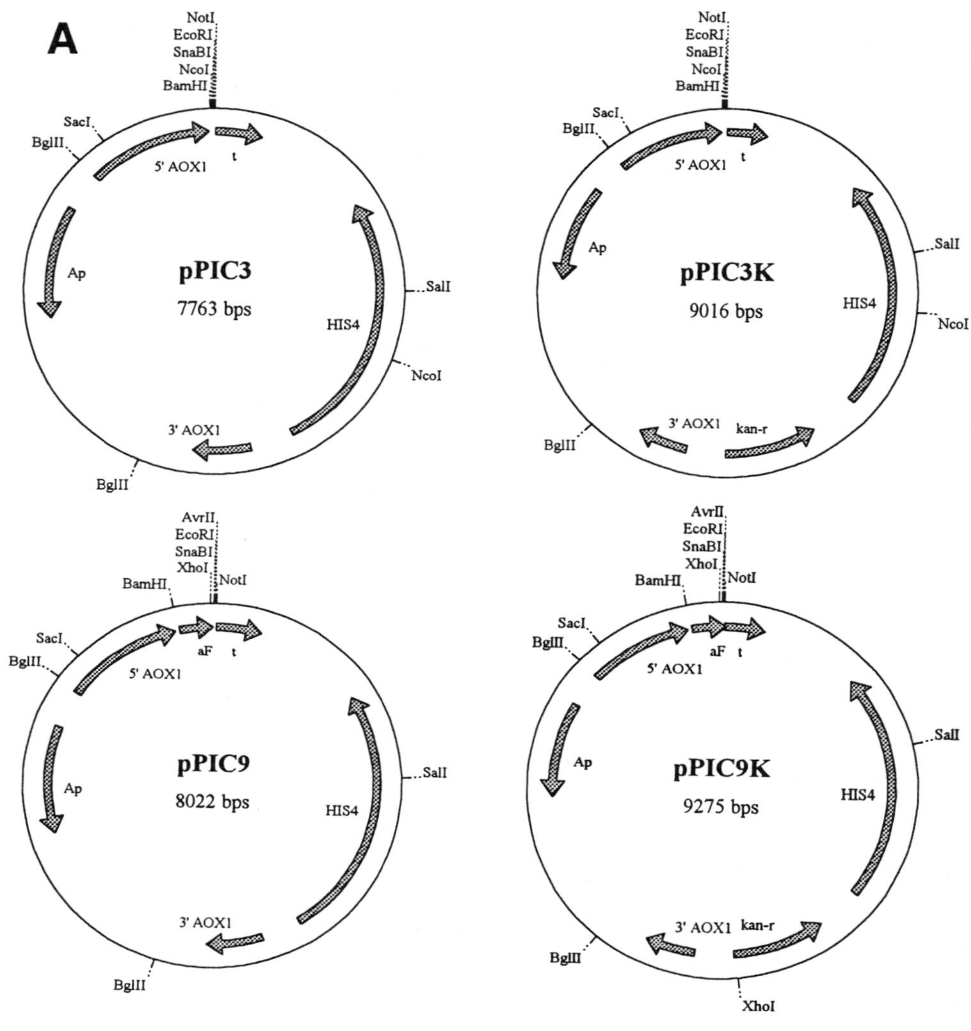

B

pPIC3/3K MCS *Bam*HI *Nco*I *Sna*BI *Eco*RI *Avr*II *Not*I

GGATCCAAACC**ATG**GAAAATACGTAGAATTCCCTAGGGCGGCCGC

pPIC9/9K MCS *Xho*I *Hin*dIII *Sna*BI *Eco*RI *Avr*II *Not*I

CTCGAGAAAAGAGAGGCTGAAGCTTACGTAGAATTCCCTAGGGCGGCCGC

L E K R E A E A

Mut^s transformants. However, expression studies showed an unexpected and extreme clonal variation in yields ***(3–5)***. A detailed analysis ***(5)*** showed that:

1. A high proportion of "transformants" contained no vector and probably represented conversion of the *his4* allele to wild type by recombination with the plasmid-borne *HIS4* gene;
2. Only 5–30% were true transplacements (Mut^s); the remainder had integrations at *AOX1* or *HIS4* and were Mut$^+$;
3. 1–10% of Mut^s transformants had up to 30 integrated copies of the transplacing fragment as tandem head-to-tail repeats; and
4. Multicopy transformants were also found among the Mut$^+$ transformants.

The diversity of events was explained by a mechanism involving in vivo ligation of the *Bgl*II fragment (**ref.** ***5*** and Chapter 13). Although transplacement usually requires the laborious spheroplast transformation technique because of low frequencies, the highly divergent population of transformants it yields can be useful for detailed optimization studies. Also, transplacement may be the only way of obtaining the very high copy numbers that have yielded enormously high levels of intracellular products: 10–30% of cell protein (>4–12 g/L in high-density fermentations) in several examples ***(3–6)***.

1.2. Strategies for Copy Number Optimization

The number of types of recombinant strains that can be generated can make the *P. pastoris* expression system appear complex. However, in the majority of cases, it is not necessary to consider all the options. For routine intracellular expression, we recommend generating *AOX1* integrants, e.g., using a *Sac*I-digested vector and isolating transformants with high vector copy numbers (>5). This can be achieved simply by drug selection using vectors, such as pPIC3K and pPIC9K (**Fig. 2**), that contain the *Tn903 kan*^{r} gene, which confers dose-dependent resistance to the antibiotic G418 in *P. pastoris* ***(8)***.

In some cases, it may be considered that there is further scope for improvement or that a more detailed optimization is required, e.g., prior to scale-up for commercial production of a protein. For intracellular proteins, we would then suggest using transplacement to isolate clones with very high copy numbers (10–30 copies). In cases where the Mut phenotype may also have an effect on yield, e.g., in secretion, transplacement can be used to generate a population of

Fig. 2. *(opposite page)* **(A)** Maps of *P. pastoris* polylinker expression and secretion vectors with and without *kan*^{r} genes for G418-selection. Shown in the maps are the ampicillin resistance (Ap) and kanamycin-resistance (*kan*^{r}) genes, 5' *AOX1* region including promoter, *AOX1* terminator (t), 3' *AOX1*, *HIS4* gene, and *S. cerevisiae* α-factor leader (αF). **(B)** Sequences of polylinkers in pPIC3 and pPIC3K, and pPIC9 and pPIC9K, showing restriction sites and α-factor leader peptide sequence (pPIC9 and 9K).

Table 2
Types of *P. pastoris* Transformant

Host strain	Vector digest[a]	Resulting His$^+$ transformants	Comments
GS115 (Mut$^+$)[a]	*Sac*I (*AOX1* integration)	Vector integrated 5' to genomic *AOX1* gene, i.e., Mut$^+$ phenotype	High-frequency transformation, using either spheroplasting or electroporation; ideal for routine use; multicopy integrants (up to 10 copies) arise at low frequency
	*Sal*I, *Stu*I (*HIS4* integration)	Vector integrated within genomic *his4* locus; Mut$^+$ phenotype	High-frequency transformation, using either spheroplasting or electroporation; note potential to generate His$^+$ pop-outs lacking foreign gene;[b] multicopy integrants (up to 10 copies) arise at low frequency
	*Bgl*II (*AOX1* transplacement)	*Bgl*II fragment replaces genomic *AOX1* gene, generating Mut^s phenotype	Low-frequency transformation, spheroplast method preferable; best method for multicopy clones: 1–10% frequency and up to 30 copies; generates a heterogeneous pool of Mut^s and Mut$^+$ transformants, including some nonexpressers
KM71 (Mut^s)	*Sac*I or *Sal*I/*Stu*I	All transformants Mut^s since *AOX1* already disrupted	Higher transformation frequency than GS115, especially with electroporation

[a]These sites are common to all vectors and can be used generally unless present in the foreign gene.

[b]This problem does not often occur in small-scale cultures.

transformants varying in both copy number and Mut phenotype for detailed comparisons. The best method for screening the resulting transformants is by rapid semiquantitative DNA dot-blot of whole-cell lysates *(6)*, since this can differentiate the very high copy "jackpot" clones, unlike G418 selection.

For secreted proteins, initial studies can be carried out with single-copy transformants in order to assess the efficiency of secretion and authenticity of the product. However, for yield optimization, it is preferable to test transformants with a range of copy numbers, which can be isolated by G418 selection *(8)* or semiquantitative DNA dot-blot screening *(6)*. These can then be tested for expression empirically, or the precise vector copy numbers can first be determined in order to identify a series with progressively increasing copy numbers for a rigorous optimization. An alternative to isolating multicopy transformants is to use a plasmid that can generate tandem copies of the expression cassette (up to 8) in vitro *(10,14)*. This method has the potential advantage of predetermining the foreign gene copy number prior to transformation (although recombination can occur to alter the number of copies), but it also has the potential disadvantage that several sequential DNA cloning steps are required and these become increasingly difficult.

In cases where a good assay is available for the secreted protein, optimization can be carried out empirically by initial high-throughput expression screening of transformants without any prior knowledge of copy number *(15)*. For a more thorough recent review of factors affecting expression efficiency and optimization strategy, *see* **ref. *16***.

1.3. Overview of the Methods in This Chapter

This chapter will describe two general methods for identifying multicopy transformants: G418 selection and semiquantitative DNA dot-blot screening. G418 selection can very rapidly yield transformants with >5 integrated vector copies, which are usually sufficient for high-level expression. This method is most valuable when used with electroporation, but it cannot differentiate transformants with copy numbers of >5.

Semi-quantitative dot-blot screening can be applied following any type of transformation; we have found it very useful in screening for high copy number transformants following spheroplast transformations (which yield the highest proportion of multicopy clones). In this method, several hundred transformants are picked and cultured in 96-well plates. Fixed volumes of cells are then filtered onto nitrocellulose and lysed prior to DNA hybridization.

The other methods described here are for the purification of DNA from *P. pastoris* transformants, and its analysis by Southern blotting or quantitative DNA dot-blot hybridization in order to determine vector copy number and structure of the integrated DNA.

2. Materials

2.1. *P. pastoris* Strains

The host strains that are most commonly used are GS115 (*his4*) and KM71 (*his4 aox1::ARG4*). Transformation frequencies are two- to fourfold higher for KM71 by all methods used. The more recently available protease-deficient strains (e.g., SMD1168; *his4 pep4*) are finding increasing use in reducing proteolytic cleavage of secreted proteins. KM71 (*his4 aox1*) and other strains that have an inactive *AOX1* gene have an Mut^s phenotype and are fundamentally different in that they grow very slowly on methanol as a carbon source, e.g., during induction.

2.2. *P. pastoris* Vectors

P. pastoris has no stable episomal vectors and, therefore, integrating vectors are used. Most use *HIS4* as a selectable marker and have the same general organization. **Figure 2** illustrates the typical intracellular and α-factor secretion vectors pPIC3 and pPIC9, respectively, and their counterparts pPIC3K and pPIC9K for G418 selection. pPIC3 and pPIC3K polylinkers are designed for ligation of the 5'-end of the foreign gene to the *Bam*HI or *Nco*I (ATG) site, and the 3'-end to one of the remaining sites. The secretion vectors pPIC9 and pPIC9K comprise a *S. cerevisiae* α-factor leader sequence with an engineered *Xho*I site just 5' to the DNA encoding the KEX2 cleavage site. Foreign genes are ligated to this site, but the KEX2 cleavage site must be reconstructed using a synthetic oligonucleotide. The *Xho*I site in pPIC9 is unique, but in pPIC9K, there is an additional site located in the kan^r gene. Versions of pPIC9K are now available in which the *Xho*I cloning site is unique. By cloning at the *Hin*dIII site (not unique) of the polylinker, the GluAla spacers of the native prepro-α-factor peptide may also be included.

2.3. Secondary G418 Selection

1. Hemocytometer.
2. Geneticin, G418 sulfate, Gibco. Stock = 50 mg/mL in H_2O, filter-sterilized and stored at –20°C.
3. YPD agar plates containing different concentrations of G418, from 0.25–2.0 mg/mL. Prepare by adding stock G418 solution to molten YPD agar (1% yeast extract, 2% peptone, 2% glucose, 1.5% agar; autoclave and cool to 50°C) prior to pouring the plate. Store plates at 4°C for up to 2 mo.

2.4. Spheroplast Transformation

1. Standard reagents for *P. pastoris* spheroplast transformation and subsequent selection for prototrophs (Invitrogen kit or *see* Chapter 3).
2. Twenty micrograms of digested vector DNA (e.g., digested with *Bgl*II or *Sac*I) for each transformation. Phenol-extract or purify using a proprietary resin, ethanol-precipitate and redissolve in 5 µL of TE buffer.

3. Sterile 50-mL centrifuge tubes.
4. Sterile H_2O.
5. Hemocytometer.
6. YNBD agar plates (1.34% yeast nitrogen base with ammonium sulfate without amino acids, 4 × 10^{-5}% biotin, 1% glucose, 1.5% agar).

2.5. Picking, Growth and Storage of Transformants in Microtiter Wells

1. Sterile toothpicks or loops.
2. Sterile 96-well microtiter plates with lids (Falcon, Fisher Scientific, Pittsburgh, PA).
3. YPD broth.
4. 75% (v/v) glycerol: Autoclave and store at room temperature.
5. Multichannel pipeters (20 and 200 μL max volume).

2.6. Determination of Mut Phenotype

1. Sterile 96-well microtiter plates with lids (Nunclon).
2. Multichannel pipeters (20 and 200 μL max volume).
3. 15-cm Disposable plastic Petri dishes.
4. 15-cm YNBD agar plates: 1.34% yeast nitrogen base with ammonium sulfate without amino acids, 4 × 10^{-5}% biotin, 1% glucose, 1.5% agar. Air-dry in sterile cabinet before use.
5. 15-cm YNB*M* agar plates: 1.34% yeast nitrogen base with ammonium sulfate without amino acids, 4 × 10^{-5}% biotin, 0.5% methanol, 1.5% agar. Air-dry in sterile cabinet before use.

2.7. Cell Lysis and Filter Hybridization

1. 96-Well dot-blot manifold (Schleicher and Schuell "Minifold," Keene, NH).
2. Vacuum pump.
3. Sterile 96-well microtiter plates with lids (Falcon).
4. Multichannel pipeters (20 and 200 μL max volume).
5. YPD broth (1% yeast extract, 2% peptone, 2% glucose)
6. Sterile toothpicks or loops.
7. Nitrocellulose sheets cut to fit dot-blot manifold (e.g., Schleicher and Schuell).
8. Solution I: 50 m*M* EDTA, 2.5% 2-mercaptoethanol, pH 9.0.
9. Solution II: 3 mg/mL zymolyase 100T. Make up fresh.
10. Solution III: 0.1 *M* NaOH, 1.5 *M* NaCl.
11. Solution IV: 2X SSC, stock 20X SSC + 3 *M* NaCl, 0.3 *M* sodium citrate, pH 7.0.
12. Whatman 3M*M* paper, cut to the same size as the nitrocellulose.
13. Hybridization probe for foreign gene. Label to high specific activity by ^{32}P by random-primed synthesis.

2.8. Preparation of Chromosomal DNA

1. YPD broth.
2. Sterile 250-mL conical flasks.

3. SCED: SCE containing 10 m*M* dithiothreitol. Prepare by adding 1 *M* dithiothreitol stock (store at –20°C) to SCE buffer (1 *M* sorbitol, 10 m*M* sodium citrate, 10 m*M* EDTA; autoclave and store at room temperature) immediately before use.
4. Zymolyase-100T, 3 mg/mL. Make fresh.
5. 1% Sodium dodecyl sulfate (SDS).
6. 5 *M* potassium acetate, pH 8.9.
7. TE buffer: 10 m*M* Tris-HCl, pH 7.4, 0.1 m*M* EDTA.
8. Ribonuclease A (DNase-free), 10 mg/mL. Store at –20°C.
9. Isopropanol.

2.9. Southern Blot Analysis

1. Standard equipment and materials for DNA agarose electrophoresis, Southern transfer, hybridization, and autoradiography.
2. Random-primed, ^{32}P-labeled *HIS4* hybridization probe (e.g., 0.6-kb *Kpn*I fragment from pPIC3). Optional: *AOX1* probe (1.4-kb *Cla*I fragment from pPIC3).
3. PhosphorImager and cassettes.

2.10. Quantitative DNA Dot-Blot Hybridization

1. 96-Well dot-blot manifold (Schleicher and Schuell "Minifold" or similar).
2. Vacuum pump.
3. Nitrocellulose or nylon filters cut to fit dot-blot manifold.
4. 1 *M* NaOH.
5. 1 *M* HCl.
6. 20X SSC stock solution.
7. Random-primed, ^{32}P-labeled hybridization probes for the foreign gene and for *ARG4* (0.75-kb *Bgl*II fragment, e.g., from plasmid pYM32).
8. PhosphorImager and cassettes.

3. Methods

3.1. G418 Selection Of Multicopy Transformants

G418 selection greatly facilitates the rapid isolation of multicopy transformants for routine laboratory use *(8)*. Because selection is from a large pool of transformants, it can be used with electroporation, which gives a low frequency of multicopy clones. This avoids the need for the much more laborious spheroplast transformation method (which gives a much higher frequency of multicopy clones). However, G418 cannot readily be used for direct selection because the level of resistance shows a dependence on plating density above 10^5 cells/plate; cells must be selected at $<10^5$ cells/plate. Therefore, His$^+$ colonies are first selected from an electroporation (total ~10^8 cells) on minimal medium, and these are then pooled for a secondary selection on G418.

In practice, the electroporation/G418-selection method works best using *AOX1* integration (e.g., with *Sac*I-digested vector) in the strain KM71 because of the two- to fourfold higher transformation frequencies (e.g., 1000–2000 colonies/μg) achieved with this strain. Using electroporation, we have found the frequency of transplacement, i.e., transformation using *Bgl*II-cut vector, to be 20-fold lower than this value.

1. Use standard electroporation protocol (Chapter 3) to transform KM71 or GS115 with digested vector.
2. Pool His^+ colonies selected on YNBD plates by resuspending them in 2–3 mL sterile YPD using a spreader.
3. Determine the cell density of the resuspended transformants using a hemocytometer (typical values may be 5–50 × 10^8 cells/mL). Alternatively measure the A_{600} (1 U ~ 5 × 10^7 cells/mL). Dilute the cell suspensions to a final density of ~10^6 cells/mL (typically 1000-fold).
4. Spread 100 μL of cell suspension (~10^5 cells) on each 9-cm YPD agar plate containing G418. Use a range of concentrations of G418, e.g., 0.5, 1.0, 1.5, or 2.0 mg/mL.
5. Incubate at 30°C. Resistant colonies should take 2–5 d to appear. The numbers of resistant colonies drop off steeply above 0.5–1.0 mg/mL G418.
6. Pick several colonies for further analysis, e.g., a total of six including colonies selected on the highest G418 concentration and a range of colonies from lower G418 concentrations (*see* **Note 1**). Store as glycerols stocks at –70°C.
7. Analyze protein production from these representative transformants.
8. Determine the vector copy number in each transformant using Southern blot analysis and/or quantitative DNA dot-blot hybridization (optional, *see below*).

3.2. Semiquantitative DNA Dot-Blot Screening of Lysed Cells for Multicopy Transformants

In order to identify rare multicopy integrants, we have used a semiquantitative DNA dot-blot technique applied to whole-cell lysates to screen several hundred transformants grown in microtiter wells ***(6)***. This method involves many more manipulations than G418 selection, but it does give a good initial indication of the vector copy number. Therefore, we have found it most useful in screening spheroplast transformants, which contain the highest proportion of multicopy clones and, in particular, for distinguishing the very high copy number "jackpot" clones (up to 30 copies) that can arise during transplacement.

3.2.1. Spheroplast Transformation

1. Use the standard spheroplast transformation protocol (*see* Chapter 3) to transform GS115 with 20-μL digested vector (*see* **Note 2**).
2. Resuspend the regenerated His^+ spheroplasts. Carefully remove the soft agarose top layer containing the colonies using a sterile scalpel. Place in a 50-mL sterile tube and vortex vigorously, adding 5–10 mL of sterile H_2O until the agarose is well broken. Allow the agarose to settle, and remove the supernatant containing liberated cells. Determine the cell density using a hemocytometer or by spectrophotometry (1 A_{600} = 5 × 10^7 cells/mL).

3. Replate the transformants to single colonies on a YNBD agar plate (500–1000 colonies/plate). Incubate at 30°C for 2–3 d until colonies appear (*see* **Note 3**).

3.2.2. Picking, Growth, and Storage of Transformants in Microtiter Wells

1. Pick individual colonies of His$^+$ transformants from YNBD agar, using sterile toothpicks or loops, into individual wells containing 200 μL YPD broth in a 96-well microtiter (*see* **Note 4**).
2. Culture for 2 d in a 30°C static incubator.
3. Resuspend the cells that have settled to the bottom of each well by pipeting up and down using a multichannel pipeter. Subculture by transferring 5 μL of each individual microculture using a multichannel pipeter into 200 μL fresh YPD in a second microtiter dish. Incubate for 1 d at 30°C and keep as a working stock (*see* **Note 5**).
4. Use the original microtiter well cultures to prepare frozen stocks (*see* **Note 6**). Add 50 μL of sterile 75% glycerol to each well using a multichannel pipeter, and mix by pipeting up and down. Tape down the lid of the microtiter plate, and freeze by gradual cooling (overnight at –20°C and then store at –70°C).

3.2.3. Determination of Mut Phenotype of Gridded Transformants

We have found that the following method of determining Mut phenotype gives very low rates of misclassification compared to replica plating. It is also extremely convenient once the transformants are available as cultures in microtiter wells.

1. Resuspend the cells in the microtiter well cultures by repeated pipeting up and down using a multichannel pipeter.
2. Withdraw 2 μL of resuspended culture from each well, and spot on 15-cm Petri dishes containing YNBD and YNB*M* agar. Recreate the original 96-well grid on each plate. This volume of culture should be absorbed into the agar medium (*see* **Note 7**).
3. Incubate the plates at 30°C for 2 d. All transformants should show an even circular patch of growth on YNBD, but only Mut$^+$ transformants should show significant growth on methanol as carbon-source (YNBM; *see* **Note 8**).

3.2.4. Cell Lysis and Filter Hybridization

1. Resuspend the working microcultures thoroughly by pipeting up and down using a multichannel pipeter, and then remove 50-μL aliquots and filter onto nitrocellulose held in a 96-well filter manifold (*see* **Note 9**). Air-dry the filter.
2. In trays, set up stacks of four sheets of Whatman 3MM paper saturated with each of solutions I–IV. Place each nitrocellulose filter sequentially on the stacks as follows:
   a. Solution I, 15 min at room temperature;
   b. Solution II, 3–4 h at 37°C (cover tray to minimize evaporation; *see* **Note 10**);
   c. Solution III, 5 min at room temperature; and
   d. Solution IV, 2 × 5 min at room temperature.
3. Air-dry the filters and bake under vacuum at 80°C.

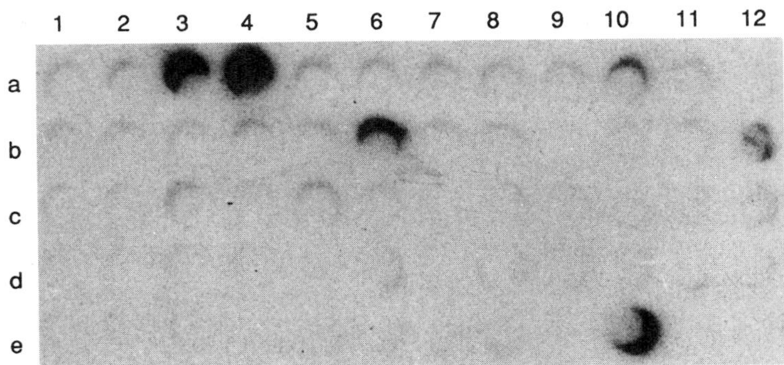

Fig. 3. DNA dot-blot screen for multicopy transformants. The result of a screen of GS115 Muts transformants using pPIC3 expressing the pertactin gene is shown ***(6)***. Faint crescents are positive signals arising from single- or low-copy number transformants, whereas rare transformants with very high copy numbers (12–30 copies) gave a much more intense signal. GS115 (e11) was included as a negative hybridization control.

4. Hybridize to a probe consisting of the foreign gene DNA labeled with ^{32}P using the random-priming method. Multicopy integrants can be identified using autoradiography by their stronger hybridization signal compared with the majority of single-copy integrants (*see* **Fig. 3**). They arise typically at a frequency of 1–10% (*see* **Note 11**).

3.3. Quantitative Determination of Vector Copy Number Using Purified DNA

In order to determine the absolute vector copy number, it is necessary to isolate total DNA from transformed strains (*see* **Note 12**). Southern blot analysis is then used to determine the chromosomal structure of integrated vector DNA (site of integration, *AOX1* or *HIS4*, vector copy number, and whether *AOX1* gene replacement has occurred). However, for accurate determination of absolute copy number in multicopy strains, the quantitative DNA dot-blot method is preferable. This method requires a known single-copy transformant, identified by Southern blot analysis, as a control.

3.3.1. Preparation of Total DNA from P. pastoris

1. Prepare an overnight culture of each transformant (including a GS115 or KM71 control) by inoculating 50 mL of YPD broth and incubating at 30°C.
2. Harvest the cells by low-speed centrifugation (1500*g* for 10 min).
3. Wash the cells by resuspending in 10 mL sterile H_2O and repelletting.
4. Resuspend the cells in 5 mL SCED. Add 100 µL of 5 mg/mL zymolyase, and incubate for 1 h at 37°C to spheroplast the cells.
5. Add 5 mL 1% SDS, mix gently by inversion, and place on ice for 5 min.

6. Add 3.75 mL 5 *M* potassium acetate, pH 8.9, and mix gently.
7. Remove the precipitate by centrifugation (10,000*g* for 20 min).
8. Add 2 vol of ethanol to the supernatant and mix. Leave at room temperature for 15 min, and then harvest by centrifugation (10,000*g* for 20 min).
9. Drain the pellet well then redissolve in 3 mL TE; this may require gentle agitation for several hours.
10. Remove undissolved material by centrifugation (10,000*g* for 10 min).
11. Add 15 μL 10 mg/mL RNase and incubate at 37°C for 30 min to digest RNA.
12. Add 1 vol of isopropanol slowly to form a separate layer. Mix gently so that the DNA forms a fibrous precipitate at the interface.
13. Remove the DNA by "spooling" around a glass rod; then place into 250 μL TE (*see* **Note 13**). Store at 4°C.

3.3.2. Southern Blot Analysis

Different combinations of restriction enzyme digestions and hybridization probes can be used. However, for standard *P. pastoris* vectors, it is always informative to digest the chromosomal DNA with *Bgl*II and hybridize with a *HIS4*-specific probe. The resulting Southern blots usually show two bands: a 2.7-kb band corresponding to the *HIS4* gene and a larger band equivalent to the *HIS4*-containing *Bgl*II fragment of the expression vector (**Fig. 4**). A direct comparison of the intensity of these two bands gives the copy number of the foreign gene; the majority of transformants are expected to be single-copy and should give two bands of similar intensity. (This is not an ideal method for absolute copy number determination, since the efficiency of transfer of DNA depends on the fragment size, and in some transformants, the *Bgl*II sites can be lost so that the larger fragment has a much larger size than expected; *see* **Fig. 4**.) For a detailed example of the analysis of transformants using *Bgl*II-cut DNA, *see* **ref. 5** and Chapter 13.

1. Digest 20 μL of DNA from each transformant with *Bgl*II in a 50-μL reaction. Include a control digest of GS115 DNA.
2. Load each digest onto a 0.6% agarose gel alongside radiolabeled markers, separate by electrophoresis, and transfer to a nylon membrane using a standard Southern blotting protocol.
3. Prepare a ^{32}P-labeled *HIS4* probe (e.g., the 0.6-kb *Kpn*I fragment of *HIS4*) by random primed labeling.
4. Hybridize using a standard protocol and visualize by autoradiography.
5. In order to determine the vector copy number, compare the intensity of the native *HIS4* gene fragment (2.7 kb) to that of the vector *Bgl*II fragment using a PhosphorImager (*see* **Note 14**).

3.3.3. Quantitative Dot-Blot Hybridization Using Purified DNA for Vector Copy Number

This method compares the hybridization signal for the foreign gene in equal amounts of chromosomal DNA from different transformants. It is totally

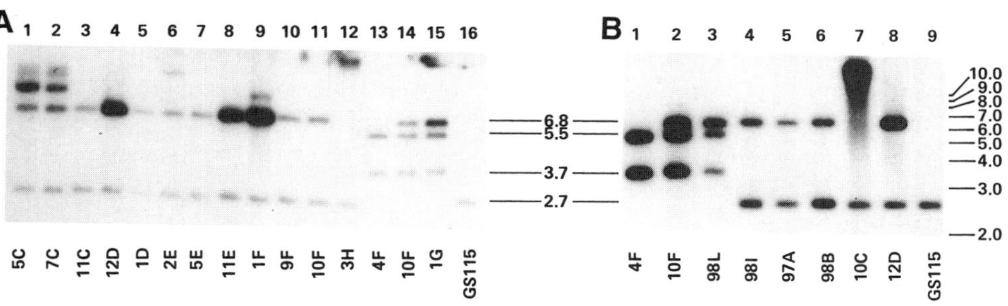

Fig. 4. Southern blot analysis of chromosomal DNA from different *P. pastoris* Mut^s **(A)** or Mut$^+$ **(B)** integrant clones expressing tetanus toxin fragment C from a transplacement transformation. The DNA was cut with *Bgl*II, and the filter hybridized with an *HIS4* probe. For the Mut^s transformants, this gave the expected two bands: a 2.7-kb band containing the chromosomal *his4* gene and a 6.8-kb band containing the fragment C gene and the wild-type *HIS4* gene. In single-copy transformants (A, lanes 3, 5, 7, 11), these have the same intensity, whereas in multicopy transformants, the intensity of the 6.8-kb band is greater. Densitometer scanning of the blot or quantitative dot-blot hybridization analysis was used to determine the precise copy numbers. (In some clones [A, lanes 1, 2, 6, 9, 12; B, lane 7] bands larger than 6.8 kb can be seen, which are consistent in size with loss of one or more *Bgl*II sites, possibly owing to exonucleolytic trimming prior to integration.) Identical patterns are usually seen for Mut$^+$ transformants from "transplacements," except that some clones have vector fragments integrated at *his4*, so that the 2.7-kb fragment is increased in size (B, lanes 1, 2, 3). For further analysis of these clones, *see* Chapter 13.

dependent on having an accurate control for the amount of DNA, e.g., hybridization to the *ARG4* gene (any other chromosomal DNA could be used), and having a known single-copy transformant to calibrate the copy number in other transformants.

1. Denature 10 μL of chromosomal DNA by adding 2.5 μL 1 *M* NaOH and incubating for 15 min. Place on ice, neutralize by adding 2.5 μL of 1 *M* HCl, and then dilute with 175 μL 10X SSC.
2. For each transformant under test, apply three 30-μL aliquots to three separate wells of a dot-blot manifold fitted with a nitrocellulose filter (presoaked in 10X SSC), and use vacuum to draw the samples through. Wash through with 300 μL 2X SSC, remove the filter, air-dry, and bake under vacuum at 80°C for 2 h. Repeat this entire procedure with a separate filter in order to prepare a duplicate.
3. Generate two probes: one specific to the foreign gene and one to *ARG4* (e.g., 0.75-kb *Bgl*II fragment from pYM32), which is used as an internal control for the amount of chromosomal DNA. Radiolabel each probe by incorporation of α-[^{32}P]-ATP by random-primed synthesis.

4. Following hybridization and washing under standard conditions, quantitate the radiolabel in each spot using a PhosphorImager (*see* **Note 17**), averaging the triplicate spots. Determine the ratio of foreign gene/*ARG4* counts for each transformant. Divide the ratio for each transformant by that for the known single-copy strain in order to determine the absolute foreign gene copy number.

4. Notes

1. Although G418 selection is effective in isolating multicopy transformants, experience has shown that it is not possible to make comparisons of copy number for transformants selected at the same G418 concentration in different experiments. The reason for this is not clear. In addition, we find that transformants with five to seven copies can be selected at the highest practical concentrations of G418 (2–4 mg/mL), so it is not possible to differentiate higher copy numbers.
2. We have found the proportion of multicopy transformants to be variable, although it is likely to correlate with the amount of DNA used in transformations. We use 10 or 20 μg/transformation. The number of transformants is typically: 5000 for *Sac*I-cut DNA, 1000 for *Bgl*II-cut DNA; the same amount of the replicating vector pHILA1 typically yields 10,000–20,000 transformants. However, spheroplast transformation of *P. pastoris* requires some skill, and frequencies can be much lower. Critical factors are the degree of spheroplasting and the source of polyethylene glycol used (we have found better results with Fisher Scientific grade PEG 3350 than with other sources).
3. If applying G418 selection following spheroplast transformation, then plate out on G418-containing YPD agar as described in **Subheading 3.3.**
4. We test from one to four 96-well plates/transformation. Ideally, reserve wells for the following strains: untransformed GS115, a Mut^s strain (e.g., KM71), and a known single-copy transformant. These will be useful as controls in subsequent analysis of the Mut phenotype and copy number.
5. Working stock cultures in 96-well plates remain viable for 2 wk at 4°C. However, note that only freshly grown cells are readily digested by zymolyase.
6. This is a very convenient way of storing the transformants in gridded format; clones can remain viable for several years.
7. Plates should be well dried so that the spotted aliquot of cells is rapidly absorbed.
8. Mut^s transformants will also show growth on prolonged incubation.
9. For this procedure, it is essential that the cultures are both of uniform density and sufficiently fresh to be readily lysed using zymolyase prior to dot-blot hybridization. Older cells are not readily lysed. Therefore, either use fresh 96-well cultures, or subculture again by inoculating 200 μL of YPD with 5 μL of stored culture and incubating overnight.
10. Failure to lyse the cells can occur at this step if lower concentrations of zymolyase (we use 3 mg/mL) are used. If necessary, the filters can be incubated overnight.
11. Transformations using *Bgl*II-cut DNA appear to give a higher proportion of multicopy transformants compared to, for example, *Sac*I-cut DNA. A significant proportion (~25% of total) of the His$^+$ colonies are not true transformants and do

not express foreign protein, but prescreening for the Mut^s phenotype eliminates all nonexpressers. Since Mut^s transformants comprise 5–25% of the starting number, and the frequency of "jackpot" clones among them is typically 1–10%, as few as 1/1000 of the original number of His$^+$ colonies are multicopy Mut^s clones. Very high copy number "jackpot" clones also occur in the Mut$^+$ transformants. We have limited evidence that cotransformation with uncut plasmid increases the frequency of multicopy clones among the Mut^s population.

12. A potential rapid alternative method for copy number determination is described in Chapter 13. This method measures the degree of G418 sensitivity of a *kan*^{r} vector transformant in a growth-inhibition assay. We found that sensitivity correlated accurately with vector copy number in a random selection of transformants, allowing us to identify rapidly a complete 1–12 copy number series for optimization of HIV-1 ENV expression ***(8)***. Note that this is different from G418 selection, which did not give this clear correlation.
13. The spooling technique, which yields very pure DNA, may not work if the DNA is too dilute. As an alternative, extract with phenol/chloroform, then chloroform/isoamyl alcohol after RNase treatment, and then ethanol-precipitate as usual.
14. An alternative to a PhosphorImager is to excise the band or spot, and quantitate by scintillation counting, or else to autoradiograph and then quantitate by densitometry.

References

1. Cregg, J. M. and Madden, K. N. (1987) Development of transformation systems and construction of methanol-utilization-defective mutants of *Pichia pastoris* by gene disruption, in *Biological Research on Industrial Yeast*, vol. II (Stewart, G. G., Russell, I., Klein, R. D., and Hiebsch, R. R., eds.), CRC, Boca Raton, FL, pp. 1–18.
2. Cregg, J. M., Tschopp, J. F., Stillman, C., Siegel, R., Akong, M., Craig, W. S., Buckholz, R. G., Madden, K. R., Kellaris, P. A., Davies, G. R., Smiley, B. L., Cruze, J., Torregrossa, R., Velicelebi, G., and Thill, G. P. (1987) High-level expression and efficient assembly of hepatitis B surface antigen in the methylotrophic yeast *Pichia pastoris*. *Bio/Technology* **5,** 479–485.
3. Sreekrishna, K., Nelles, L., Potenz, R., Cruze, J., Mazzaferro, P., Fish, W., Fuke, M., Holden, K., Phelps, D., Wood, P., and Parker, K. (1989) High-level expression, purification, and characterization of recombinant human tumor necrosis factor synthesized in the methylotrophic yeast *Pichia pastoris*. *Biochemistry* **28,** 4117–4125.
4. Sreekrishna, K., Potenz. R. B., Cruze, J. A., McCombie, W. R., Parker, K. A., Nelles, L., Mazzaferro, P. K., Holden, K. A., Harrison, R. G., Wood, P. J., Phelps, D. A., Hubbard, C. E., and Fuke, M. (1988) High level expression of heterologous proteins in methylotrophic yeast *Pichia pastoris*. *J. Basic Microbiol.* **28,** 265–278.
5. Clare, J. J., Rayment, F. B., Ballantine, S. P., Sreekrishna, K., and Romanos, M. A. (1991) High-level expression of tetanus toxin fragment C in *Pichia pastoris* strains containing multiple tandem integrations of the gene. *Bio/Technology* **9,** 455–460.

6. Romanos, M. A., Clare, J. J., Beesley, K. M. Rayment, F. B., Ballantine, S. P., Makoff, A. J., Dougan, G., Fairweather, N. F., and Charles, I. G. (1991) Recombinant *Bordetella pertussis* pertactin (P69) from the yeast *Pichia pastoris*: high-level production and immunological properties. *Vaccine* **9,** 901–906.
7. Clare, J. J., Romanos, M. A., Rayment, F. B., Rowedder, J. E., Smith, M. A., Payne, M. M., Sreekrishna, K., and Henwood, C. A. (1991) Production of mouse epidermal growth factor in yeast: high-level secretion using *Pichia pastoris* strains containing multiple gene copies. *Gene* **105,** 205–212.
8. Scorer, C. A., Clare, J. J., McCombie, W. R., Romanos, M. A., and Sreekrishna, K. (1994) Rapid selection using G418 of high copy number transformants of *Pichia pastoris* for high-level foreign gene expression. *Bio/Technology* **12,** 181–184.
9. Siegel, R. S., Buckholz, R. G., Thill, G. P., and Wondrack, L. M. (1990) Production of epidermal growth factor in methylotrophic yeast cells. International Patent Application No. WO 90/10697.
10. Brierley, R. A., Davis, G. R., and Holtz, G. C. (1994) Production of insulin-like growth factor-1 in methylotrophic yeast cells. US Patent No. 5,324,639.
11. Thill, G. P., Davis, G. R., Stillman, C., Holtz, G., Brierley, R., Engel, M., Buckholtz, R., Kinney, J., Provow, S., Vedvick, T., and Seigel, R. S. (1990) Positive and negative effects of multi-copy integrated expression vectors on protein expression in *Pichia pastoris*, in *Proceedings of the 6th International Symposium on Genetics of Microorganisms*, vol. II (Heslot, H., Davies, J., Florent, J., Bobichon, L., Durand, G., and Penasse, L., eds.), Société Française de Microbiologie, Paris, pp. 477–490.
12. Scorer, C. A., Buckholz, R. G., Clare, J. J., and Romanos, M. A. (1993) The intracellular production and secretion of HIV-1 envelope protein in the methylotrophic yeast *Pichia pastoris*. *Gene* **136,** 111–119.
13. Romanos, M. A., Scorer, C. A., and Clare, J. J. (1992) Foreign gene expression in yeast: a review. *Yeast* **8,** 423–488.
14. Cregg, J. M., Vedvick, T. S., and Raschke, W. C. (1993) Recent advances in the expression of foreign genes in *Pichia pastoris*. *Bio/Technology* **11,** 905–910.
15. Laroche, Y., Storme, V., De Meutter, J., Messens, J., and Lauwereys, M. (1994) High-level secretion and very efficient isotopic labeling of tick anticoagulant peptide (TAP) expressed in the methylotrophic yeast, *Pichia pastoris*. *Bio/Technology* **12,** 1119–1124.
16. Romanos, M. A. (1995) Advances in the use of *Pichia pastoris* for high level gene expression. *Curr. Opinion Biotechnol.* **6,** 527–533.

6

Isolation of Nucleic Acids

David J. Miles, Katie Busser, Christine Stalder, and David R. Higgins

1. Introduction

Thorough characterization of recombinant strains of *Pichia pastoris* requires analysis of genomic DNA and RNA. The protocols in this chapter describe methods for isolation of high-quality genomic DNA suitable for hybridization experiments, crude genomic DNA suitable for use as a PCR template, and total RNA suitable for a wide range of transcriptional analyses. They are based on standard protocols developed for *Saccharomyces cerevisiae* and have been modified by ourselves or others for *P. pastoris*.

Southern analysis of genomic DNA has been relied on to determine details of recombinant *P. pastoris* strains. Using appropriate probes and restriction enzymes, the presence and genomic location of the gene of interest can be verified. The number of copies of the gene integrated can be estimated by an increase in molecular weight of the DNA fragment released by restriction enzymes with sites flanking the integration site or by increase in signal intensity by slot-blot hybridization. In addition, it is possible to distinguish strains in which the gene of interest has been incorporated into the genome via a single integration event transplacing host sequences or by gene replacement removing host sequences (e.g., replacement of *AOX1* coding sequences with the gene of interest to produce a strain with an Mut^{s} phenotype).

PCR is a powerful tool that can be used to screen many clones rapidly at once for the presence of heterologous sequences as well as to verify strain phenotypes (e.g., Mut^{+} vs Mut^{s} strains) and has become the method of choice for selection of clones following transformation for subsequent expression analysis.

RNA hybridization (Northern) experiments may be helpful in understanding why some constructions do not express a heterologous protein in *P. pastoris*.

From: *Methods in Molecular Biology, Vol. 103:* Pichia *Protocols*
Edited by: D. R. Higgins and J. M. Cregg © Humana Press Inc., Totowa, NJ

For example, higher eukaryotic genes frequently contain AT-rich regions that may be recognized by 3'-RNA processing enzymes in *P. pastoris* resulting in truncation of the mRNA *(1)*. Such aberrant mRNA processing sites can be identified by Northern blotting and corrected by introducing silent changes in the DNA to break up AT-rich regions.

2. Materials

2.1. Isolation of High-Quality Genomic DNA

1. YEPD liquid medium: 1% yeast extract, 2% peptone, 2% dextrose. Medium should be autoclaved for 20 min and is stable at room temperature for months.
2. Deionized, water sterilized by autoclaving for 20 min.
3. SZB: 2.3 mL of sterile, deionized water, 12.5 mL of 2 *M* sorbitol, 2.5 mL of 1 *M* sodium citrate, and 7.5 mL of 0.2 *M* EDTA. Sterilize by autoclaving for 20 min and store at room temperature. Just prior to use, add 15 mg of Zymolyase 60,000 (Seikagaiku America, Inc., Rockville, MD) and 0.2 mL 2-mercaptoethanol.
4. Triton X-100.
5. SDS-TE: 5 mL of 10% SDS, 2.5 mL of 1 *M* Tris-HCl, pH 8.0, 0.125 mL of 0.2 *M* EDTA, and 17.4 mL of sterile, deionized water. This solution is stable at room temperature for several months.
6. 5 *M* potassium acetate sterilized by autoclaving for 20 min.
7. 5 *M* ammonium acetate sterilized by autoclaving for 20 min.
8. 100% Isopropanol chilled to –20°C.
9. 80% Ethanol chilled to –20°C.
10. 10 mg/mL RNaseA.
11. 5 *M* sodium chloride acetate sterilized by autoclaving for 20 min.
12. 20 mg/mL pronase (Sigma, St. Louis, MO).
13. Buffer-saturated phenol *(2)*: Melt phenol crystals in a 68°C water bath. Hydroxyquinoline may be added to 0.1% to slow oxidation, partially inhibit RNases, and act as a metal ion chelator. Adjust the pH of the phenol to >7.8 by repeated extraction with an equal volume of 0.1 *M* Tris-HCl, pH 8.0 (measure pH using pH paper). When the phenol is equilibrated to >pH 7.8, add 0.1 vol of 0.1 *M* Tris-HCl, pH 8.0, containing 0.2% 2-mercaptoethanol. The equilibrated phenol is stable at 4°C in a dark container for ~1 mo.
14. A 25:24:1 mixture of buffer saturated phenol, chloroform, and isoamyl alcohol.
15. 100% Ethanol chilled to –20°C.
16. TE: 10 m*M* Tris-HCl, pH 8.0, 1 m*M* EDTA sterilized by autoclaving for 20 min.
17. 65°C Water bath.
18. 37°C Water bath.
19. A microscope and slides.

2.2. Preparation of Crude Genomic DNA for PCR

1. Sterile, deionized water.
2. 1 mg/mL Zymolyase 60,000 (Seikagaiku America, Inc.) in sterile, deionized water. This solution should be prepared fresh just prior to use.

3. 30°C water bath.
4. 10X buffer for PCR (usually provided by vendor with *Taq* DNA polymerase).
5. 25 m*M* magnesium chloride sterilized by passing through 0.22-μ filter.
6. 100 m*M* dNTP mix: Combine equal volumes of 100 m*M* solutions of dATP, dCTP, dGTP, and TTP to make a final solution that is 25 m*M* with respect to each dNTP.
7. Primers for analytical PCR: In general, these do not need to be purified by HPLC or gel electrophoresis. They are maintained in sterile, deionized water at 100 ng/μL.
8. Mineral oil may be required depending on the type of thermocycler used for PCR.
9. *Taq* DNA polymerase diluted to 0.16 U/μL in 1X reaction buffer.
10. 30°C Water bath.
11. Thermocycler.

2.3. Isolation of Total RNA

1. YEPD liquid medium.
2. AE buffer: 50 m*M* sodium acetate, pH 5.3, 10 m*M* EDTA. This buffer should be sterilized before use by autoclaving for 20 min.
3. 10% SDS.
4. Buffer-saturated phenol (*see* **Subheading 2.1., item 13**).
5. A dry-ice/ethanol bath.
6. A 1:1 mixture of buffer-saturated phenol and chloroform.
7. 3 *M* sodium acetate, pH 5.3, sterilized by autoclaving for 20 min.
8. 100% Ethanol chilled to –20°C.
9. 80% Ethanol chilled to –20°C.
10. Diethyl pyrocarbonate- (DEPC) treated water: add 0.1% DEPC to deionized water, and stir at room temperature overnight. Inactivate the DEPC by autoclaving for 20 min.
11. 250-mL sterile shake flasks with cap or stopper.
12. Sterile microcentrifuge tubes.
13. 65°C Water bath.

3. Methods

3.1. Isolation of High-Quality Genomic DNA

The protocol described here for the isolation of high-quality genomic DNA from *P. pastoris* was developed originally for *S. cerevisiae* ***(3)***. The typical yield is between 30 and 100 μg from a 10-mL culture sample. **Figure 1A** shows 3 μg of undigested, genomic DNA electrophoresed on a 1% agarose gel (lane 1). Nuclease contamination and shear are factors that can reduce the quality of genomic DNA. *See* **Note 1** for a discussion.

1. Inoculate 50 mL of liquid YEPD with a single colony in a 250-mL sterile, shake flask. Incubate overnight at 30°C with vigorous shaking. By morning, the culture should have reached stationary phase and have a density of ~10^8 cells/mL (~5–10 OD_{600} U/mL).

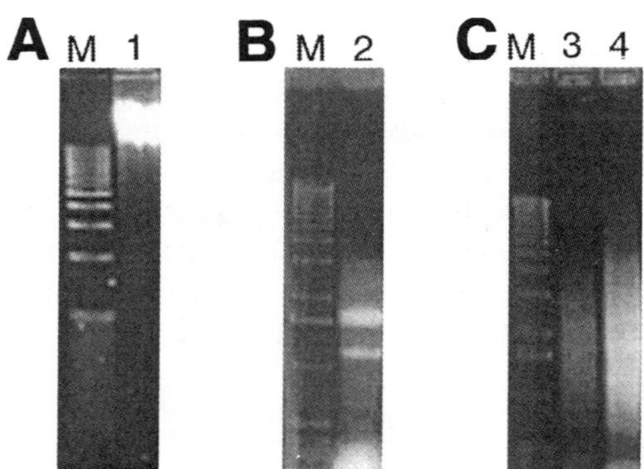

Fig. 1. Lane A1, 3 μg of undigested genomic DNA was electrophoresed through a 1% agarose-TAE gel. Lane B2, 25% of the total RNA isolated from a 10-mL culture was electrophoresed through a 1% agarose-TAE gel. Lanes C3 and C4, polyadenylated RNAs isolated using a commercial kit (Invitrogen Corporation, Carlsbad, CA) were electrophoresed through a 1% agarose-TAE gel. M lanes contain DNA mol-wt markers.

2. Transfer 10 mL of the overnight culture to a disposable centrifuge tube (e.g., Falcon 2059 [Fisher Scientific, Pittsburgh, PA]) and centrifuge at 2500*g* (usually about 3500 rpm in a tabletop centrifuge) for 5 min.
3. Resuspend cells in 1 mL sterile, deionized water and transfer them to a sterile microcentrifuge tube.
4. Remove the cell walls to generate spheroplasts as follows. Centrifuge the resuspended cells for 20–30 s in a microcentrifuge and resuspend in 0.15 mL freshly prepared SZB. Incubate in a 37°C water bath for 30–40 min, shaking occasionally. After 20 min, check the cells for spheroplasting by adding 5 μL of the sample to a drop of 5% Triton X-100 on a microscope slide and check for lysed cells. These will appear to be like "ghosts" relative to unlysed cells. When >80% of the cells are spheroplasted (i.e., lyse in Triton X-100), proceed to the next step.
5. Lyse the spheroplasts by adding 0.15 mL of the SDS-TE solution. Vortex briefly and incubate in a 65°C water bath for 5–10 min.
6. Precipitate proteins and cellular debris by adding 0.15 mL of 5 *M* potassium acetate. Vortex briefly and incubate on ice for 30–45 min.
7. Centrifuge the precipitated material away from nucleic acids in solution in a microcentrifuge at maximum speed for 15 min. Transfer 0.3–0.4 mL of the supernatant to a fresh microcentrifuge tube. Discard the pellet.
8. Precipitate the nucleic acids by adding 0.2 mL of 5 *M* ammonium acetate and 1 mL of 100% isopropanol chilled to –20°C. Hold at –20°C for 10 min or overnight. This is a good point to hold the procedure if necessary.
9. Centrifuge the nucleic acids at maximum speed for 5 min at 4°C. Pour off and discard the supernatant, recentrifuge for a few seconds, and remove trace supernatant with a micropipet. Dry the pellet lightly under a vacuum.

10. Redissolve the pellet in 200 µL of sterile, deionized water. Remove most of the RNA by adding 2 µL of 10 mg/mL RNaseA. Incubate in a 37°C water bath for 10 min.
11. The next two steps help remove any remaining proteins associated with the genomic DNA. Add 4 µL of 5 *M* sodium chloride and 2 µL of 20 mg/mL pronase. Incubate in a 37°C water bath for 10 min.
12. Extract the sample by adding 200 µL of buffer-saturated phenol and vortex for 20 s. Separate the aqueous and organic phases by centrifugation at maximum speed for 5 min. Transfer the upper aqueous phase to a fresh microcentrifuge tube.
13. Extract the sample again with 200 µL of 25:24:1 phenol:chloroform:isoamyl alcohol. Vortex for 20–30 s, and centrifuge at maximum speed for 5 min. Again, carefully transfer the upper, aqueous phase to a fresh microcentrifuge tube. Leave any apparent interface behind with the organic phase.
14. Add 0.1 vol (20 µL) of 3 *M* sodium acetate, pH 5.3, and 2.5 vol (500 µL) of 100% ethanol chilled to –20°C. Hold at –80°C for 30 min or overnight at –20°C. Centrifuge the DNA at maximum speed for 10 min at 4°C. Rinse the pellet with 80% ethanol chilled to –20°C, and centrifuge again for 5 min. Discard the supernatant, and centrifuge again for a few seconds. Remove trace amounts of the supernatant with a micropipet. Allow to air-dry for a few minutes at room temperature.
15. Dissolve the DNA pellet in 25–50 µL of TE, and store at 4°C (*see* **Note 2** for comments on storage of genomic DNA).

3.2. Preparation of Crude Genomic DNA Template for PCR

This method, first described by Linder et al. ***(4)***, is particularly useful because single colonies or liquid culture can be used as starting material. It relies on treatment with zymolyase to weaken the cell wall followed by a freeze-and-heat cycle to release the genomic DNA in a crude lysate. This template is suitable for PCR using "hot-start" conditions where the reaction components are heated to 95°C before addition of the thermostable DNA polymerase. In addition, the procedure does not require an extraction step, so the investigator is not exposed to phenol or chloroform and can process many samples in parallel.

1. Streak cells for single colonies on selective agar medium. With a sterile toothpick or pipet tip, pick a single, well-isolated colony, and resuspend the cells in 10 µL sterile, deionized water in a microcentrifuge tube (*see* **Note 3** for use of liquid cultures as starting material).
2. Add 5 µL of 1 mg/mL Zymolyase 60,000 to each cell suspension. Vortex briefly and incubate in a 30°C water bath for 10 min.
3. Freeze the samples at –80°C for 10 min or immerse closed tubes containing the samples in liquid nitrogen for 1 min.
4. Prepare a hot-start PCR in a tube appropriate for the thermocycler to be used. A 50-µL reaction should include 5 µL of 10X reaction buffer, 5 µL of 25 m*M* magnesium chloride, 1 µL of 100 m*M* dNTP mix, 1 µL of each primer (100 ng/µL), and 27 µL of sterile, deionized water.

5. Add 5 μL of each crude lysate sample to the reaction mixture, vortex briefly, and overlay with 20 μL of mineral oil if required for the thermocycler.
6. Place the PCR samples into the thermocycler, and heat them to 95°C for 5 min.
7. Add 5 μL of 0.16 U/μL of *Taq* DNA polymerase to each reaction.
8. Use a temperature cycling program that repeats the following steps 30 times: 95°C for 1 min to denature the template, 54°C for 1 min to anneal the primers and template, 72°C for 1 min to extend the primers along the template. Add a final extension step to the thermocycler program to incubate the samples at 72°C for 10 min (*see* **Note 4** for cycling parameters for longer templates).
9. Analyze 10 μL of each PCR sample by agarose gel electrophoresis. These samples may be stored at 4°C for a few days.

3.3. Isolation of Total RNA

This total RNA isolation method was originally described by Sherman et al. *(5)* and is a standard method for *S. cerevisiae*. Glass beads are used to break open the yeast cells in the presence of phenol to inactivate ribonucleases (RNases). Schmitt et al. *(6)* modified this protocol substituting heating and freezing for the glass beads to break open the cells. This rapid protocol is suitable for isolation of total RNA from many samples in parallel. Typical yields of RNA range from 5–10 mg from a 10 mL culture. In **Fig. 1B**, 25% of the RNA isolated was separated on a 1% agarose gel. Total RNA may be used for isolation of polyadenylated RNA with a variety of commercially available kits (*see* **Note 5**). As a rule, the yield of polyadenylated RNA is between 1 and 3% of the total RNA.

1. Inoculate 50 mL of liquid YEPD with a single colony in a 250-mL sterile shake flask. Incubate overnight at 30°C with vigorous shaking. By morning, the culture should have reached stationary phase and have a density of ~10^8 cells/mL (~5–10 OD_{600} U/mL).
2. Transfer 10 mL of the overnight culture to a disposable centrifuge tube (e.g., Falcon 2059 [Fisher Scientific, Pittsburgh, PA]) and pellet the cells by centrifugation at 2500*g* (usually about 3500 rpm in a tabletop centrifuge) for 5 min.
3. Resuspend the pelleted cells in 400 μL AE buffer, and transfer them to a sterile microcentrifuge tube.
4. Add 40 μL of 10% SDS and vortex for ~20 s.
5. Add 500 μL of buffer-saturated phenol and vortex for ~20 s.
6. Place the mixture in a 65°C water bath for 4 min.
7. Quickly transfer the mixture from the 65°C water bath to a dry-ice/ethanol bath for 1 min or until phenol crystals begin to appear.
8. Separate the aqueous and organic phases by centrifugation for 2 min in a microcentrifuge at maximum speed.
9. Transfer the upper aqueous phase to a fresh tube.
10. Add an equal volume of 1:1 phenol-chloroform and vortex for ~20 s.
11. Again, separate the aqueous and organic phases by centrifugation for 2 min in a microcentrifuge at maximum speed.

12. Transfer the upper aqueous phase to another fresh tube.
13. Precipitate the RNA by adding 0.1 vol (40 μL) of 3 *M* sodium acetate, pH 5.3, and 2.5 vol (1 mL) of 100% ethanol chilled to –20°C. Mix by inverting several times.
14. Centrifuge the precipitated RNA at 4°C in a microcentrifuge for 15 min at maximum speed.
15. Remove the ethanol, and wash the pellet with 80% ethanol chilled to –20°C.
16. Centrifuge for 5 min at maximum speed, and discard the ethanol.
17. If polyA RNA is to be purified from this total RNA using a commercial kit, resuspend the pellet in 100 μL of the buffer recommended for that kit. For total RNA, resuspend in 20 μL of DEPC-treated water. Total RNA may be stored indefinitely at –80°C (*see* **Note 6**).

4. Notes

1. Significant sources of DNases are reagents, containers, water baths, and the hands of the investigator. Use of ultrapure or molecular biology-grade reagents and sterile, deionized water will minimize DNase contamination in reagents. When possible, use fresh plastic-ware instead of glassware as containers, and make sure that samples are closed tightly when placed in water baths. The investigator should wear gloves at all times to avoid contaminating the samples. In addition, high-quality genomic DNA should be treated very gently to minimize damage by shearing forces. The product should never be vortexed and wide-bore pipets should be used whenever possible.
2. Repeated freeze–thaw cycles will break genomic DNA reducing the overall quality of hybridization studies, such as Southern analysis. DNA prepared with this procedure is quite stable when stored at 4°C.
3. Liquid cultures grown to stationary phase in YEPD or a selective medium may be used as a starting material for this procedure. Add 1 μL of the liquid culture to 9 μL sterile, deionized water in a microcentrifuge tube and continue at **Subheading 3.2., step 2**. Colonies on plates may be stored at 4°C for several weeks before analysis. In contrast, cells in liquid cultures stored at 4°C without shaking must be used within 3 d.
4. The cycling parameters given are adequate for targets up to 2000 bp. It is necessary to increase the extension time at 72°C for longer template targets.
5. Total RNA isolated with this procedure makes an excellent starting material for many commercially available kits for isolation of polyadenylated RNA. An example of polyadenylated RNA isolated from total RNA is shown in **Fig. 1C**. Samples in lanes 3 and 4 were isolated from total RNA using Invitrogen's FastTrack™ Kit.
6. Long-term storage at –80°C minimizes the activity of RNases that may remain after the procedure. Repeated freeze-thaw cycles should be avoided but generally are not as damaging to RNA as they are to large, genomic DNA.

References

1. Scorer, C. A., Buckholz, R. G., Clare, J. J., and Romanos, M. A. (1993) The intracellular production and secretion of HIV-1 envelope protein in the methylotrophic yeast *Pichia pastoris*. *Gene* **136,** 111–119.

2. Sambrook, J., Fritsch, E. F., and Maniatis, T. (1989) *Molecular Cloning. A Laboratory Manual.* Cold Spring Harbor Laboratory, Cold Spring Harbor, NY.
3. Strathern, J. N. and Higgins, D. R. (1991) Recovery of plasmids from yeast into *Escherichia coli* shuttle vectors, in *Guide to Yeast Genetics and Molecular Biology* (Gutherie, C. and Fink, G. R., eds.), Academic, San Diego, pp. 319–329.
4. Linder, S., Schliwa, M., and Kube-Granderath, E. (1996) Direct PCR screening of *Pichia pastoris* clones. *BioTechniques* **20,** 980–982.
5. Sherman, F., Fink, G. R., and Hicks, J. B. (1983) *Methods in Yeast Genetics Laboratory Manual.* Cold Spring Harbor Laboratory, Cold Spring Harbor, NY.
6. Schmitt, M. E., Brown, T. A., and Trumpower, B. L. (1990) A rapid and simple method for preparation of RNA from *Saccharomyces cerevisiae. Nucleic Acids Res.* **18,** 3091,3092.

7

Generation of Protease-Deficient Strains and Their Use in Heterologous Protein Expression

Martin A. G. Gleeson, Christopher E. White, David P. Meininger, and Elizabeth A. Komives

1. Introduction

In optimizing expression of heterologous protein, the issue of proteolysis is often an important factor, since many peptides and proteins are susceptible to degradation by proteases produced in the host organism. In such cases even if the protein product is expressed at high levels, overall yield can be drastically reduced through proteolysis, which not only reduces the amount of intact material but also complicates the recovery process. The use of protease-deficient strains has been shown to be a successful approach to improve the yield of fully active, expressed proteins in both *Saccharomyces cerevisiae* and *Escherichia coli* *(1,2)*. Proteolysis can occur during expression or during the first stages of purification. Certainly, expression of intracellular proteins, which requires cell lysis for purification, will result in their exposure to proteases. Similarly, secreted proteins will be exposed to vacuolar proteases through cell lysis that occurs during growth in the fermentor. Consequently, production of many proteins, particularly heterologously expressed proteins, depends on the amount of proteolytic activity they are exposed to during expression and purification, and the use of strains that are deficient in proteases can significantly improve overall yields.

The proteolytic activities of the yeast *S. cerevisiae* have been characterized in detail *(3)*, and they appear to be similar in *Pichia pastoris*. Most proteolytic activities in the yeast cell are contained within membrane-bound vacuoles. Proteinase A, a vacuolar, aspartyl protease, is encoded by the *S. cerevisiae PEP4* gene, which is capable of self-activation, as well as subsequent activation of additional vacuolar proteases, such as carboxypeptidase Y and proteinase B.

From: *Methods in Molecular Biology, Vol. 103:* Pichia *Protocols*
Edited by: D. R. Higgins and J. M. Cregg © Humana Press Inc., Totowa, NJ

Carboxypeptidase Y appears to be completely inactive prior to proteinase A-mediated proteolytic activation. Proteinase B, encoded by the *PRB1* gene in *S. cerevisiae*, is approx 50% bioactive in its precursor form, prior to activation by proteinase A. Consequently, a strain deficient in protease A is also lacking carboxypeptidase Y activity as well as partially deficient in protease B activity. In generating the *P. pastoris* protease-deficient strains, comparisons showed that *P. pastoris* had a similar genotype and phenotype with regard to the *PEP4* gene. Comparing *pep4* strains of *S. cerevisiae* to *P. pastoris*, similar reduction in the activities of carboxypeptidase Y and proteinase A are seen ***(4)***. As with the protease-deficient strains of *S. cerevisiae*, the protease-deficient strains of *P. pastoris* are not as robust as the wild-type strains, and require greater care in growth and storage.

In this chapter, two protocols are described. The first is a protocol for re-engineering an expression strain, converting it to a *pep4* genotype. An example of the utility of this approach is given in Chapter 11, where the overall productivity of an IGF-expressing strain was significantly enhanced in a *pep4* strain. The second protocol describes expression of secreted proteins from the *pep4* strain, SMD1168. For expression of thrombomodulin fragments, utilization of the *pep4* strain of *P. pastoris* has proven to be the only method for large-scale production of active protein. No thrombomodulin activity was detected when expressed in a wild-type host. The C-terminal "tail" of the protein is essential for activity, and it is thought that carboxypeptidase Y activity is responsible for loss of activity in wild-type cultures. Improvements in yield of 5–10-fold for proteins expressed in the *pep4* strain include nitrate reductase (N. Crawford, unpublished) and a fusion protein composed of coagulation factor X and the cellulose-binding protein (M. Guarna, unpublished). In other recently described work, Weiss et al. utilized the SMD1163 strain to produce the 5-HT5A receptor. These researchers report improvements in yield of 28-fold ***(5)***. Thus, the utility of the *pep4* strain for production of protease sensitive proteins is well established.

2. Materials

2.1. Yeast Strains

1. *P. pastoris*: GS115 (*his4*) and SMD1168 (*his4 ura3 pep4*::*URA3*). Both are available from Invitrogen (Carlsbad, CA).
2. *S. cerevisiae*: DBY747 (control *PEP4* strain) (a *his3-D1 leu2–3 leu2–112 ura3–52 trp1–289 gal^s, can1 CUP*; 20B12 (control *pep4* strain) (a *pep4–3 trp1*).
3. Both are available from the yeast genetic stock Culture Collection.

2.2. Plasmids

The expression plasmid pPIC9K(tm45) consists of the gene for a thrombomodulin fragment cloned into the expression plasmid pPIC9K ***(6)***. The disruption plasmid pDR421 contains an internal portion of the *P. pastoris PEP4* gene, which, when introduced into the *P. pastoris* genome at the *PEP4* locus, results in two incomplete and nonfunctional copies of the *PEP4* gene. To construct pDR421, the *URA3* gene of *P. pastoris* was cloned into a pUC19 vector carrying a 450-base pair internal portion of the *PEP4* gene (**Fig. 1**). This plasmid contains a bacterial origin of replication and an ampicillin resistance gene for routine subculture and maintenance in *E. coli*. Selection for transformation in *P. pastoris* is provided by the *URA3* gene. There is a unique *Bgl*II site in the middle of the *PEP4*, fragment which is convenient for linearizing the plasmid to direct integration into the *PEP4* locus of *P. pastoris*.

2.3. Media and Culture Vessels

1. YPD: 1% yeast extract, 2% peptone, 2% dextrose. For solid media, add 2% agar.
2. SD: 0.67% yeast nitrogen base (without amino acids), 2% dextrose, 2% agar.
3. 5-FOA plates: 0.67% yeast nitrogen base (without amino acids), 2% dextrose, 750 mg/L 5-fluoroorotic acid (PCR Inc., Gainsville, FL), 48 mg/L uracil, 2% agar.
4. MM (minimal growth medium): 20 m*M* potassium phosphate, pH 6.0, 10% casamino acids, and 10% yeast nitrogen base (both from Difco Laboratories, Detroit, MI), and 0.4 μg/L biotin from Fisher Biotech (Pittsburgh, PA). The medium is supplemented with 1–2% glycerol or 2% methanol as carbon source. A rich medium containing 1% yeast extract, 2% peptone, and 2% glucose is supplemented with 15% glycerol for frozen storage of yeast strains. All carbon sources are from Fisher Chemicals.
5. Shake-flask cultures are grown in 0.2–1-L volumes in 4-L baffled flasks for maximum aeration. Small cultures are grown in borosilicate glass culture tubes (25 × 150 mm) covered with four layers of cheesecloth (all from Fisher).
6. Fermentation is in a BioFlo 3000 fermentor (New Brunswick Scientific, Edison, NJ).
7. Basal salts medium: 13.3 mL phosphoric acid, 14.3 g potassium sulfate, 2.3 g calcium sulfate·$2H_2O$, 11.7 g magnesium sulfate·$7H_2O$, 40 mL glycerol, 3.9 g potassium hydroxide/L, 4 mL/L PTM salts (*see* **step 8**), and 4 mL/L biotin solution (25 mg biotin to 100 mL of 0.02 *M* NaOH). Ammonium hydroxide, which is used as the base as well as the nitrogen source in the fermentation, is from Fisher Chemicals.
8. PTM salts: 2 g cupric sulfate·$5H_2O$, 0.08 g sodium iodide, 3 g manganese sulfate, 0.2 g sodium molybdate·$2H_2O$, 0.02 g boric acid, 0.5 g cobalt chloride, 7 g zinc chloride, 22 g ferrous sulfate·$7H_2O$, and 5 mL sulfuric acid/L of H_2O (*see* **Note 1**).
9. The fermentation requires periodic addition of small amounts of Antifoam 289 (Sigma, St. Louis, MO).

2.4. Protein Purification

Purification of the protein is by adsorption onto QAE Sephadex A-120 (Sigma) equilibrated with 50 m*M* morpholinesulfonic acid (Fisher Chemicals).

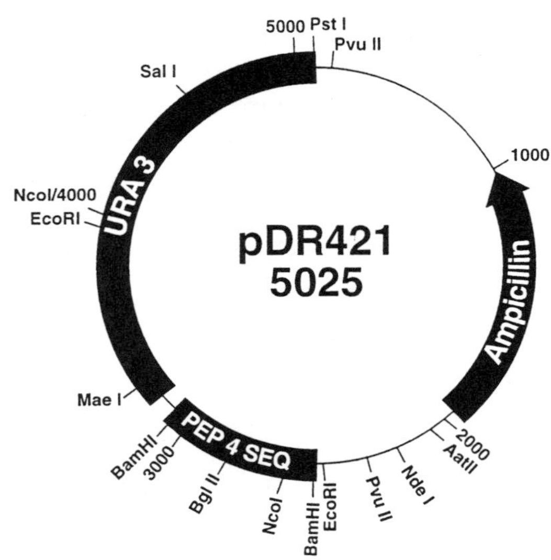

Fig. 1. Restriction map of the *PEP4* disruption vector pDR421.

2.5. Transformation

1. The transforming DNA (15 µg of fresh maxi prep) is linearized using 15 U of the restriction enzyme, *Bgl*II, from New England Biolabs (Beverly, MA) and purified with GENECLEAN (Bio 101 Inc., Vista, CA).
2. Spheroplast transformation requires Zymolyase T20 from ICN Chemicals (Costa Mesa, CA) and D-sorbitol and dithiothreitol, both from Fisher Chemicals.

2.6. Southern Blot

1. 10X TE buffer (for cell breakage): 10 m*M* Tris-HCl, 10 m*M* EDTA, pH 7.4.
2. Cells are placed in 12 × 75-mm glass tubes containing 0.5 g of 425–600 mm acid-washed glass beads from Sigma.
3. Preparation of the probe and detection is done using the GENIUS kit (Boehringer–Mannheim, Indianapolis, IN).

2.7. Carboxypeptidase Y APNE Overlay Assay

1. YPD plates.
2. Overlay reagent: 40% dimethylformamide, 0.6% agar, 1.2 mg/mL *N*-acetyl D/L-phenylalanine β-naphthyl ester (APNE) from Sigma Chemicals. Prepare fresh and hold molten agar solution at 50°C.
3. Fast Garnet GBC salt (5 mg/mL) from Sigma in 100 m*M* Tris-HCl, pH 7.4.

2.8. Protease A Assay

1. 1% acid denatured hemoglobin: Dissolve 1.25 g hemoglobin (Sigma) in 50 mL water; dialyze against water 3 × 3 L over ~24 h at 4°C (use dialysis tubing with a

molecular weight cutoff of 12,000–13,000). Add 1 *N* HCl to pH 1.8, incubate for 1 h at 35°C, adjust to pH 3.2 with 1 *M* NaOH, and bring the volume to 62.5 mL with water (aliquots of this 2% solution can be stored at –80°C). On day of use, mix with 62.5 mL 0.2 *M* glycine-HCl, pH 3.0.

2. 1 *N* perchloric acid.
3. 0.5 *M* NaOH.
4. 2% Na_2CO_3 in 0.1 *M* NaOH.
5. 1% $CuSO_4 \cdot 5H_2O$.

3. Methods

3.1. Re-Engineering of an Expression Strain by Disruption of the PEP4 Locus

3.1.1. Growth and Maintenance of pep4 Strains

P. pastoris strains and transformants are propagated in rich medium (YPD) with rapid shaking at 30°C. The *pep4* strains of *P. pastoris* are not stable when stored for more than a few weeks at 4°C on a plate or more than a few months on a slant. However, *pep4* strains are stable for several years when stored in rich medium with 15% glycerol at –70°C.

3.1.2. Isolation of Spontaneous ura3 Mutants

The approach taken to disrupt the *pep4* gene of an established expression strain is to first isolate a spontaneously arising *ura3* mutant of the strain, and then transform this variant with a *pep4* disruption vector. Isolation of spontaneously arising *ura3* mutations is facilitated by selecting for resistance to 5-fluoro-orotic acid (5FOA). 5FOA is an analog of a uracil, biosynthetic pathway intermediate. When metabolized it is converted to 5-fluoro uracil which is toxic to the cell *(7)*. Cells that carry an inactivating mutation at the *URA3* locus are resistant to 5FOA as well as having an absolute requirement for uracil. A vector that carries the *URA3* gene can be used for functional complementation and transformation of the strain. In this way a strain already developed to express a heterologous protein at high levels can be re-engineered using a *pep4*-disruption vector to reduce the effects of proteases. The steps involved in identifying a suitable spontaneous *ura3* mutant include selection of 5FOA-resistant colonies, identification of stable uracil auxotrophs, crossing strains to designate complementation groups, and transformation.

1. Inoculate a fresh colony of the desired strain in 10 mL of YPD in a 50-mL flask with shaking overnight.
2. Centrifuge and resuspend cells in sterile water to a volume of 1 mL.
3. Spread ~5×10^7 cells (typically 100 µL) onto 5FOA plates. After 5–6 d incubation at 30°C, colonies can be picked and streaked onto YPD plates.

4. Single colony clones from YPD plates are transferred to a patch on a YPD master plate. Once the patches have grown (~12–24 h), they are replica-plated to two SD plates, one supplemented with and the other without uracil.
5. Colonies are scored for uracil auxotrophy after 2 d growth at 30°C. Suitable uracil auxotrophs will not grow on media lacking uracil (i.e., will not display leaky growth on the selective media) and will have a low reversion frequency. This is estimated by plating a known amount of cells onto selection plates and determining the frequency of prototrophs arising (*see* **Note 2**).
6. Mutations in two genes, *URA3* and *URA5*, result in auxotrophs resistant to 5FOA. To distinguish these two classes, perform complementation analysis using standard crossing techniques (*see* Chapter 2).
7. An expedient approach to identifying which of the two complementation groups represents mutations in the *URA3* gene is to directly transform a candidate strain from each complementation group with the pDR403 vector ***(4)***. The *ura3* host strain will generate prototrophic transformants with the vector, whereas a strain from the other complementation group will not.

3.1.3. Transformation with Disruption Vector

To direct integration of the disruption vector pDR421 to the *PEP4* locus, 10 μg of the vector is linearized by digestion with *Bgl*II, which cuts the vector once in the middle of cloned *pep4* fragment **(Fig. 1)**. The recombinogenic ends of the linearized DNA stimulate homologous recombination at the *PEP4* locus, resulting in disruption of the *PEP4* gene. Transformation by either spheroplast or whole-cell methods can be used (*see* Chapter 3 for details). Colonies arising after 4–7 d are picked and screened for CPY activity.

3.1.4. Screening for Disrupted PEP4 Clones

3.1.4.1. APNE Overlay Assay

A satisfactory plate assay that directly measures protease A (PrA) activity in colonies is not available. However, since loss of PrA activity results in a failure to generate a mature functional form of carboxypeptidase Y (CpY), the disruption of *PEP4* gene can be monitored by following CpY activity in the colonies using a simple plate overlay assay ***(8)***. Cells are permeabilized by dimethylformamide in the overlay. The APNE is specifically cleaved by CpY in the cells to produce β-naphthol, which reacts with Fast Garnet salt solution to give an insoluble red dye. Colonies in which active CpY is present will be red relative to the pink color of colonies in which the CpY is inactive.

1. Replica plate strains to thick YPD plates and grow for 3 d at 30°C.
2. Add 2.5 mL of APNE solution to 4 mL of molten agar in a 15-mL tube, mix thoroughly, and pour over the surface of the colonies.
3. Once the Agar has solidified, carefully flood the surface of the agar with 5 mL of Fast Garnet GBC solution. Watch for the color to develop (typically <5 min) and pour off the solution.

3.1.4.2. Proteinase A Assay

The assay for PrA activity measures the release of amino acid residues from acid-denatured hemoglobin. At acid pH, PrA is the only protease to carry out this reaction. This protocol is adapted from the method of Jones et al. ***(9)***.

1. The assay reaction mix consists of 50 µL of cell-free extract added to 400 µL of 1% hemoglobin in an Eppendorf tube. This is incubated at 37°C in a shaking water bath for 90 min.
2. The reaction is stopped by adding 200 µL of 1 *N* perchloric acid.
3. The perchloric acid insoluble material is removed by centrifugation for 5 min at full speed in a mini-centrifuge.
4. A negative control reaction is made by adding 200 µL of 1 *N* perchloric acid to a 50-µL sample of cell-free extract prior to adding the substrate; then add substrate and spin down.
5. A 200-µL sample of the acidic supernatant is neutralized by mixing with 200 µL of 0.31 *N* NaOH.
6. The perchloric acid soluble product is assayed using the Bradford protein assay. A 40-µL sample of the neutralized supernatant is mixed with 760 µL of distilled water. To this is added 200 µL of concentrated Bradford reagent (Bio-Rad). Mix well and read the OD_{595}.
7. Generate a standard curve with 0–20 µg of bovine serum albumin (BSA). By subtracting the OD_{595} of the blank from the OD_{595} of the sample, the amount of BSA equivalents can be determined.
8. Determine the protein concentration of the cell-free extract. Typically 2 µL of cell extract is required. Definition of specific activity: one unit of activity corresponds to 1 µg of BSA equivalent/min. For a 90-min incubation, 40-mL sample, the total units in the sample equal the number of milligrams of BSA × 0.361 divided by the number of milligrams of protein in a 50-µL cell-free extract (U/mg protein).

3.2. Expression of a Thrombomodulin Fragment in a pep4 Strain

This protocol describes the production by shake flask growth and by fermentation of large quantities of thrombomodulin fragment from the SMD1168 strain of *P. pastoris* transformed with multiple copies of the pPIC9K expression vector containing the gene for the thrombomodulin fragment. The SMD1168 strain was chosen for transformation after proteolytic sensitivity tests (**Subheading 3.2.5.**) using culture broth from GS115 cells showed proteolytic sensitivity.

3.2.1. Transformation of pep4 Strains

1. To obtain multicopy insertion of the gene of interest, transform *P. pastoris* strain SMD1168 with linearized DNA of pPIC9K vector as described in Chapters 5 and 13 ***(6)***. The DNA is linearized with *Bgl*II and purified with Gene-Clean. Purification of the linearized DNA with phenol or transformation with unpurified DNA results in a drastic reduction of transformation efficiencies (*see* **Note 3**).

2. After 4–7 d, harvest and pool transformed His$^+$ yeast colonies from the top agar. Plate diluted portions on MD plates, incubate at 30°C for 2 d, and then replica plate to selection plates (YPD medium plates containing 2 g/L G418).

3.2.2. Screening for Expression

Typically, 5–10 G418^r cells are picked from the G418 selection plate for further analysis.

1. To test for protein expression, cells of each G418^r strain are grown overnight in 10 mL of minimal medium supplemented with 1% glycerol in borosilicate glass culture tubes (25 × 150 mm) covered with four layers of cheesecloth, and shaken rapidly to maximize aeration (*see* **Note 5**).
2. After 2 d, the cells are collected by centrifugation and resuspended in 2 mL minimal medium supplemented with 2% methanol and grown as before.
3. After 1–2 d more, the medium is tested for thrombomodulin cofactor activity using an enzymatic assay for activation of protein C ***(10)***. Expression is maximal after 2 d. The amount of thrombomodulin cofactor activity in cultures of G418^r cells indicated expression levels of 1–4 mg/L. When the expression vector is used to transform wild-type cells, or when single copy transformants are assayed, no measurable thrombomodulin cofactor activity is observed.

3.2.3. Screening for Mut Phenotype

Since the expression vector can insert into the *P. pastoris* genome at several loci, one of which is the *AOX1* gene, the transformants producing the highest levels of thrombomodulin activity are screened for their ability to grow on methanol as a sole carbon source. This test defines the transformants as methanol utilization slow (Mut^s), indicating that the expression vectors have inserted into the *AOX1* gene, or Mut$^+$, indicating that the *AOX1* gene is intact and functional (*see* **Note 6**).

To ascertain the Mut phenotype, it is necessary to do liquid culture growths.

1. Inoculate each test strain into 10 mL of minimal medium containing only 0.1% glycerol.
2. After 24 h, the culture will have used up the glycerol and attained an OD_{600} of 5–10. At this time, the cells are centrifuged and resuspended in 10 mL of minimal medium containing 2% methanol. The OD_{600} is determined immediately and again after 24 h.
3. A 300-μL aliquot of 50% methanol is added to each culture and the OD_{600} is determined again at 36 and 48 h. The Mut^s cultures typically attain an OD_{600} of 25 and the Mut$^+$ cultures attain an OD_{600} of 40.

3.2.4. P. pastoris *DNA Extraction and Southern Blot Analysis*

1. To purify genomic DNA from transformed *P. pastoris* cells, each G418^r transformant is grown from 10 μL of frozen culture in a 10-mL culture in mini-

mal medium containing 1% glycerol. As a control, a culture is also grown from a transformant previously determined to contain one copy of the pPIC9K insert. Each culture is grown for 2 d at 30°C and 300 rpm, and 100 OD_{600} of cells are collected by centrifugation at 1500*g*.

2. The cells are washed once by centrifugation with 15 mL H_2O, then with 15 mL of 10X TE buffer. The cell pellet is resuspended in 400 mL of 10X TE and transferred to glass tubes (12 × 75 mm) containing 0.5 g of 425–600 μ acid-washed glass beads. The mixture is vortexed for 4 min, then the broken cells are transferred to 1.5-mL Eppendorf tubes. The glass beads are washed with another 400 μL 10X TE, which is added to the broken cells. RNase is added to 10 μg/mL, and samples are incubated for 15 min at room temperature. NaCl is then added to 100 m*M*, and phenol-chloroform extraction of DNA is performed with 400 μL phenol and 400 μL chloroform. The mixture is vortexed for 15 s and centrifuged at high speed in a minicentrifuge for 5 min, and the top supernatant is divided into equal aliquots and transferred to two tubes containing 800 μL ethanol each. After 30 min storage at –20°C, the tubes are centrifuged at high speed for 15 min, the ethanol is removed by aspiration, and the DNA is resuspended in 50 μL of TE buffer. The concentration is determined by reading the absorbance at 260 nm.
3. For Southern blot analysis, 15 μg of the purified DNA is digested with 20 U of *Bgl*II restriction enzyme overnight at 37°C. Since this is genomic DNA, it is important to let the reaction proceed overnight. The DNA is then transferred onto nitrocellulose paper and crosslinked. Prepare a labeled probe using the GENIUS kit reagents and the 700-bp *Sac*I–*Bam*HI fragment of pPIK9K. After hybridization, unbound probe is washed off, and permanent color generation is attained using reagents for alkaline phosphatase colorimetric detection of the digoxigenin-labeled probe provided in the kit.

3.2.5. Test for Proteolytic Stability of Foreign Protein from Wild-Type and Protease-Deficient Strains

To determine whether the protein of interest is unstable under expression conditions because of proteolysis, an experiment to test protein stability may be performed.

1. Proteolytic stability of the protein of interest is tested by comparison of the stability of the protein in culture broth from both the wild-type (GS115) and the *pep4* strain. The culture broth used in this experiment differs depending on whether the protein will be expressed intracellularly or extracellularly, and whether fermentation will be the ultimate production method. If the protein is expressed intracellularly, a cell extract is prepared by growing the *P. pastoris* to stationary phase, washing 20 OD_{600} units of cells with 120 m*M* sodium azide, and then lysing in 400 μL of 100 m*M* Tris, pH 7.5, by vortexing the cells with acid-washed beads for 1 min. If the protein is secreted, the supernatant from a culture grown to ~20 OD_{600} U is used instead. If fermentation is to be the ultimate production method, cells or broth from a fermented culture of *P. pastoris* is used in this experiment.

2. The culture extract or broth is tested for proteolytic activity towards the protein of interest. Measurement of enzymatic activity of the protein of interest is preferable since there are many other proteins present in both culture extracts and culture supernatants of *P. pastoris*. If an enzymatic assay is not available, analysis on sodium dodecyl sulfate–polyacrylamide gel electrophoresis (SDS-PAGE) or western blots also works. If the culture supernatant is from a fermentation, western blotting is necessary because many other proteins are present in the fermentation supernatants. A solution of approx 1% protein is prepared by mixing 350 μL of buffer containing purified protein for which proteolytic sensitivity is being assessed to 50 μL of cell extract. Alternatively, the protein is added to 400 μL of culture supernatant (which is more dilute in proteases than the culture extract). Two controls, one without the protein of interest and the other without the culture extract or supernatant, are also prepared. A 40-μL portion of the reaction is removed at the beginning of the reaction, and again at 2, 4, 6, 12, and 24 h. The samples are analyzed for the presence of the protein of interest using enzymatic assay, SDS-PAGE, or western blotting.

3.2.6. Protein Production in Shake-Flask Cultures

1. The transformed SMD1168 strain that produces the highest levels of the protein of interest in small cultures is used for all large scale growths. Typically, 10 μL of the frozen stock is used to inoculate 2 × 10 mL of BMGY medium. The culture is grown with rapid shaking at 30°C for 2 d. These cultures are then used to inoculate 2 × 1 L of BMGY in two 4-L baffled shake flasks that are grown for 2 d with rapid shaking (300 rpm) and covered with four layers of cheesecloth for maximal aeration. The cell density at the end of this glycerol growth phase is typically 50–60 OD_{600}/mL.
2. To start the induction phase, the cells are collected by centrifugation and resuspended in 2 × 200 mL of BMMY (2% methanol, pH 6.0) and grown in the same manner for an additional 2 d. Methanol is again added to a final concentration of 2% after 24 h. For the thrombomodulin fragments, equivalent amounts of protein were produced by both Mut^s and Mut^+ strains from shake-flask cultures.

3.2.7. Fermentation

1. Production of large amounts of thrombomodulin EGF ***(4,5)*** by fermentation can be performed using a BioFlo 3000 fermentor equipped with a 5.0-L capacity bioreactor using the protocol described in Chen et al. ***(11,12)*** or in Chapter 9.
2. On the first day of the fermentation protocol, a 10-mL culture (25 × 150-mm tubes) in minimal medium supplemented with 2% glycerol is started from 10 μL of a frozen stock. The culture is allowed to grow at 30°C in a rotary shaker at 300 rpm. On the second day, 5 mL of the culture is added to 200 mL of minimal medium again supplemented with 2% glycerol in a 500-mL disposable corning Erlenmeyer flask, and allowed to continue to grow at 30°C, 300 rpm.
3. The fermentation vessel is also prepared on the second day. Before autoclaving the fermentation vessel, calibrate the pH probe and check the dissolved oxygen

(DO) probe membrane for leaks (proper DO control is essential to a good fermentation of *P. pastoris*). The fermentation vessel is prepared with 3 L of basal salt medium. At the time of autoclaving, two 1-L bottles with tubing caps loosely attached are also prepared; one of the bottles is empty, and the other contains 1 L of 50% glycerol. The vessel and bottles are autoclaved on liquid cycle for 30 min with slow exhaust. It is important to check to make sure liquid levels do not drop significantly, and to let the vessel cool slowly before attaching it to the Bioflow 3000 console; otherwise damage to the DO probe membrane may be sustained. After cooling, the vessel is connected to the BioFlo 3000 console, the agitation set to 500 rpm, and the temperature is set to 30°C. The DO monitor is turned on and set to 30 for the entire run. DO levels are maintained at 30 during the entire run by sparging with pure O_2 during periods of heavy oxygen uptake. The concentrated (30%) ammonium hydroxide feed is begun so that the pH can equilibrate to 5.0 overnight. Once the pH begins to approach 5.0, a precipitate will form, but this will not affect the success of the fermentation.

4. On the third day, the accuracy of the pH probe is checked by removing a sample of the medium and determining the pH on an external meter. The DO probe is also calibrated by disconnecting the probe and setting the zero point and then sparging the solution with oxygen and setting the 100% point. Add 12 mL of sterile filtered PTM salts and 12 mL of biotin solution. The vessel is then inoculated with the 200 mL culture. Antifoam is added via a syringe through the top of the addition port after 3–6 h fermentation. The DO level begins to drop after about 8–10 h.
5. A wet cell weight is determined. Close to the end of the glycerol batch phase as the glycerol becomes exhausted (evident by a sudden and steady rise in the DO value), typically 24–30 h post inoculation, determine the wet cell weight (~120 g/L).
6. When the glycerol batch phase is completed, begin the glycerol limited fed-batch phase immediately by attaching to a second feed pump the bottle containing 1 L of 50% glycerol (autoclaved or sterile-filtered) supplemented with 12 mL PTM salts and 12 mL biotin solution. The DO almost instantaneously dropped back down to 30. The glycerol should be kept growth-rate limiting and not fed too fast, with the feed lasting 12–24 h and using approx 500–1 L of glycerol. At this point, the wet cell weight should be approx 350 g/L.
7. Once the glycerol fed-batch phase is completed, begin the methanol induction. It is important to start the methanol induction with 50% methanol supplemented with 6 mL/L PTM salts and 6 mL/L biotin solution at a feed rate of 1.8 mL/h. The feed is doubled each hour for about 5–6 h. During this period, the DO decreases from ~100 to ~70%. If the methanol induction phase is to begin at night, increase the 50% methanol feed to 20 mL/h overnight. Once the cells seem to be growing at a setting of 20 mL/h for at least 1 h, the feed is switched to 100% methanol supplemented with 12 mL/L PTM salts and 12 mL/L biotin solution. The 100% feed begins at 10 mL/h, and is increased every few hours with each step preceded by a DO spike (*see* Chapter 9 for definition and description of DO spike). Depending on the strain (Mut^+ or Mut^s), the maximum setting is from 18 (Mut^s)

to 50 (Mut$^+$) mL/h. The time intervals between increases of the methanol feed are also longer for the Mut^s strains. If the strain is Mut$^+$, the wet cell weight will continue to rise to about a maximum of 425–450 g/L and the cells will utilize a significant amount of base during the induction phase. For a Mut^s strain, the wet cell weight should remain about the same as when methanol induction begins.

8. Both Mut$^+$ and Mut^s strains of the transformed *pep4 P. Pastoris* SMD1168 give high expression levels. A graph of a typical fermentation run is shown in **Fig. 2**. The cells are typically harvested 48–72 h after induction. Induction times may vary for different proteins. Cells are removed from the fermentor by decanting its contents or applying pressure to the vessel and collecting the cells from the sample port. The cells are separated from the culture medium by centrifugation at 2000*g* for 1 h.
9. The supernatant may be stored in 1-L portions at –70°C until use. The thrombomodulin fragment is further stabilized against proteolysis by addition of 10 m*M* EDTA prior to freezing the supernatant. The first step of purification is to concentrate the protein by anion-exchange chromatography. For the thrombomodulin fragment, the frozen culture supernatant is thawed quickly by constant mixing under a stream of 55°C H_2O. In the first purification step, the culture supernatant is diluted to 2 L with H_2O containing 10 m*M* EDTA and loaded onto a 300 mL column of QAE Sephadex A-120 (Sigma) equilibrated with 50 m*M* morpholinesulfonic acid (Fisher) buffer, pH 6.5 (buffer A) at 4°C. The thrombomodulin fragment is eluted with a linear gradient of 750 mL buffer A to 750 mL buffer A containing 1 *M* NaCl. Further purification is afforded by Mono Q chromatography on an FPLC in Tris-HCl, pH 7.5, followed by reverse-phase HPLC. Typical crude yields are 150–200 mg of thrombomodulin/L of culture supernatant, and purified yields are approx 50%. Further details of the protein activity and purification are reported elsewhere ***(10)***.

4. Notes

1. If the PTM salts are sparged with N_2 gas before sterile filtering, and the biotin solution is kept separately, the PTM salts last for several months without discoloration or precipitation.
2. A reversion frequency of $<1 \times 10^{-8}$ is ideal.
3. Spheroplasts of the SMD1168 strain are especially sensitive to breakage, and vortexing significantly reduces the yield of spheroplasts and subsequently the transformation yield. Therefore, resuspension of the spheroplasts is accomplished by gentle rocking of the culture tubes.
4. The transformation efficiency using the spheroplast method is lower for the *pep4* strain than for the GS115 strain by 50–80%. Typically, 50–100 His$^+$ colonies per plate are obtained from transformation of the *pep4* strain.
5. It is important to note that once *P. pastoris* cells have been induced, they are not recoverable, so a frozen stock of each culture should be made from a portion of the culture prior to induction.
6. In order to know the amount of methanol to feed during fermentation, and the rate at which to begin feeding, it is essential to know the Mut phenotype of the strain.

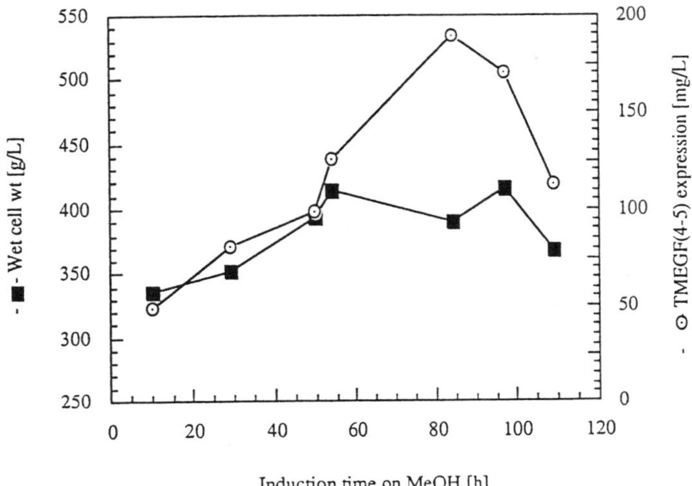

Fig. 2. High-cell-density fermentation of *P. pastoris* SMD1168-thrombomodulin-EGF(4–5) Mut^s strain. The plot shows the expression level of thrombomodulin EGF(4–5) in mg/L vs the wet cell weight in g/L during the induction phase. (■) Wet cell weight, (●) mg/L thrombomodulin fragment based on thrombomodulin cofactor activity assay.

References

1. Sander, P., Grunewald, S., Bach, M., Haase, W., Reilander, H., and Michel, H. (1994) Heterologous expression of the human D2S dopamine receptor in protease-deficient *Saccharomyces cerevisiae* strains. *Eur. J. Biochem.* **226**, 697–705.
2. Meerman, H. J. and Georgiou, G. (1994) Construction and characterization of a set of *E. coli* strains deficient in all known loci affecting the proteolytic stability of secreted recombinant proteins. *Biotechnology* **12**, 1107–1110.
3. Van Den Hazel, B. H., Kielland-Brandt, M. C., and Winther, J. B. (1996) Review: biosynthesis and function of yeast vacuolar proteases. *Yeast* **12**, 1–16.
4. Patent Application 50848.
5. Weiss, H. M., Haase, W., Michel, H., and Reilander, H. (1995) Expression of functional mouse 5-HT5A serotonin receptor in the methylotrophic yeast *Pichia pastoris*: pharmacological characterization and localization. *FEBS Lett.* **377**, 451–456.
6. Scorer, C. A., Clare, J. J., McCombie, W. R., Romanos, M. A., and Sreekrishna, K. (1994) Rapid selection using G418 of high copy number transformants of *Pichia pastoris* for high-level foreign gene expression. *Biotechnology* **12**, 181–184.
7. Boeke, J., LaCroute, F., and Fink, G. R. (1984) The *Saccharomyces cerevisiae* genome contains functional and nonfunctional copies of transposon Ty1. *Mol. Gen. Genet.* **197**, 345–349.
8. Jones, E. W. (1977) Proteinase mutants of *Saccharomyces cerevisiae*. *Genetics* **85**, 23–33.

9. Jones, E. W., Zubenko, G. S., and Parker, R. R. (1982) *PEP4* gene function is required for expression of several vacuolar hydrolases in *Saccharomyces cerevisiae*. *Genetics* **102**, 665–677.
10. White, C. E., Hunter, M. J., Meininger, D. P., White, L. R., and Komives, E. A. (1995) Large scale expression, purification and characterization of the smallest active fragment of thrombomodulin: the roles of the sixth domain and of methionine-388. *Protein Eng.* **8**, 1177–1187.
11. Chen, Y. L., Cino, J., Hart, G., Freedman, D., White, C. E., and Komives, E. A. (1996) High protein expression in fermentation of recombinant *Pichia pastoris* by fed batch process. *Process Biochem.* **32**, 107–111.
12. Chen, Y. L., Cino, J., White, C. E., and Komives, E. A. (1996) Continuous production of thrombomodulin from a *Pichia pastoris* in fermentation. *J. Chem. Technol. Biotechnol.* **67**, 143–148.

8

Glycosylation Profiling of Heterologous Proteins

José A. Cremata, Raquel Montesino, Omar Quintero, and Rossana García

1. Introduction

The methylotrophic yeast *Pichia pastoris* has been widely exploited for its high-level expression of heterologous proteins by recombination of gene sequences of interest with the methanol-inducible alcohol oxidase gene (*AOX1*) promoter ***(1–3)***. Secreted and cytoplasmic expression of heterologous proteins at levels equivalent to *Escherichia coli* and significantly higher than in *Saccharomyces cerevisiae* has been achieved ***(4)***. In addition, *P. pastoris* cultures can be easily scaled up to high cell densities, and as a result, yields are also high on a volumetric basis (e.g., 12 g/L, for tetanus toxin fragment C ***[5]***, 2.5 g/L for invertase ***[6]***, 2.5 g/L for α-amylase ***[7]***). Secreted products can comprise more than 80% of the protein in the medium ***(6)***. However, secretion is a complex process that is not only dependent on gene dosage, but also on other factors, such as signal sequence recognition and processing, proteolysis, and glycosylation.

With regard to glycosylation, heterologous proteins secreted from *S. cerevisiae* are often hyperglycosylated with outer chains of mannose units of up to 50–150 residues added at *N*-asparagine-linked sites, which makes such glycoproteins highly antigenic and, therefore, unsuitable for use as human therapeutic drugs ***(8)***. To investigate glycosylation in *P. pastoris*, Tschopp et al. expressed the *S. cerevisiae SUC2* gene in *P. pastoris* and examined in detail the structure of N-linked oligosaccharides added to its product, invertase ***(6)***. They observed that *S. cerevisiae* invertase secreted from *P. pastoris* is not hyperglycosylated, but contains outer chains of only 8–14 mannose residues (Man_{8-14}), compared to an average length of >50 when the same enzyme is secreted from *S. cerevisiae* ***(9)***. The *P. pastoris*-secreted invertase resembles in size the endoplasmic reticulum core-glycosylated form of invertase seen in *S.*

From: *Methods in Molecular Biology, Vol. 103:* Pichia *Protocols*
Edited by: D. R. Higgins and J. M. Cregg © Humana Press Inc., Totowa, NJ

cerevisiae sec18 mutants at nonpermissive temperature ***(9)***. The core-like structure of the oligosaccharides was confirmed by Trimble et al. using NMR techniques ***(10,11)***. They found that 75% of the oligosaccharides on *P. pastoris* secreted invertase are Man_{8-9}, and most of the remaining oligosaccharides are Man_{10-11}. The structures are all typical of core glycosylation structures on fungal proteins ***(11)***. Of potential importance to the use of *P. pastoris* glycoproteins as human pharmaceuticals, the oligosaccharides in *P. pastoris*-secreted invertase lack terminal α1,3-mannose residues that are commonly found on proteins secreted from *S. cerevisiae* and are known to be highly antigenic ***(10,11)***. However, the suitability of foreign glycoproteins expressed in *P. pastoris* for pharmaceutical use remains problematic, since the lower eukaryotic structure of *P. pastoris* oligosaccharides is significantly different than that of mammalian cells.

Unfortunately, little information exists on oligosaccharide structures on other glycoproteins secreted from *P. pastoris*. Although some proteins appear to have only short oligosaccharide structures like invertase, others appear to be hyperglycosylated ***(4,12)***. The variability of carbohydrate structures on glycoproteins secreted from *P. pastoris* makes their analysis an important aspect of protein characterization in this yeast. As a step in this analysis, oligosaccharide "profiling" methods have been developed in recent years. One of them, called FACE for fluorophore-assisted carbohydrate electrophoresis, is based on the reductive amination of glycans by 8-amino-1,3,6-naphthalene trisulfonic acid (ANTS) and their separation by polyacrylamide gel electrophoresis ***(13)***. More recently, a second method based on the resolution of the same ANTS–oligosaccharide derivatives by HPLC analysis has been described ***(14)***. By a two-dimensional analysis process involving HPLC retention times in one dimension and relative migration index (RMI) during electrophoresis in the other, oligomannoside structures can readily be determined. These profiling methods, which are described in this chapter, greatly simplify the work of characterizing carbohydrates present on natural and recombinant glycoproteins.

2. Materials

2.1. Chemicals and Enzymes

All reagents should be of analytical grade. Recommended sources of specialty chemicals and enzymes include: ANTS from Molecular Probes (Eugene, OR); *N*-glycosidase F (PNGase F) from either Boehringer Mannheim (Indianapolis, IN) or New England BioLabs (Beverly, MA); and α-mannosidase from *Aspergillus saitoi* (an exoglycosidase) from Oxford GlycoSystems (Abingdon, UK).

2.2. Buffers, Stock Solutions, and Equipment

1. Enzyme incubation buffer: 0.2 m*M* sodium phosphate buffer, pH 8.6.
2. Denaturing solution: 2.5% SDS, 1.5 *M* β-mercaptoethanol.

3. ANTS solution: 0.15 *M* ANTS in an acetic acid/water solution (3:17 v/v). Gentle warming in a 60°C bath is required to dissolve the ANTS completely. This solution may be stored at –70°C.
4. Sodium cyanoborohydride solution: 1.0 *M* in dimethyl sulfoxide (DMSO). This solution must be made fresh daily.
5. Gel electrophoresis apparatus: Mighty Small SE250, Hoeffer Scientific Instruments (San Francisco, CA).
6. Acrylamide stock solution: an aqueous solution of 60% (w/v) acrylamide, 1.6% (w/v) *N*, *N'*-methylene-bis-acrylamide.
7. 4X stock of gel buffer: 1.5 *M* Tris-HCl, pH 8.5.
8. 10X stock of electrophoresis buffer: 1.92 *M* glycine, 0.25 *M* Tris base, pH 8.5.

3. Methods

3.1. Release of Oligomannosides from Proteins

1. Dialyze a 50–250-μg sample of glycoprotein against distilled water, and then dry the sample in a centrifugal vacuum evaporator (*see* **Note 1**).
2. Dissolve the dried glycoprotein sample in 30 μL of enzyme incubation buffer.
3. Add 2 μL of denaturing solution, and heat at 100°C for 5 min.
4. Cool the sample to room temperature and add 5 μL of a 7% (v/v) solution of Nonidet P-40.
5. Add 5 μL of PNGase F, mix well, and centrifuge briefly to bring the entire sample to the bottom of the tube.
6. Incubate the sample at 37°C for 16 h.
7. Centrifuge the sample at 10,000*g* for 5 min to clear the solution. Discard the pellet.
8. Add 130 μL of cold ethanol, and incubate for 1 h in an ice bath or for 20 min at –20°C.
9. Centrifuge at 10,000*g* for 3 min.
10. Carefully separate the supernatant, which contains the oligosaccharide pool, from the pellet.
11. Dry the protein pellet from step 9 in a centrifugal vacuum evaporator for 10 min. Resuspend it in 50 μL of water, and add 3 vol of cold ethanol. Incubate for at least 1 h in an ice bath or for 20 min at –20°C.
12. Centrifuge at 10,000*g* for 3 min, and add the supernatant from **step 10** (*see* **Note 2**).
13. Dry the oligosaccharide pool in a centrifugal vacuum evaporator for at least 1 h with low heating. Do not exceed 45°C during evaporation. If the ANTS derivatization reaction is not to be performed immediately, store the pellet at –20°C.

3.2. ANTS Derivatization of Oligosaccharides

1. Add 5 μL of ANTS solution to the dry oligosaccharide pool sample described in **Subheading 3.1., step 13**.
2. Add 5 μL of the cyanoborohydride solution, and mix well. Be sure that all the oligosaccharides are in solution.
3. Centrifuge briefly to bring sample to the bottom of the tube, and incubate the sample for 16 h at 37°C or 2 h at 45°C (*see* **Note 3**).

4. Dry the reaction product in a centrifugal vacuum evaporator for at least 1 h with low heat (<45°C).
5. Add 10 μL of water, and store at –70°C until HPLC or electrophoresis steps.

3.3. Fluorophore-Assisted Carbohydrate Electrophoresis

1. For each 12 mL of the resolving gel solution, mix 6 mL of the 60% stock acrylamide solution, 3 mL of gel buffer, 70 μL of a freshly prepared ammonium persulfate solution (10% w/v), 3 mL of water, and 10 μL of TEMED.
2. As soon as the TEMED is added, pour the resolving gel solution into the gel mold. Ensure there are no bubbles in the gel solution before it sets.
3. Allow the polymerization reaction to proceed (~15 min).
4. Prepare the stacking gel solution by mixing 0.63 mL of 60% stock acrylamide solution, 1.25 mL of stock gel buffer, and 70 μL of a freshly prepared ammonium persulfate solution (10% w/v). Add 6.3 mL of water and 10 μL of TEMED.
5. Pour the stacking gel solution over the resolving gel in the mold.
6. Immediately insert a sample well-forming comb containing the desired number of "teeth" (typically 8–11 wells for a gel that is 8 cm in width).
7. Allow the polymerization reaction to proceed (~30 min), and then remove the comb gently.
8. Place the gel-containing glass sandwich into the electrophoresis apparatus, and fill the anode and cathode reservoirs with 1X electrophoresis buffer solution prepared by dilution of the 10X stock (*see* **Note 4**).
9. Load 2–3 μL of the samples (from **Subheading 3.2., step 5**) into the wells with the aid of a microliter syringe.
10. For a gel that is 0.5 × 80 × 80 mm, apply a constant current of 15 mA for 1.5 h at 4°C (*see* **Note 4**).
11. After electrophoresis, view the electrofluorogram by irradiating the gel with a hand-held UV lamp (*see* **Note 5** and **Fig. 1A**). Photograph using Polaroid 53 film. The excitation wavelength is 353 nm, and the emission wavelength is 535 nm. (**Note:** The typical DNA UV lamps are not at the appropriate wavelength.)

3.4. HPLC Analysis of ANTS–Oligosaccharide Derivatives

1. Adjust the HPLC equipment for gradient mode with two pumps and an automatic controller. Use a fluorometric detector set at an excitation wavelength of 353 nm and an emission wavelength of 535 nm.
2. Prepare the following buffer systems:
   a. Glacial acetic acid (6% v/v) in a 70:30 (v/v) mixture of acetonitrile–water titrated to pH 5.5 with triethylamine.
   b. Glacial acetic acid (6% v/v) titrated to pH 5.5 with triethylamine.
3. Stabilize the NH_2–HPLC column (Nucleosil 5S NH2, 4.6 × 250 mm, Knauer, Germany) with 95% buffer A and 5% buffer B for at least 30 min, and then perform an initial "blank" HPLC run without sample.
4. Inject a 1–3-μL sample of ANTS-derivatized oligosaccharide, and immediately initiate the gradient flow. Gradient conditions are: 95% buffer A and 5% buffer B to 50% each in 90 min at a flow rate of 0.7 mL/min. After the HPLC run, measure

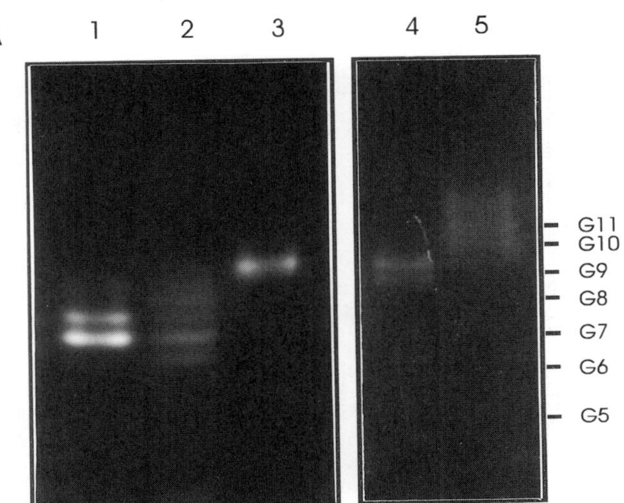

Fig. 1. Glycosylation profiles of recombinant proteins expressed in *P. pastoris*. Periplasmic invertase and secreted dextranase were obtained from the expression of the SUC2 gene from *S. cerevisiae* and the *DEX1* gene from *Penicillium minioluteum* under the control of the *P. pastoris AOX1* promoter, respectively. Both proteins were directed into the secretory pathway by the invertase signal sequence, which was fused to the amino-terminus of each. The *S. cerevisiae HIS3* gene was used as a selection marker for transformation of a *P. pastoris his3* host strain ***(15,16)***. For secretion of the catalytic subunit of bovine enterokinase (EK_L), plasmid pAO815-S, which contains the *S. cerevisiae* α-MF prepro secretion signal and the *P. pastoris HIS4* gene, was used. The *P. pastoris HIS4* gene was the selection marker for transformation of the EK_L vector into the *P. pastoris his4* host GS115 ***(12)***. **(A)** FACE profiles of ANTS–oligosaccharide pools derived from: lane 1, ribonuclease B; lane 2, invertase; lanes 3 and 4, dextranase; and lane 5, EKL. Markers designated G5–G11 denote the positions of oligosaccharides from the malto oligosaccharide series of varying degrees of polymerization (GLYKO) *(continued on next page)*.

the retention times of each peak relative to the $Man_7GlcNAc_2$–ANTS standard (Oxford GlycoSystems) (*see* **Note 6** and **Fig. 1B**).

3.5. Exoglycosidase Digestion Using A. saitoi α1,2-Mannosidase

1. Reconstitute one vial of α1,2-mannosidase from *A. saitoi* in 10 μL of water. Enzyme will be at a concentration of 1 μU/μL in 100 m*M* sodium acetate, pH 5.0.
2. Add 2.6 μL of the reconstituted enzyme to 40 pmol of dry oligomannoside sample. To estimate 40 pmol of samples, compare fluorescence intensities of samples to that of the G4 oligosaccharide standard (50 pmol) from the malto oligosaccharide series (GLYKO, Novato, CA). The final glycan concentration will be 15 μ*M*, and the final enzyme concentration will be 1 mU/mL.
3. Incubate the sample in enzyme for 16–24 h at 37°C. The sample is then ready for further analysis (e.g., FACE or HPLC) (*see* **Note 8** and **Fig. 2**).

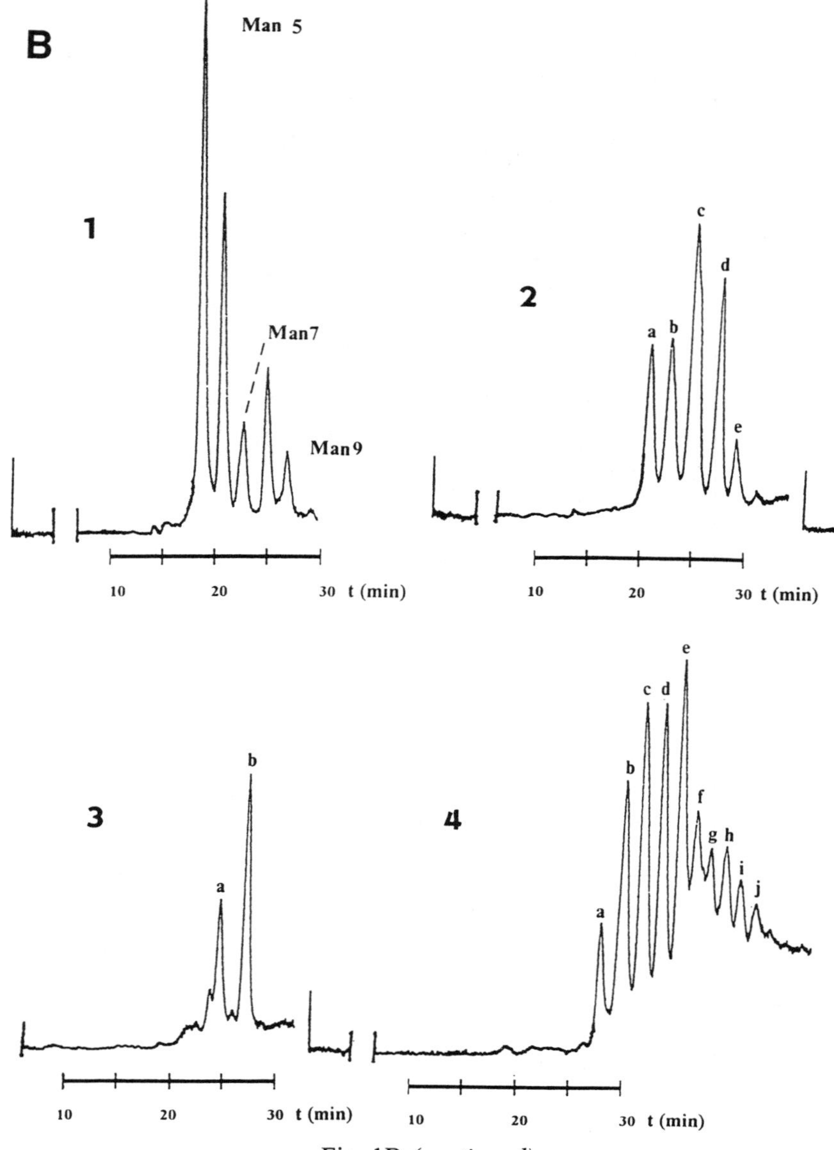

Fig. 1B *(continued)*.

4. Notes

1. Protein solutions should be at a low salt concentration. Although the initial steps of deglycosylation are not particularly sensitive to salt, glycan derivatization with ANTS is strongly inhibited by the presence of even low salt concentrations. Sev-

C

Suggested structures	Ribonuclease B	Invertase	Dextranase	Enterokinase
$Man_5\,GlcNAc_2$	0.81			
$Man_6\,GlcNAc_2$	0.90	a = 0.92		
$Man_7\,GlcNAc_2$	1.00	b = 1.01		
$Man_8\,GlcNAc_2$	1.10	c = 1.11	a = 1.12	
$Man_9\,GlcNAc_2$	1.18	d = 1.23	b = 1.23	a = 1.23
$Man_{10}\,GlcNAc_2$		e = 1.30		b = 1.33
$Man_{11}\,GlcNAc_2$				c = 1.40
$Man_{12}\,GlcNAc_2$				d = 1.48
$Man_{13}\,GlcNAc_2$				e = 1.57
$Man_{14}\,GlcNAc_2$				f = 1.63
$Man_{15}\,GlcNAc_2$				g = 1.70
$Man_{16}\,GlcNAc_2$				h = 1.76
$Man_{17}\,GlcNAc_2$				i = 1.83
$Man_{18}\,GlcNAc_2$				j = 1.90

Fig. 1 *(cont.)*. **(B)** Oligosaccharide profiles of the samples in (A), but analyzed by NH_2–HPLC. Profiles are oligosaccharides derived from: 1, ribonuclease B; 2, invertase; 3, dextranase; and 4, EK_L. Note the higher sensitivity and resolution achieved by HPLC analysis. Comparison of the $Man_5GlcNAc_2$ to $Man_9GlcNAc_2$ oligomannoside commercial standards derived from ribonuclease B demonstrates a broad range of oligomannoside structures on invertase that are mainly of relatively small size. EK_L contains oligosaccharides of up to $Man_{18}GlcNAc_2$, but on dextranase, more than 90% of the oligosaccharides correspond to $Man_{8–9}GlcNAc_2$ species. The inserted table in (B) shows the retention times of the same samples relative to the $Man_7GlcNAc_2$ standard. Note that the $Man_9GlcNAc_2$ in the recombinant samples differs in retention time from that of the commercial standard, a consequence of the presence of one α1,2-linkage instead of one α1,6-linkage on carbohydrates from *P. pastoris* recombinant proteins (*see* **Note 7**). **(C)** Comparison of the relative retention times with the corresponding standards of ribonuclease B. From $Man_{10}GlcNAc_2$ up to $Man_{18}GlcNAc_2$, the retention times were estimated.

eral different conditions for enzymatic release of oligosaccharides with PNGase F have been described, all with pH values between 7.4 and 8.7.

2. It has been reported that a portion of the oligosaccharides in samples may coprecipitate with proteins on addition of 3 vol of cold ethanol. Our experience is that the amount that coprecipitates is dependent on the properties of the protein. Thus, recovery of oligosaccharides after the first precipitation step is variable. To

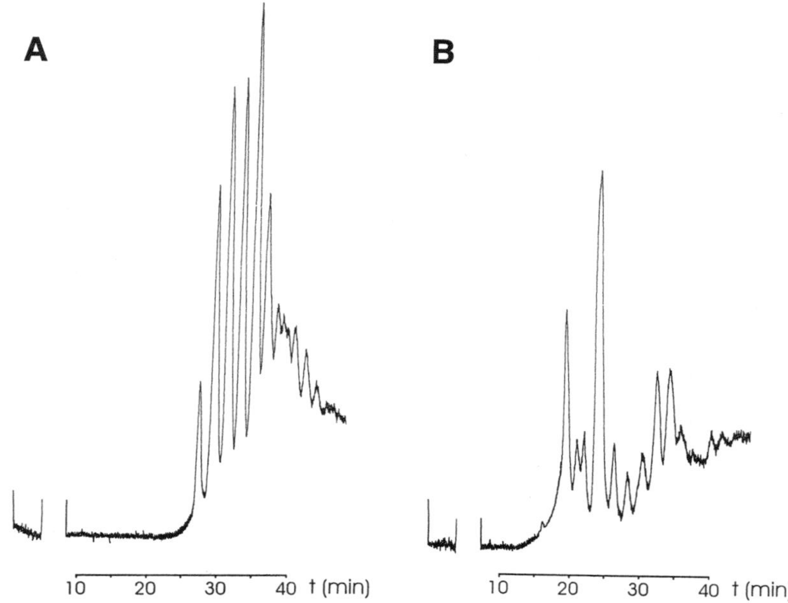

Fig. 2. Exoglycosidase digestion using α1,2-mannosidase from *A. saitoi*. Oligosaccharides derived from EK_L were digested with the exoglycosidase. As a result, virtually all carbohydrate peaks are shifted to lower retention times. This result excludes the possibility of α1,3-terminal-mannose residues on these oligosaccharides. Residual amounts of undigested oligosaccharides are also present. $Man_6GlcNAc_2$, which is theoretically the final product of specific α1,2-mannosidase digestion of $Man_9GlcNAc_2$, showed the same retention time as the smallest oligosaccharide obtained from digestion of EK_L. This result demonstrates that a terminal α1,2-mannose linkage on $Man_9GlcNAc_2$ carbohydrates from *S. cerevisiae* is changed to an α1,6-linkage in *P. pastoris*. Moreover, the latter oligosaccharide has a slightly higher retention time than the corresponding commercial standard $Man_9GlcNAc_2$ derived from ribonuclease B (*see* insert table in **Fig. 1**). **(A)** Profile from EK_L oligosaccharides not treated with exoglycosidase. **(B)** Profile of EK_L oligosaccharides after specific α1,2-mannosidase digestion.

avoid inefficient recovery of oligosaccharides, repeat the protein reconstitution and ethanol precipitation steps at least twice. Protein remaining in oligosaccharide samples can interfere with the resolution of oligosaccharides during electrophoresis. To reduce protein in samples, ethanol precipitations should be repeated until a transparent oligosaccharide pellet is obtained.

3. Although, the derivatization reaction can be performed at either 37°C for 16 h or 45°C for 2 h, in our experience, 37°C for 16 h has resulted in the highest percentage of labeled oligosaccharides.

4. Electrophoresis should be performed at 4°C to avoid self-degradation of the fluorophore-labeled oligosaccharides. Most electrophoresis apparatuses allow for cooling of only one side of the gel. To avoid cooling problems, perform electrophoresis with a precooled chamber using precooled electrophoresis buffer. The apparatus designed for FACE by GLYKO (Novato, CA) is designed so that both sides of the gel are submerged in electrophoresis buffer, thereby providing a more uniform temperature over both gel surfaces. To recover samples after electrophoresis, the desired bands are excised, placed in an Eppendorf tube with 100 μL of water, and allowed to stand for 3 h at 4°C. The supernatants are then dried in a centrifugal vacuum evaporator.
5. For better sensitivity and precision in determining RMI values, an automated system composed of a CCD camera coupled to a personal computer with appropriate software is available from GLYKO.
6. Since the buffer system used for HPLC was designed for conditions of ion suppression, small variations in its preparation may introduce changes in the absolute retention times of ANTS–oligosaccharide derivatives. The magnitude of this effect depends on the oligosaccharides that are being analyzed, i.e., whether they are neutral or charged. $Man_7GlcNAc_2$ was selected as a standard to compare relative retention times, since it is not affected by changes in ion suppression. Analysis of other oligomannosides with additional mannose residues showed an increase of 0.1 in relative retention time per added mannose, relative to the retention time of the $Man_7GlcNAc_2$ standard.
7. The RMI values determined by FACE analysis are typically dependent on the mass:charge ratio. However, with neutral oligosaccharides, such as oligomannosides on yeast glycoproteins, RMI values are dependent only on the mass. For example, the change in linkage configuration when one α1,2-mannose linkage on $Man_9GlcNAc_2$ is changed to an α1,6-mannose linkage is not distinguishable by FACE. However, this change is detected by NH_2–HPLC analysis of the same ANTS derivative (*see* **Fig. 1A,B**). The relative retention time of the $Man_9GLcNAc_2$ structure is increased owing to the presence of the additional α1,6-linkage in place of an α1,2-linkage as demonstrated by comparison of isomers derived from *P. pastoris* invertase and ribonuclease B.
8. The use of a specific α-mannosidase facilitates structural analysis of oligosaccharides from heterologous proteins expressed in P. pastoris. Addition of α1,3-mannose residues on oligosaccharides starting at Man_{11} is typical of oligomannosides from *S. cerevisiae*, whereas α1,2-extensions characterize *P. pastoris* oligosaccharides. This difference is clearly observed in **Fig. 2**, which contains the reaction products of digestion with α-mannosidase from *A. saitoi*, an enzyme that is highly specific for α1,2-linkages relative to α-mannosidases from jack bean or other sources. These structural features are not directly observable by simple inspection of FACE or HPLC results with non-α-mannosidase-treated samples.

References

1. Cregg, J. M., Tschopp, J. F., Stilman, C., Siegel, R., Akong, M., Craig, W. S., Buckholz, R. G., Madden, K. R., Kellaris, P. A., Davis, G. R., Smiley, B. L., Cruze, J., Torregrossa, R., Velicelebi, G., and Thill, G. P. (1987) High level expression and efficient assembly of hepatitis B surface antigen in the methylotrophic yeast *Pichia pastoris*. *Bio/Technology* **5,** 479–485.
2. Cregg, J. M., Vedvick, T. S., and Raschke, W. C. (1993) Recent advances in the expression of foreign genes in *Pichia pastoris*. *Bio/Technology* **11,** 905–910.
3. Cregg, J. M. and Madden, K. R. (1988) Development of the methylotrophic yeast *Pichia pastoris* as a host for the production of foreign proteins. *Dev. Ind. Microbiol.* **29,** 33–41.
4. Romanos, M. A., Scorer, C. A., and Clare, J. J. (1992) Foreign gene expression in yeast: a review. *Yeast* **8,** 423–488.
5. Clare, J. J., Rayment, F. B., Ballantine, S. P., Sreekrishna, K., and Romanos, M. A. (1991) High level expression of tetanus toxin fragment C in *Pichia pastoris* strains containing multiple tandem integrations of the gene. *Bio/Technology* **9,** 455–460.
6. Tschopp, J. F., Sverlow, G., Kosson, R., Craig, W., and Grinna, L. (1987) High level secretion of glycosylated invertase in the methylotrophic yeast *Pichia pastoris*. *Bio/Technology* **5,** 1305–1308.
7. Paifer, E., Margolles, E., Cremata, J. A., Montesino, R., Herrera, L., and Delgado, J. M. (1994) Efficient expression and secretion of recombinant α-amylase in *Pichia pastoris* using two different signal sequences. *Yeast* **10,** 1415–1419.
8. Romanos, M. (1995) Advances in the use of *Pichia pastoris* for high level gene expression. *Curr. Opinion Biotechnol.* **6,** 527–533.
9. Grinna, L. S. and Tschopp, J. F. (1989) Size distribution and general structural features of N-linked oligosaccharides from the methylotrophic yeast *Pichia pastoris*. *Yeast* **5,** 107–115.
10. Trimble, R. B. and Atkinson, P. H. (1986) Structure of yeast external invertase Man_{8-14} GlcNAc processing intermediates by 500-megahertz ^{1}H NMR spectroscopy. *J. Biol. Chem.* **261,** 9815–9824.
11. Trimble, R. B., Atkinson, P. H., Tschopp, J. F., Townsend, R. R., and Maley, K. (1991) Structure of oligosaccharides on *Saccharomyces SUC2* invertase secreted by the methylotrophic yeast *Pichia pastoris*. *J. Biol. Chem.* **266,** 22,807–22,817.
12. Vozza, L. A., Wittwer, L., Higgins, D. R., Purcell, T. J., and Bergseid, M. (1996) Production of a recombinant bovine enterokinase catalytic subunit in the methylotrophic yeast *Pichia pastoris*. *Bio/Technology* **14,** 77–81.
13. Jackson, P. (1990) The use of polyacrylamide gel electrophoresis for the high resolution separation of reducing saccharides labeled with fluorophore 8-aminonaphthalene-1,3,6-trisulfonic acid. *Biochem. J.* **270,** 705–713.
14. Montesino, R., Quintero, O., García, R., and Cremata, J. A. (1995) N-linked glycan profiling of native and recombinant glycoproteins using a modification of the fluorophore-assisted carbohydrate electrophoresis methodology. *Glycoconjugate J.* **12,** 406.

15. Herrera, L., Yong, V., Margolles, E., Delagado, J., Morales, J., Silva, A., Paifer, E., Ferbeyre, G., Sosa, A., Martínez, V., Aguiar, J., Seralena, A., González, T., Montesino, R., Cremata, J. A., Villareal, A., and González, B. (1991) Method for the expression of heterologous genes in the yeast *Pichia pastoris*, expression vectors and transformed microorganisms. European Patent Application No. 0 438 200 A1.
16. Roca, H., García, B. M., Margolles, E., Curbelo, D., Delgado, J., Cremata, J. A., González, M. E., Rodríguez, E., García, R., Fernández, C., and Morera, V. (1995) Dextranase enzyme, method for its production and DNA encoding the enzyme. European Patent Application No. 0 663 443 A1.

9

High Cell-Density Fermentation

Jayne Stratton, Vijay Chiruvolu, and Michael Meagher

1. Introduction

The purpose of this chapter is to provide sufficient instruction for the reader to implement fermentation strategies to produce recombinant proteins in the yeast *Pichia pastoris*. *P. pastoris* utilizes the highly efficient alcohol oxidase 1 gene *(AOX1)* promoter for high-level expression of foreign genes. The *AOX1* promoter is induced by methanol, but repressed in the presence of excess glycerol. Glycerol is readily metabolized by *P. pastoris*, and high-cell densities can be obtained utilizing this carbon source. The standard fermentation protocol is to grow *P. pastoris* in excess glycerol to repress expression followed by induction of protein production by the addition of methanol after the glycerol is exhausted.

With regard to methanol-utilizing ability, three phenotypes of expression strains are available. The first is termed Methanol Utilization positive (Mut^+), which has functional *AOX1* and *AOX2* genes and grows on methanol at the wild-type rate. The second is Methanol Utilization Slow (Mut^s), in which the *AOX1* gene is disrupted. Methanol metabolism in this strain is dependent on the transcriptionally weaker *AOX2* gene. The third is called Methanol Utilizing negative (Mut^-). In this strain, both *AOX* genes are disrupted and, as a result, the strain cannot metabolize methanol at all.

Expression with all three *P. pastoris* Mut strains is performed in fermentor cultures using three-step procedures. The first two fermentation steps are the same for all three phenotypes of strains. The first step is a glycerol batch phase, which serves to generate cell mass. The second step is a fed-batch phase, where glycerol is added at a growth-limiting rate. Limiting glycerol during this phase derepresses the methanol metabolic machinery and allows the cells to transition smoothly from glycerol to methanol growth ***(1)***. The third

From: *Methods in Molecular Biology, Vol. 103:* Pichia *Protocols*
Edited by: D. R. Higgins and J. M. Cregg © Humana Press Inc., Totowa, NJ

step is the methanol induction phase and differs depending on the phenotype of the expression host. However, a common feature of this third phase is that methanol levels in the fermentor must be carefully monitored either directly by gas chromatography or indirectly by dissolved oxygen (DO) concentration. The following sections describe how to perform fermentations with each of the Mut phenotype strains. In addition, a description is presented of a low-density fermentation method for Mut$^+$ strains that does not require the addition of oxygen. This method is convenient for the production of foreign proteins using standard unmodified fermentors.

1.1. Mut$^+$ Strains

Mut$^+$ strains of *P. pastoris* are sensitive to high residual methanol concentrations *(2)*. In the presence of excess oxygen and methanol, formaldehyde, the first product of methanol metabolism, rises to toxic levels and "pickles" the cells *(2–5)*. Thus, for successful cell growth and protein expression with Mut$^+$ strains, it is critical to maintain a low methanol concentration in the fermentor. The advantage of Mut$^+$ strains for expression relative to *AOX*-defective strains is their much faster methanol growth (and foreign protein production) rate. However, the concentration of methanol must be tightly controlled.

1.2. Mut^s Strains

Mut^s (*AOX1*-deletion) strains can be generated during transformation by gene replacement events in which the *AOX1* gene is replaced by the expression cassette *(6)*. Alternatively, a *P. pastoris* strain, such as KM71, which contains a disrupted *AOX1* gene, can be utilized as host with expression vectors inserted by single crossover type integration. As mentioned above, these strains grow slowly on methanol because of their reliance on the *AOX2* gene for alcohol oxidase *(7)*. The fermentation strategy for Mut^s strains is the same as with Mut$^+$ strains, except for a lower methanol feed rate to maintain methanol at a concentration between 0.2 and 0.8% in the fermentor *(7)*. An advantage of Mut^s strains is that the culture is not as sensitive to residual methanol in the fermentor media relative to Mut$^+$ strains and hence the process of scale-up can be easier.

An alternative induction strategy for Mut^s strains is to use a mixed glycerol:methanol feed. Brierley et al. *(1)* investigated mixed feed ratios of 4:1 and 2:1 (g glycerol/h:g methanol/h). They also described a mixed feeding strategy in which feeding was begun at a 2:1 ratio, and gradually the proportion of methanol in the feed was increased until methanol began to accumulate. The maximum methanol uptake rate was achieved when a mixed feed ratio between 1:1 and 1:3 was observed. This method of slowly increasing the proportion of methanol during the mixed feed resulted in the highest recombinant protein expression levels and shortened the induction phase from 175 to 45 h. However, mixed feeding at high rates can also result in the production of inhibitory levels of ethanol ***(1)***.

Another induction feeding method for Mut[s] strain fermentations is a batch methanol mode in which a series of methanol additions are carried out over the length of the induction period ***(8)***. The first addition of methanol is initiated 30 min after exhaustion of glycerol from the first fed-batch phase.

1.3. Mut⁻ Strains

A Mut⁻ strain of *P. pastoris* with disruptions in both the *AOX1* and *AOX2* genes can be utilized for expression ***(9)***. This strain is totally defective in alcohol oxidase and therefore cannot utilize methanol at all. However, methanol will still induce transcription of the *AOX1* promoter and expression of genes regulated by the promoter ***(10)***. The inability of the strain to grow on methanol requires the use of an alternate carbon source, such as glycerol, for growth and protein production. However, excess glycerol can cause repression of the *AOX1* promoter, thus affecting expression of foreign genes. The feeding strategy for Mut⁻ expression strains is similar in concept to the mixed feed strategy for Mut[s] strains. Glycerol is fed under growth-limiting conditions and methanol is maintained at 0.5% during the induction phase ***(10)***. It is important to monitor the methanol level every 5–6 h and to periodically add methanol to replace that lost to evaporation from the fermentor. The optimum glycerol feed rate was determined to be 0.7 g glycerol/g of dry cell weight for maximum β-galactosidase protein production with minimal cell mass and ethanol production ***(10)***.

The following sections describe how to perform each of these fermentations. In addition, we describe a moderate-density fermentation procedure for Mut⁺ strains that does not require oxygen supplementation.

2. Materials

2.1. Media

2.1.1. Shake-Flask Media

MGY medium: Autoclave 700 mL of DI water. After autoclaving, add 100 mL of sterile 1 *M* potassium phosphate buffer, pH 6.0, 100 mL of filter sterilized solution of 10X (13.4 g/100 mL) yeast nitrogen base without amino acids, 100 mL of a filter-sterilized solution of 10X (10 g/100 mL) glycerol in water, and 2 mL of a filter sterilized solution of 500X biotin (20 mg/100 mL DI water).

2.1.2. Fermentor Media

1. *P. pastoris* grows well in defined media, reaching cell densities of 75–100 g dry cell wt/L. Two basic media formulations, basal salts and FM22, consistently have given good results and are described below. Fermentation media are typically added to and autoclaved in the fermentor.

2. Basal salts medium (per liter): 26.7 mL 85% H_3PO_4, 0.93 g $CaSO_4 \cdot 2H_2O$, 18.2 g K_2SO_4, 14.9 g $MgSO_4 \cdot 7H_2O$, 4.13 g KOH, 40 g glycerol. The medium will have a pH of 1.0–1.5 after autoclaving. After the medium reaches temperature (30°C) in the fermentor, the pH is adjusted to pH 5.0 with 30% NH_4OH. NH_4OH also serves as a source of nitrogen during fermentation.
3. FM22 medium (per liter): 42.9 g KH_2PO_4, 5 g $(NH_4)_2SO_4$, 1.0 g $CaSO_4 \cdot 2H_2O$, 14.3 g K_2SO_4, 11.7 g $MgSO_4 \cdot 7H_2O$, 40 g glycerol and adjust to pH 4.5 with KOH. The pH should be 4.5 after cooling (*see* **Note 1**).

2.1.3. Trace Element Solutions

Once the fermentor is attached to the control systems and the temperature of the medium has cooled to 30°C in the fermentor, trace salts are added. The primary difference between the two trace salts solutions, PTM1 and PTM4, is the level of ferrous sulfate and zinc chloride, with PTM1 containing approximately three times the amount of these two chemicals (*see* **Note 2**).

1. PMT1 (for use with Basal salts medium) (per liter): 6.0 g $CuSO_4 \cdot 5H_2O$, 0.8 g KI, 3.0 g $MnSO_4 \cdot H_2O$, 0.2 g $Na_2MoO_4 \cdot 2H_2O$, 0.2 g H_3BO_3, 0.5 g $CaSO_4 \cdot 2H_2O$, 20 g $ZnCl_2$, 65 g $FeSO_4 \cdot 7H_2O$, 0.2 g biotin, 5 mL conc. H_2SO_4. Filter-sterilize and add 2 mL/L of Basal salts medium. The trace salt solution is also added to the methanol feed supply at 2 mL/L.
2. PMT4 (for use with FM22 medium) (per liter): 2.0 g $CuSO_4 \cdot 5H_2O$, 0.08 g NaI, 3.0 g $MnSO_4 \cdot H_2O$, 0.2 g $Na_2MoO_4 \cdot 2H_2O$, 0.02 g H_3BO_3, 0.5 g $CaSO_4 \cdot 2H_2O$, 0.5 g $CoCl_2$, 7 g $ZnCl_2$, 22 g $FeSO_4 \cdot 7H_2O$, 0.2 g biotin, 1 mL conc. H_2SO_4. Filter-sterilize and add 1 mL/L of FM22 medium. The trace salt solution is also added to the methanol feed supply at 4 mL/L.

2.2. Fermentors

This section describes how three different fermentors are set up for *P. pastoris* high cell density fermentation. The three fermentors have working volumes of 1, 4, and 60 L and discussion will refer to each of the fermentors based on their working volume. The first part of this section will describe the standard attributes of the fermentors and specific modifications that were made to each fermentor. Later there will be a section on generic changes to the fermentors that were necessary to accommodate the high oxygen demand of the *P. pastoris* fermentations.

2.2.1. One-Liter Fermentor

For 1 L working volume (1.5 L total volume) benchtop fermentor, the Multigen manufactured by New Brunswick Scientific (NBS) Co., Inc. (Edison, NJ) or other suitable fermentor can be used. The unit should be integrated for control of temperature and agitation. Several of these small fermentors can be conveniently located side-by-side and connected to one computer for data acquisition. pH control is accomplished through base addition from external controllers, such as the pH 4000 controller from NBS, which includes a dual pump module for control of two different fermentors. (The pH 4000 pump module is designed to add acid and base, but the acid pump can be configured to add base on the second channel of the pH 4000 controller.) Alternatively, the Series 500 pH controller from Valley Instrument Co. (Exton, PA) can be used along with a 101F fixed-speed peristaltic pump from Watson-Marlow (Wilmington, MA).

Sterile air is introduced into the vessel through a sparger and is metered through a flow meter by a needle valve. To accommodate high oxygen demands, a gas-blending panel is used and is described in **Subheading 2.2.5.1.** Because substantial volumes of glycerol and methanol are added during fermentation, the starting volume of medium for batch growth phase should be 0.4–0.5 L. With the Multigens, the maximum working volume is approx 1.0–1.1 L.

A second modification to the Multigen fermentor is needed to improve mixing. The MultiGen vessels are factory equipped with two-baffle mixers mounted to the headplate. These baffles must be replaced with a four-baffle cage, similar to the baffle cage used in the BioFlo III fermentors (*see* **Fig. 1**). A third modification is necessary for temperature control at high cell densities. At optical densities (OD_{600nm}) above 300, *P. pastoris* cultures generate a substantial amount of heat. The "cold finger" heat exchangers that are supplied with the Multigen fermentor units extend only two-thirds of the distance into the culture. To correct this problem, the "cold fingers" should be modified to extend the full length of the fermentor vessel. A flow rate of 0.1–0.2 gallons/min of cold water (~13°C) is used to regulate temperature.

2.2.2. Four-Liter Fermentor

For 4-L working volume (5 L total volume) fermentor, a BioFlo III or the BioFlo 3000 (NBS) **(Fig. 1)** or other suitable fermentor can be used. (The rest of this chapter primarily refers to equipment and specifications for the BioFlo III.) The BioFlo III comes equipped with a microprocessor for control of pH, DO, agitation, temperature, nutrient feed rate, and foam control, and a serial communication port for supervisory computer control and data acquisition. Sterile air is introduced into the vessel through a ring sparger and is metered through a flow meter by a needle valve located on the front panel of the unit. However, for maximum control of air and oxygen flow rates, the gas blending package described in **Subheading 2.2.5.1.** is recommended and can be mounted on top of the BioFlo III (*see* **Fig. 1**).

The 5-L BioFlo III vessel has a maximum working volume under high agitation conditions of 4.0–4.5 L. An initial volume of 3 L is recommended for short induction periods (<40 h), whereas longer induction periods require an initial volume of 2.0 or 2.5 L. Initial volumes of 2.0–2.5 L require longer (320 mm) dissolved oxygen and pH probes. The standard probes (220 mm) require a minimum volume of 3.0 L.

Fig. 1. Picture of BioFlo III fermentor setup for a *P. pastoris* fermentation.

2.2.3. Equipment Common to Both One- and Four-Liter Fermentors

2.2.3.1. pH Control

Ammonium hydroxide is used to control pH and serves as the nitrogen source for the fermentation. Because of the volatility of NH_4OH, silicone or even Marprene tubing is not recommended for the entire run of tubing from the feed bottle to the fermentor. The diffusivity of NH_4OH through the tubing is sufficient to cause air gaps in the tubing. Teflon tubing (Western Analytical Prod., Temecula, CA) is used from the feed bottle to the fermentor, except for a short length of 0.8-mm bore Marprene tubing (Watson-Marlow) that is just long enough to fit in the peristaltic pump head. The Marprene tubing is connected to the Teflon tubing using peristaltic tubing adapters (UpChurch Scientific, Oak Harbor, WA). One end of the Teflon tubing is connected with an adapter to a 25-gage needle and inserted into the fermentor through a septum. The cap on the base-containing bottle is fitted with two openings, one for an air exhaust vent filter and the other for the Teflon tubing, which should extend down into the liquid base for feeding. The suction side of the pump (i.e., feed bottle to pump) is 1/8-in.-diameter (i.d.) Teflon tubing, whereras the supply side (i.e., pump outlet to the fermentor) is 1/16-in. i.d. tubing.

2.2.3.2. Methanol Feed Control

The same bottle and tubing configuration that is used for base addition is also used for the methanol feed because of the same diffusion problems. If silicone tubing is used from the feed bottle to the fermentor, gaps of air in the tubing will appear as a result of diffusion of the methanol through the tubing wall. This will cause irregular feeding of methanol, resulting in irregular DO profiles. Methanol is fed to the fermentor using a variable speed pump, such as the Watson-Marlow 101U/R. This pump is capable of delivering liquid volumes at rates as low as 0.001 mL/min (using 0.5-mm bore tubing) and as high as 53 mL/min (using 4.8-mm bore tubing). With this wide range of flow rates, the pump can be used for feeding methanol to fermentors 1–80 L in volume.

It is critical to set up the methanol feed bottle to allow for a continuous flow of liquid. The needle on the end of the feed line must be inserted at a 90° angle, without touching the sides of the fermentor headplate port to allow methanol to flow freely. If the needle is inserted at an angle, methanol pools at the headplate and an inconsistent delivery occurs. The pooling and release of large drops of methanol results in an inconsistent feed of methanol and an erratic DO profile, possibly resulting in a failed fermentation (*see* **Note 3**). Another alternative is to use a feed tube that is completely submerged, allowing methanol addition directly into the fermentor broth.

2.2.4. Larger Fermentors

The setup required for larger fermentors is generally similar to that for 4-L fermentors, only scaled up in volume. The main difference is that the amount of medium, methanol, and base used will increase. The initial and final volumes must still be calculated just as with the 4-L fermentation, only on a larger scale. Feeding strategies remain the same. We have the most experience with an 80-L (60-L working volume) Mobile Pilot Plant (NBS) and, therefore, will describe equipment and methods for this vessel. This bioreactor has a control panel and features similar to the BioFlo III, including data logging and computer control. However, like most large fermentation vessels, it is sterilizable-in-place and possesses a bottom-drive double mechanical agitator seal. The Watson-Marlow 101U/R variable-speed pump equipped with 0.8 mm-bore tubing is used for methanol feeding.

2.2.5. Modifications for High Cell Density Fermentation

2.2.5.1. Gas Blending Package

For high-cell density fermentations of *P. pastoris*, fermentors must be modified for addition of oxygen. To avoid oxygen limitation, the air supply must be supplemented with pure oxygen. For this, a gas-blending package can be purchased from Valley Instrument. The gas-blending package consists of an oxygen, air, and totalizing flow meter mounted on a stainless-steel panel. The oxygen flow meter for the Multigen has a range of 0–500 mL/min, whereas the BioFlo III uses 0–2000 mL/min. The air and totalizing flow meters for the Multigen and BioFlo III are 0–2000 and 0–5000 mL/min, respectively. Each of the gas-blending packages is equipped with an oxygen regulator, essential to ensure a steady flow of O_2. The O_2 supply pressure is regulated to 60 psig (pounds-per-square-inch gage) and the individual fermentor regulators are set to 20 psig each. A diagram of the Multigen setup, including the blending package, is presented in **Fig 2**. To ensure proper blending of air and O_2, the supply-side O_2 regulator should be set 5 psig higher than the air regulator, which ensures that O_2 will displace the air as the O_2 flow is increased. The DO controller for the Multigens (Valley Instrument Model 516M) controls a proportioning valve that meters oxygen automatically to maintain the DO at the desired set point.

2.2.5.2. Air Exhaust System

The next suggested modification concerns the air exhaust system. A standard air filter on the exhaust line can become plugged with condensate or medium and cells during lengthy fermentations. Instead of using a filter on the air exhaust line, a length of Marprene tubing can be attached to the end of the condenser and the end of the tubing placed in a 500-mL medium bottle containing 250 mL of 1 *N* NaOH. This medium bottle is then placed into a 4-L plastic beaker and covered in foil. This arrangement will contain cells and culture ejected during fermentation in the event of foaming.

2.2.5.3. Computer Monitor

Although not essential, DO, pH, and temperature can be logged electronically using such software as the NBS BioCommand through a Universal I/O converter also manufactured by NBS. The converter has multichannel, analog-to-digital/digital-to-analog signal conversion capabilities that can convert up to 10 independent analog inputs to NBS digital outputs, and vice-versa for data archiving and graphics display of fermentor outputs.

3. Methods

3.1. Inoculum Preparation

3.1.1. One- and Four-Liter Fermentors

1. Start inoculum from frozen cell stock cultures. Thaw frozen cells (1 mL) at room temperature and transfer to a 250-mL shake flask containing 50 mL of MGY medium. (One 50-mL shake flask is used to inoculate a 1-L fermentor and three are needed to inoculate a 4-L fermentor.).
2. Incubate flasks with shaking at 30°C for 20–24 h. The culture should be at an OD_{600nm} of between 2 and 5.
3. Add inoculum aseptically to the fermentor through a septum using a sterile 60-cc syringe with an attached needle. Another method is to set up the shake flasks with stoppers that contain a vent filter and inoculating tubing (stainless steel or glass) connected to sterile tubing of appropriate size and length. The inoculating end of the tubing can be attached to a needle that is then inserted into the septum on the fermentor. A peristaltic pump is then used to add the inoculum.

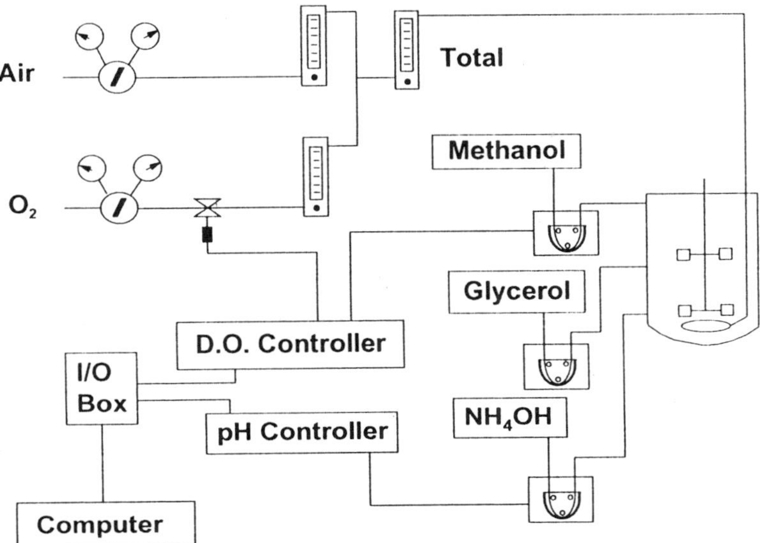

Fig. 2. Schematic diagram of setup for 1-L Multigen fermentor.

3.1.2. 60-Liter Fermentor

1. The inoculum train for a 60-L fermentation follows the same steps outlined in **Subheading 3.1.1.** for the 4-L fermentor. The 4-L fermentor is used to inoculate the 60-L fermentor. The medium used for the 4-L fermentor is the same as for the 60-L fermentor. The 4-L fermentor is prepared and autoclaved with a transfer line attached to the harvest port. The transfer line consists of sterile tubing of appropriate size and length that can be threaded through a peristaltic pump, such as 4.8-mm-i.d. Marprene tubing (Watson-Marlow), slipped over the end of the harvest port on the 4-L vessel. The other end is attached to a 16-gage needle and tie-wrapped to prevent leakage.
2. The needle is then punched through a septum on the 60-L fermentor, allowing the inoculum to be transferred by a peristaltic pump. If a peristaltic pump is not available, air pressure can be used to transfer the culture. The 4-L fermentor is operated in batch mode (i.e., no additional carbon source is added to the fermentor once the fermentor is inoculated). The OD_{600nm} in the 4-L fermentor should reach 40–50 before inoculating the 60-L fermentor. The high cell density inoculum will shorten the initial glycerol batch phase from 20–24 h to 16–18 h.

3.2. Fermentation

3.2.1. General Fermentation Procedure for Glycerol Batch and Fed-Batch Steps

1. The following is the procedure for a typical 4-L fermentation of a *P. pastoris* expression strain but can be scaled up or down in volume. For all fermentation

processes, whether working with Mut^+, Mut^s, or Mut^- phenotype strains, the strategy is to grow the yeast to high cell density on glycerol followed by induction with methanol. The three different phenotypes are distinguished by their growth rates on methanol—Mut^+ (0.14 h^{-1}), Mut^s (0.04 h^{-1}), and Mut^- (0.0 h^{-1})—and their induction procedures are described in this section.

2. The temperature, pH, and DO should be set at 30°C, 5.0, and 35%, respectively, and each should be set to automatic control. The air flow rate is set at 1.25 vvm (2.5 L) and is not changed during the fermentation.
3. The initial glycerol concentration is 4% (w/v).
4. The pH is controlled using a 30% by volume NH_4OH solution.
5. The 4-L fermentor (containing ~2 L of glycerol medium) is inoculated with the three 250-mL shake flasks, each containing 50 mL of culture as described above. This initial glycerol batch phase usually lasts 20–24 h. Automatic DO control is set at 35% during the glycerol batch phase until the agitation rate reaches 800 rpm. At this point, the DO control is turned off and agitation is controlled manually (800 rpm for the 4-L fermentor; 450 rpm for the 60-L fermentor), adjusting the flow of O_2 into the fermentor. When the glycerol is completely exhausted, the DO reading will rise sharply. The OD_{600nm} at this point should be 60–70.
6. Begin the glycerol fed-batch phase. The glycerol fed-batch phase further increases the cell density, initiates derepression of the *AOX1* promoter, and provides for a smooth transition to the methanol fed-batch stage. The feed pump with glycerol (50% w/v) is set to deliver 15 mL/L h based on the starting volume of 2.0 L. For the BioFlo III fermentor, the nutrient feed pump is used to feed glycerol. This phase will last for 4 h. During the glycerol fed-batch phase, the DO is controlled manually by setting the agitation to maximum, the total air flow to constant, and varying the amount of oxygen to keep the DO steady at 30–35% (*see* **Note 4**). During the glycerol fed-batch phase, perform "DO spikes" at 0.5, 1.0, 2.0, and 4.0 h to ensure that glycerol is not accumulating. A DO spike consists of shutting off the glycerol feed pump and timing how long it takes for the dissolved oxygen reading to rise 10%. Once the DO has increased 10%, the feed pump is turned back on. A final DO spike should also be performed at the end of the glycerol fed-batch phase just prior to starting the methanol feed. The OD_{600nm} at this point of the fermentation should be between 100 and 120.
7. Add antifoam if necessary to control excess foaming. The BioFlo III is equipped with sensors and peristaltic pumps for automatic addition of antifoam. A solution containing 5% Struktol JA 673 antifoam (Struktol Company of America; Stowe, OH) in 100% methanol is used to control foaming. This solution can also be added through a syringe filter. Undiluted antifoam (1–2 drops) may also be added to the fermentor through a 0.45-μ syringe filter to control excessive foaming.

3.2.2. Methanol Feeding of Mut^+ Strains

There are two key points to the fed-batch strategy for Mut^+ strains. The first is to make sure that all the glycerol is exhausted before beginning methanol feeding, and the second is to make sure that methanol does not accumulate in the medium during the induction phase.

1. Turn off the glycerol feed pump. The DO should rapidly rise to 100% or greater.
2. Set the methanol feed rate at 3.5 mL/L·h. Adjust pH for optimum protein production (*see* **Note 5**). This low initial methanol feed rate allows the culture to adapt to the new carbon source. This period may last anywhere from 2 to 5 h. During adaptation, the DO will steadily decrease from 100 to 40% over a 1–3-h period. Eventually the DO will stabilize and then begin to increase. An increase in DO means that the culture is beginning to starve for methanol and is ready to accept a faster methanol feed rate. Perform a DO spike before changing either the methanol or O_2 feed rate (*see* **Note 6**). Once the DO reaches 60%, the methanol feed rate can be increased.
3. Increase the methanol feed rate to 5 mL/L·h. After 2 h, perform DO spike. Continue this feed rate until the DO spike time is 30 s or less. Once this occurs proceed to the next step.
4. Increase the methanol feed rate by 1 mL/L·h. After 1 h check the status of the culture by performing a DO spike. If the spike time is <30 s, increase the feed rate another 1 mL/L·h. If the spike time exceeds 35 s, maintain that feed rate until the DO spike time drops below 30 s.
5. Continue the incremental increases at ~1-h intervals until an optimal feed rate of 11–12 mL/L·h is achieved. If DO spike times become longer than 1 min, reduce the methanol feed rate by 1–2 mL/L·h. Ideally, the feed rate should be brought to the maximum as quickly as possible. The methanol induction period may be as short as 18 h or as long as 96 h. The best way to determine the optimum induction period is to run a fermentation for the maximum induction period, examining time-course samples throughout and empirically determining the optimum time of harvest. The harvest should take place at the point in time when the most product has accumulated. If you have the capability to measure methanol concentration, keep the concentration above 0.4% and below 3% ***(11)***. A residual methanol level of 1% provides the best growth conditions ***(11)***. The final OD_{600nm} may reach 400 or higher. After centrifugation of methanol-induced cultures, the cell pellet should occupy 30–40% of the culture volume.

3.2.3. Methanol Feeding of Mut^s Strains

The fermentation process for Mut^s strains is identical to Mut$^+$ strains for steps described in **Subheading 3.2.1.** (glycerol batch and fed-batch steps). At the end of the glycerol fed-batch step, the methanol feed is started, and methanol concentration is maintained at <1% throughout the methanol-induction phase. Start the methanol feed rate at 1 mL/L·h and increase the rate over 6–8 h to 6 mL/L·h. Maintain that rate during the length of the induction period.

3.2.4. Methanol Induction of Mut$^-$ Strains

The fermentation process for Mut$^-$ expression strains is similar to that for *Escherichia coli* strains expressing proteins under control of the *lacZ* promoter.

In both, a carbon source must be provided along with the nonmetabolizable inducing agent. For Mut^- strains, the carbon source is glycerol. The batch and fed-batch steps are identical to those for Mut^+ strains (**Subheading 3.2.1.**) except that the initial concentration of glycerol is 20 instead of 40 g/L. At the completion of the fed-batch glycerol phase, the induction phase is started by initiating a glycerol feed rate of 1 g/L·h, and methanol is added to a final concentration of 0.5% *(10)*. During the induction phase, the methanol concentration should be maintained at 0.5%, monitored every 6–12 h, and adjusted accordingly.

3.3. Fermentation at Moderate Cell Density

A moderate cell density fermentation with Mut^+ strains requiring no oxygen supplementation or additional cooling is also possible with *P. pastoris*. Conditions are the same as described in **Subheadings 3.2.1.** and **3.2.2.** with the exceptions described below.

1. For the Bioflo III fermentor, set for automatic DO control throughout the entire fermentation. Perform DO spikes as described in **Subheading 3.2.1., step 6**. The agitation rate should not exceed 450 rpm at any point during the fermentation.
2. Prepare batch glycerol medium with a concentration of glycerol of 0.5–1.0%. This should result in a final batch density of 4–8 OD_{600nm} units. It may be possible to increase the concentration of glycerol in the initial batch glycerol medium to 1.0 or 1.5%, as long as the oxygen transfer rate of the fermentor is not limited. Fermentations with shorter or longer methanol induction times will need to adjust the initial batch glycerol concentration and feed strategies to maximize cell mass/protein production within the limits of the fermentor.
3. Once the initial glycerol is exhausted, a 50% glycerol solution is fed at 1.5 mL/L·h until the OD_{600nm} is 15 (2–3 h).
4. At the completion of the glycerol fed-batch phase, start a methanol feed at a rate of 1 mL/L·h. Adaptation to methanol usually requires approx 2 h.
5. Once adapted, increase the methanol feed rate to 2 mL/L·h. Continue methanol feeding for 4 h at this rate, then increase the rate to 3 mL/L·h for 2 h, then further increase the rate to 4 mL/L·h and maintain this rate for the duration of the methanol growth and induction phase. A maximum agitator speed of 400–500 rpm should be sufficient to attain a final OD_{600nm} of 50 with an 80-h induction.

3.4. Analytical Methods

Methanol and ethanol concentrations in the media are determined by gas chromatography using a Hewlett Packard 5880A GC with a column packed with chromosorb W (AW) (80/100; 10% carbowax 20M-TPA + 0.1% H_3PO_4 [Supelco, Bellefonte, PA]). Conditions include using nitrogen as the carrier gas at a flow rate of 18 mL/min and hydrogen as the fuel gas. The internal standard is n-propanol. The column oven temperature should be set at 80°C, the flame-ionization detector at 350°C, and the injector at 250°C.

Glycerol concentrations are determined by HPLC using a ICE-ASI, Dionex column, and Dionex Pulsed Amperometric Detector [Platinum electrode, E1, 0.20(60); E2, 1.25(60); E3, -0.1(240)] (Dionex, Sunnyvale, CA). The isocratic solvent was 100 m*M* perchloric acid at a flow rate of 1 mL/min ***(10)***.

3.5. Summary

The purpose of this chapter is to educate the reader about the basic equipment and strategies used in fermentations of *P. pastoris* in both bench-top and pilot-scale operations. A key element in expression of foreign proteins in this yeast is the need for sufficient aeration, which is achieved by proper mixing of the media and by blending gases to control dissolved oxygen content. Automatic pH control is essential for growth and expression in *P. pastoris*. Finally, fed-batch fermentations require the use of peristaltic pumps and tubing capable of low rates of delivery for the feeding of nutrients and base. Teflon tubing and peristaltic pump adapters are recommended for fed-batch operations.

The information in this chapter should enable a reader with little or no experience to perform a high-cell-density fermentation of a *P. pastoris* expression strain. Although most procedures described here are specifically for the BioFlo III (NBS), it should be possible to achieve high expression levels with almost any good-quality fermentor, modified to accommodate this organism.

4. Notes

1. Not all chemical vendors are suitable suppliers for the chemicals for defined fermentation media. Our lab has observed problems with magnesium sulfate from one source. Media prepared with chemicals from this source consistently formed precipitates, whereas the same chemical from another did not.
2. There will be some precipitate formation in batches of both PTM trace salt solutions when added to the media. A slight haze is acceptable.
3. It is important to note that the nutrient pump on the Bioflo III fermentor cannot be used for methanol feeding for Mut^+ fermentations. This pump operates intermittently, which will cause periodic fluctuations in DO levels. DO levels are not as critical during the glycerol fed-batch phase, so this pump can be used for this purpose.
4. Once the glycerol feed and the DO/oxygen flow rates are set in balance, it should not be necessary to further adjust the O_2 flow rate during the fed-batch phase.
5. The pH during methanol induction is set for optimum protein stability. This value is best determined by running a series of fermentations at different induction pHs.
6. Mut^+ strains are very sensitive to methanol concentrations. It is critical that a DO spike is performed before making any change in methanol or O_2 feed rate. If the DO is dropping steadily and reaches <15%, turn off the methanol feed pump and allow the culture to metabolize the excess methanol, which is evident by an increase in the DO. Once the DO spike has occurred, increase the O_2 by 0.1 L/min and turn on the methanol feed pump. What is happening is that the culture is metabolizing as much methanol as there is oxygen available. Basically the cul-

ture is starved for oxygen, as is evident by the low DO value. The most immediate response would be to increase the oxygen feed rate, but this could kill the culture. If there is too much methanol present (1–2%) and suddenly the oxygen level is increased, this may result in an abundance of formaldehyde in the culture, which will "pickle" the culture.

References

1. Brierley, R. A., Bussineau, C., Kosson, R., Melton, A., and Siegel, R. S. (1990) Fermentation development of recombinant *Pichia pastoris* expressing the heterologous gene: bovine lysozyme. *Biochemical Engineering VI, Ann. NY Acad. Sci.* **589,** 350–363.
2. Couderc, R. and Barratti, J. (1980) Oxidation of methanol by the yeast *Pichia pastoris:* purification and properties of alcohol oxidase. *Agric. Biol. Chem.* **44,** 2279–2289.
3. Swartz, J. R. and Cooney, C. L. (1981) Methanol inhibition in continuous cultures of *Hansenula polymorpha. Appl. Environ. Microbiol.* **41,** 1206–1213.
4. Hazeu, W. and Donker, R. A. (1983) A continuous culture study of methanol and formate utilization by the yeast *Pichia pastoris. Biotechnol. Lett.* **5,** 399–404.
5. Gleeson, M. A. and Sudbery, P. E. (1988) The methylotrophic yeasts. *Yeast* **4,** 1–5.
6. Cregg, J. M., Tschopp, J. F., Stillman, C., Siegel, R., Akong, M., Craig, W. S., Buckholz, R. G., Madden, K. R., Kellaris, P. A., Davis, G. R., Smiley, B. L., Cruze, J., Torregrossa, R., Velicelebi, G., and Thill, G. P. (1987) High-level expression and efficient assembly of hepatitis B surface antigen in *Pichia pastoris. Bio/Technology* **5,** 479–485.
7. Cregg, J. M. and Madden, K. R. (1988) Development of the methylotrophic yeast, *Pichia pastoris*, as a host system for the production of foreign proteins. *Dev. Ind. Microbiol.* **29,** 33–41.
8. Sreekrishna, K., Nelles, L., Potenz, R., Cruze, J., Mazzaferro, P., Fish, W., Fuke, M., Holden, K., Phelps, D., Wood, P., and Parker, K. (1989) High-level expression, purification, and characterization of recombinant human tumor necrosis factor synthesized in the methylotrophic yeast *Pichia pastoris. Biochemistry* **28,** 4117–4125.
9. Cregg, J. M. (1987) Genetics of methyltrophic yeasts, in *Proceedings of the Fifth International Symposium on Microbial Growth on* C_1 *Compounds* (Duine, J. A. and van Verseveld, H. W., eds.), Marinus Nijhoff, Dordrecht, The Netherlands, pp. 158–167.
10. Chiruvolu, V., Cregg, J. M., and Meagher, M. M. (1997) Recombinant protein production in an alcohol oxidase-defective strain of *Pichia pastoris* in fed-batch fermentations. *Enzyme Microb. Technol.* **21,** 277–283.
11. Murray, D. W., Duff, S. J. B., and Lanthier, P. H. (1989) Induction and stability of alcohol oxidase in the methylotrophic yeast *Pichia pastoris. Appl. Microbiol. Biotechnol.* **32,** 95–100.

10

Use of *Pichia pastoris* as a Model Eukaryotic System

Peroxisome Biogenesis

Klaas Nico Faber, Ype Elgersma, John A. Heyman, Antonius Koller, Georg H. Lüers, William M. Nuttley, Stanley R. Terlecky, Thibaut J. Wenzel, and Suresh Subramani

1. Introduction

Many subcellular processes present in multicellular eukaryotes have been shown to reside in yeast as well. The ease with which yeasts are genetically manipulated and analyzed by biochemical and ultrastructural techniques has greatly contributed to the present knowledge of these processes. In particular, the biogenesis, maintenance, and functions of the different subcellular organelles have been intensively studied in yeast. *Saccharomyces cerevisiae* has been the yeast of choice when the organelles under study are constitutively present in the cell, such as mitochondria, nucleus, Golgi, and the endoplasmic reticulum. In contrast, peroxisome biogenesis is studied in a variety of yeast species, mainly because the abundance of this organelle is highly inducible by external factors. Several nonconventional yeast species show a much more pronounced peroxisome proliferation than does *S. cerevisiae* ***(1–6)***.

Peroxisomes are believed to develop by growth and fission of pre-existing peroxisomes. Specific carbon and nitrogen sources have been found that require the activity of peroxisomal enzymes. The initial steps of oleate and methanol metabolism have been shown to reside within the peroxisome ***(7)***. *Pichia pastoris* is one of the few yeast species that is able to metabolize both these substrates with the accompanying strong proliferation of peroxisomes ***(3,4,8)***. This chapter emphasizes the importance of studying peroxisome biogenesis in yeast to the understanding of this process in higher eukaryotes (humans) and

From: *Methods in Molecular Biology, Vol. 103:* Pichia *Protocols*
Edited by: D. R. Higgins and J. M. Cregg © Humana Press Inc., Totowa, NJ

focuses on the yeast *P. pastoris*. For a general overview of peroxisome function and biogenesis, the reader is referred to recent reviews by Rachubinski and Subramani ***(9)***, Purdue and Lazarow ***(10)***, Braverman et al. ***(11)***, Subramani ***(12)***, and Van den Bosch et al. ***(13)***.

Peroxisomes are ubiquitous organelles that house different activities depending on the organism under study and the environmental conditions applied. Peroxisomes are defined as organelles bounded by a single membrane and containing an H_2O_2-producing oxidase together with catalase, which converts the H_2O_2 into H_2O and O_2 ***(14)***. Morphologically they resemble glycosomes (in trypanosomes) ***(15)***, glyoxysomes (in plants) ***(16)***, and hydrogenosomes (in anaerobic fungi) ***(17)***, and are therefore classified as microbodies. Peroxisomes play an essential role in human physiology as dramatically demonstrated by severe disorders associated with malfunctioning peroxisomes. Complementation analysis of patient cell lines revealed that over 11 genes are essential for proper peroxisome assembly in humans ***(11,18)***.

The dramatic effects of peroxisomal disorders in humans triggered the isolation of yeast mutants with similar phenotypes, e.g., the absence of functional and/or morphologically recognizable peroxisomes (peroxisome assembly or *pex*-mutants; *see* **Subheading 1.2.**) ***(3,4,19–24)***. These mutants were isolated from pools of mutagenized cells that were screened for their inability to utilize oleate and methanol. The isolation of yeast genes encoding either metabolic enzymes or proteins essential for peroxisome biogenesis has greatly contributed to our current understanding of protein targeting to, and biogenesis of, peroxisomes. Proteins involved in peroxisome biogenesis (peroxisomal matrix protein import, peroxisomal membrane biogenesis, peroxisome proliferation, and peroxisome inheritance) are designated peroxins, with *PEX* representing the gene acronym. The reader is refered to the table published by Distel et al. ***(25)***, which links the *PEX* numbers to the previously used names.

1.1. Peroxisomal Targeting Signals

1.1.1. Peroxisomal Targeting Signal 1 (PTS1)

Most peroxisomal matrix proteins are synthesized on free polysomes, sorted post-translationally to the organelle, and imported without further modification ***(26–28)***. The peroxisomal localization of firefly luciferase when expressed in mammalian cells led to the identification of the first signal involved in the targeting of peroxisomal proteins ***(29,30)***. The Ser-Lys-Leu (SKL) sequence present at the C-terminus of this protein was both necessary and sufficient for correct sorting ***(31,32)***. Since then, a great variety of SKL-like C-terminal sequences have been placed in the first group of peroxisomal targeting signals (PTS1s). In mammalian cells, the PTS1 conforms to the consensus S/A/C–K/H/R–L/*M* ***(12)***. Heterologous expression of PTS1-containing proteins in yeast, plants, insects, and mammalian cells revealed the universal nature of the signal ***(33,34)***. Furthermore, antibodies raised against the C-terminal PTS1 specifically recognize proteins from all four different subclasses of microbodies, underscoring their evolutionary relationship ***(17,34,35)***.

1.1.2. Peroxisomal Targeting Signal 2 (PTS2)

The second peroxisomal targeting signal was first characterized as an exception to the rule of "import without modification." Rat ketoacyl CoA-thiolase is synthesized as a precursor protein for which the N-terminus (26 amino acids in thiolase A; 36 amino acids in thiolase B) is cleaved on import. The first 11 residues of thiolase A contain sufficient information to direct a reporter protein to peroxisomes ***(36,37)***. Site-directed mutagenesis and identification of similar N-terminal sequences sufficient for targeting of other peroxisomal proteins have revealed a minimal consensus (R/K)(L/I/V)–X_5–(H/Q)(L/A) for the PTS2 ***(38–41)***. Clearly, processing of the PTS2 sequence is not a prerequisite for import of this class of proteins ***(36,38,42,43)***. Furthermore, in contrast to the PTS1, the PTS2 also functions at internal positions in the protein (Faber and Subramani, unpublished data). Identification of peroxisomal proteins from yeast, kinetoplasts, plants, and mammals containing a PTS2-like sequence demonstrates that this import signal is conserved throughout the eukaryotic kingdom.

1.1.3. Other PTSs and Import of Oligomers

Additional targeting signals for matrix proteins may exist, since several proteins do not contain either a PTS1 or a PTS2, such as *Candida tropicalis* acyl-CoA oxidase ***(44,45)*** and *S. cerevisiae Pex*7p ***(46)***. However, recent data provide an alternative explanation for targeting of such proteins to peroxisomes. Glover et al. ***(47)*** and McNew and Goodman ***(48)*** showed that polypeptides containing a PTS can form oligomers with polypeptides lacking a PTS and that the resultant complex can be imported into peroxisomes. These data suggest that protein unfolding is not a prerequisite for targeting to peroxisomes, a dogma generally accepted for targeting to other organelles (except to the nucleus). Walton et al. ***(49)*** showed that even gold particles of up to 9 nm in diameter and coated with a human serum albumin–HSA–SKL conjugate are imported into peroxisomes. Future experiments are needed to resolve whether import of oligomeric peroxisomal protein complexes is of importance in vivo, and if so, whether polypeptides lacking a PTS may enter the peroxisome via a heteromultimeric interaction with a protein containing a PTS.

1.1.4. Membrane Protein Targeting Signal (mPTS)

In contrast to matrix proteins, little is known about targeting of peroxisomal membrane proteins. They do not contain signals resembling either PTS1 or

PTS2 and, therefore, may be imported via a separate pathway. Defining the targeting signal(s) of peroxisomal membrane proteins (mPTSs) has been initiated by the laboratory of Goodman ***(50–52)***. PMP47, a *Candida boidinii* peroxisomal integral membrane protein, is thought to contain six transmembrane domains (TMD). A sequence between TMD 4 and 5, which is believed to face the peroxisomal matrix, has been shown to be sufficient to target an otherwise soluble reporter protein, chloramphenicol acetyltransferase (CAT), to the peroxisomal membrane. The 20 amino acid sequence contains a stretch of positively charged amino acids, which support membrane targeting.

PpPex3p is a peroxisomal membrane protein essential for peroxisome biogenesis in *P. pastoris* ***(53)***. Homologs have been found in *S. cerevisiae* ***(54)*** and *Hansenula polymorpha* ***(55)***, which share an overall identity of approx 30%. Deletion analysis revealed that the N-terminal 40 amino acids of PpPex3p are sufficient for targeting a reporter protein to the peroxisomal membrane ***(53)***. This sequence also contains a positively charged stretch of amino acids at positions 9–15 relative to the initiating methionine. Although this sequence is not particularly conserved among the three homologs, they all contain at least four positively charged amino acids in a stretch of six. The absence of an obvious membrane-spanning domain in the sequences sufficient for targeting suggests that they interact with another peroxisomal membrane protein. Surprisingly, the N-terminal 16 amino acids of HpPex3p, which includes the positively charged residues, targets a reporter protein to the endoplasmic reticulum (ER) ***(55)***. Whether this represents aberrant sorting of the fusion protein or sorting of a peroxisomal membrane protein via the ER is unclear. Although the model of autonomous proliferation of the peroxisome has been favored, these and other recent data require a re-evaluation of a possible role of the ER in peroxisome biogenesis ***(55–57)***.

1.2. Peroxisome Assembly Mutants

1.2.1. Isolation of Peroxisome Assembly Mutants

Since peroxisomes are strongly induced by growing *P. pastoris* either on methanol or oleate, their essential contribution in metabolizing these carbon sources is most obvious. In contrast, peroxisomes are barely detectable and much fewer in number in glucose-grown cells. In fact, whether they contribute at all to cell growth and maintenance under these conditions is unclear. Apparently they are not essential for cell viability, since many peroxisome assembly mutants have been isolated that are unable to utilize methanol and/or oleate, but are still able to grow on glucose-containing media.

Initial screens to identify such *pex* mutants were based on biochemical and morphological characterization of strains unable to utilize oleate and/or methanol ***(3,4,19–21)***. A large number of mutant strains of *P. pastoris* have been isolated by the laboratories of J. M. Cregg, S. J. Gould, and S. Subramani, and are presently placed into 16 complementation groups. In most mutants, the assembly of functional peroxisomes is clearly affected. However, in many, remnants of peroxisomes are detected ***(58–66)***. These peroxisomal ghosts resemble those observed in cells from patients suffering from generalized peroxisomal disorders (e.g., Zellweger syndrome). Although very successful, these isolation procedures require both biochemical (cell fractionation) and electron microscopical analyses to determine the mutant phenotype.

New tools have been developed both for identifying *pex* mutants and for screening of import of marker proteins without the need for biochemical or electron microscopical techniques. Elgersma et al. ***(23)*** developed an elegant positive screening method in baker's yeast using resistance to phleomycin as a selective criterion for *pex* mutants. A hybrid protein was produced consisting of the bleomycin resistance protein and a C-terminal portion of luciferase containing the PTS1 sequence, SKL. If properly targeted, the yeast strain is sensitive to phleomycin, since the resistance protein is sequestered from the drug in the cytoplasm. In mutants lacking functional peroxisomes, the hybrid protein is localized in the cytosol and prevents the harmful action of the drug. Among the phleomycin-resistant *S. cerevisiae* mutants obtained after chemical mutagenesis, up to 30% were affected in the biogenesis of peroxisomes. Several new *PEX* gene complementation groups were identified among 30 new *pex* mutant strains. This suggests that the pool of *PEX* genes is not yet saturated and that new genes may be identified using alternative screening methods.

For *P. pastoris*, the phleomycin screen has also been proven to be effective ***(103)*** (more details in **Subheading 1.4.**). An overview of the complementation groups and genes involved in peroxisome biogenesis in *P. pastoris* is shown in **Table 1**.

1.2.2. Genes and Functions

The intensive search for peroxisome-deficient mutants in yeasts has now resulted in approx 16 complementation groups for which, in most cases, homologs from different yeast species are found. Since the sequence data have revealed little about the functions of the encoded proteins, we now face the challenging task of unraveling the functions of these proteins by use of genetic, biochemical, and ultrastructural analysis. *PpPEX* gene products have been shown to reside either in the cytosol and on vesicles (PpPex1p ***[58]***, PpPex6p ***[62]***), or on the peroxisomal membrane (PpPex2p ***[59]***, PpPex3p ***[54]***, PpPex4p ***[60]***, PpPex5p ***[61]***, PpPex8p ***[63]***, PpPex10p ***[64]***, PpPex12p ***[65]*** and PpPex13p ***[66]***).

PpPex1p and PpPex6p are both members of the AAA-type family of ATPases of which the *N*-ethylmaleimide-sensitive factor (NSF) is the most well known

Table 1
***P. pastoris* Genes Involved in Peroxisome Biogenesis**

P. pastoris peroxin, mol wt	Mammalian homolog	Function/features of *P. pastoris* protein	Ref.
Pex1p (129)	+	AAA-type ATPase containing two ATP binding sites; C-terminal one is essential for function; cytosolic and vesicle-associated; interacts with Pex6p	***58***
Pex2p (65)	+	Peroxisomal membrane protein; Zn-finger domain at C-terminus	***59***
Pex3p (52)		Peroxisomal membrane protein; deletion strain does not contain peroxisomal ghosts; involved in early stages of peroxisome biogenesis	***53***
Pex4p (24)		Peroxisomal membrane protein; ubiquitin-conjugating enzyme	***60***
Pex5p (65)	+	Specific impairment in import of PTS1-containing proteins; peroxisomal membrane-associated protein, faces the cytosol; some may also be cytosolic; receptor for PTS1-containing proteins; contains 7 TPR domains in C-terminal part of protein; interacts with Pex13p.	***61,66***
Pex6p (127)	+	AAA-type ATPase containing one ATP binding site; cytosolic and vesicle-associated; interacts with Pex1p.	***62***
Pex7p (42)	+	Deficient only in import of PTS2-containing proteins; receptor for PTS2-containing proteins; contains β-transducin (WD40) domains	***103***
Pex8p (81)		Peroxisomal membrane protein	***63***
Pex10p (48)		Peroxisomal membrane protein; Zn finger motif at C-terminus, facing the peroxisomal matrix; deletion strain accumulates peroxisomal membranes	***64***
Pex12p (48)	+	Peroxisomal membrane protein; Zn-finger motif at C-terminus, facing the cytosol	***65***
Pex13p (41)	+	Peroxisomal membrane protein; contains SH3 domain; proposed Pex5p docking protein	***66***

(67,68). NSF, and its *S. cerevisiae* homolog Sec18p ***(69)***, are largely cytosolic proteins involved in membrane-fusion events occurring within the secretory pathway. Although the *S. cerevisiae* homolog of *PpPex1p* was the first *PEX* gene to be identified, its subcellular localization and function remain obscure. Recent data obtained in our lab show that both *P. pastoris* proteins are nonperoxisomal proteins, with significant amounts of both proteins associated with (small) membranous structures of unknown origin. This might indicate that these proteins are involved in membane traffic and addition to the peroxisome, activities similar to those of NSF and Sec18p ***(104)***.

Three peroxisomal membrane proteins containing a Zn finger motif have been shown to be required in peroxisome biogenesis in *P. pastoris* (PpPex2p, PpPex10p, and PpPex12p) ***(59,64,65)***. The corresponding deletion strains all contain peroxisomal remnants, suggesting a role for these proteins in peroxisomal matrix protein import rather than membrane protein sorting or membrane asssembly. The C-terminal Zn finger motifs are thought to be involved in protein–protein interactions, but no interacting partners have been identified so far.

PpPex3p is a peroxisomal membrane protein. The deletion strain does not contain morphologically recognizable remnants of peroxisomes in contrast to most of the other *PpPEX*-deletion strains, suggesting that PpPex3p is involved in an early stage of peroxisome biogenesis ***(53)***. Features of the remaining characterized PpPexps are given in **Table 1**. Loss of these proteins results in a general impairment in peroxisome assembly. Their functions are now being studied by determining their localization, topology, the proteins they interact with, and the effect of controlled expression on peroxisome development and morphology.

One specific class of *pex* mutants showed a selective import defect of specific matrix proteins, providing direct indication about the possible function of the complementing gene products, which will be discussed in the next section.

1.2.3. PTS Receptors

The existence of targeting signals that direct proteins to the peroxisomes implies the presence of receptor molecules for these PTSs. Among the *pex* mutants of *P. pastoris*, one complementation group (*Pppex5*) still contained small peroxisomes, which showed a selective import of proteins with a PTS2 sequence, whereas PTS1-containing proteins were localized in the cytosol ***(61)***. This indicates that these classes of proteins follow, at least partly, separate import routes. Not surprisingly, the gene that complements this mutant encodes the receptor for PTS1 proteins ***(61,70)***. The protein product, 68-kDa PpPex5p, harbors seven tetratricopeptide repeat (TPR) motifs at its C-terminal end. The TPR motif is a loosely conserved sequence thought to be involved in inter- or intramolecular interactions, and is also observed in Tom70p, a mitochondrial outer membrane protein that is the receptor for many polypeptides imported into mitochondria ***(71,72)***. Though no transmembrane segments are predicted from the primary amino acid sequence, the PpPex5p has been biochemically localized on the peroxisomal membrane facing the cytosol, a location and topology expected for a targeting signal receptor. In addition, some of the protein is found in the cytosol. In vivo and in vitro binding assays showed directly that PpPex5p binds to peptides ending in SKL ***(70)***. Furthermore, the TPR domains contain the site of PTS1 sequence binding.

Homologs of PpPex5p have been isolated from *S. cerevisiae* ***(73)***, *H. polymorpha* ***(74,75)***, and *Yarrowia lipolytica* ***(76)***, indicating that the PTS1 sequence is recognized by similar receptor molecules in the different yeasts. However, the subcellular location of each homolog appears to be different. ScPex5p is primarily found in the cytosol, HpPex5p is present in the cytosol as well as in the peroxisomal matrix, and YlPex5p is primarily present inside the peroxisome. This apparent discrepancy in the localization has led to models where the receptor shuttles into and out of the peroxisome ***(9)***. Recently, a peroxisomal membrane protein has been identified that interacts through its C-terminal src homology 3 (SH3) domain with PpPex5p ***(66)***, implying that this protein acts as the docking site for the PTS1 receptor on the peroxisomal membrane.

The PTS2R has been described in *S. cerevisiae* only ***(45,77,78)***. Mutant strains defective in PTS2R (*pex7* mutants) are selectively impaired in the import of PTS2 proteins. Lazarow and coworkers found the gene product (ScPex7p) exclusively in the peroxisome. In contrast, Kunau and coworkers reported a cytosolic and peroxisomal membrane localization, suggestive of a shuttling mechanism between the cytosol and the peroxisomal membrane. Both groups have reported the interaction between the PTS2R and the PTS2 signal ***(46,79)***. The N-terminal sequence of the PTS2R was found to act as a PTS for a heterologous reporter protein ***(46)***. This would seem to suggest a peroxisomal localization for ScPex7p, but, as with the PTS1R, a hypothetical shuttling between cytosol and peroxisomal matrix cannot be ruled out. No interaction has been found among the proposed docking protein for the PTS1R, PpPex13p, and the PTS2R. Interestingly, PTS1R and PTS2R show an interaction in a yeast–two-hybrid screen, but this requires another protein, Pex14p (W. H. Kunau, personal communication).

1.3. Human Genes

1.3.1. The Human PTS1R

The data obtained for the yeast PTS1Rs have been instrumental in the isolation of the human homolog (HsPex5p) of these proteins. Three different approaches have resulted in the isolation of this gene.

1. Dodt et al. ***(80)*** used the predicted amino acid sequences from the yeast genes to screen a human E(xpressed) S(equence) T(ag) library.
2. Fransen et al. ***(81)*** exploited the documented interaction between the PTS1 sequence and PpPex5p in the two-hybrid system to fish the receptor molecule out of a human cDNA library.
3. Wiemer et al. ***(82)*** obtained a partial cDNA clone whose product showed significant similarity with PpPex5p.

The deduced amino acid sequence of the HsPex5p shows 34% identity with PpPex5p. Highest similarity is observed between the C-termini containing the TPR domains. As with PpPex5p, HsPex5p is able to bind the SKL sequence in vitro. The HsPex5p is unable to substitute functionally for the PpPex5p in the *pex5*-deletion strain. Remarkably, a PpPex5p–HsPex5p fusion protein consisting of the N-terminal half (lacking the TPR domains) of PpPex5p and the C-terminal half (containing the TPR domains) of HsPex5p in the *pex5*-mutant resulted in the rescue of the oleate growth defect, whereas the strain remained unable to utilize methanol. Since the TPR domains are essential for, and most probably determine the specificity of, PTS binding, the HsPex5p C-terminal fragment is probably unable to recognize and transport one or more proteins essential for methanol metabolism.

As with the yeast homologs, both cytosolic and peroxisomal membrane localizations of the gene product have been reported. Additional experiments are required to resolve this discrepancy and to test the model of receptor shuttling through the peroxisome.

1.3.2. Other Human Homologs

As the sequencing of the human genome continues, many human homologs of yeast peroxisome assembly proteins are expected to appear in the data bases. Complementation screens using mammalian cells have revealed two genes (RnPEX2 and RnPEX6) for which yeast homologs are defined. RnPex2p is a peroxisomal membrane protein with a Zn binding motif ***(83)***, and of the three different classes of Zn binding proteins involved in peroxisome biogenesis, PpPex2p shows the highest homology ***(59)***.

Human *Pex*6p is a member of the AAA family of ATPases ***(84)***. From studies in yeast, it is clear that at least two individual members of the AAA family of ATPases are involved in peroxisome biogenesis ***(58,62,85–87)***. Human Pex6p shows highest homology to PpPex6p proteins localized to the cytosol ***(62,84)***. In contrast, the ortholog from rat, RnPex6p, was localized to the peroxisomal membrane ***(88)***, and PpPex6p was found to be predominantly vesicle-associated ***(104)***. Further experiments are needed to reveal whether PpPex6p/RnPex6p shuttle between the cytosol and the peroxisomal membrane and whether they have functions resembling that of NSF.

Recently, four additional human *PEX* homologs were identified in the EST data bases by similarity to their yeast homologs, *HsPEX6*, *HsPEX7*, *HsPEX12*, and *HsPEX13*, underscoring the value of yeast as a model organism for peroxisome biogenesis ***(89)***. All of these genes show restoration of peroxisomal functions in specific cell lines derived from fibroblasts from patients with peroxisome-related diseases, implying that mutations in these proteins are the direct cause of these diseases.

1.4. In Vivo Labeling of Peroxisomes

Improved procedures have been developed for biochemical and ultrastructural analysis of organelle biogenesis and protein trafficking in eukaryotic cells. Through cell fractionation, fractions enriched for specific organelles can be obtained and used for in vitro studies; fixed cells may be processed for electron microscopical analysis to determine organelle morphology and the location of proteins. These procedures are, in general, very laborious and do not allow studies within living cells. The recent cloning and characterization of the green fluorescent protein (GFP) from the jellyfish *Aequorea victoria* have provided a tool to study organelle biogenesis and protein targeting in vivo by direct fluorescence microscopy.

1.4.1. Green Fluorescent Protein

The 27-kDa GFP emits green light (λmax = 509 nm) on excitation with blue light (λmax = 395 nm) ***(90)***. The nontoxicity of heterologously expressed GFP and the minimal photobleaching of the fluorophore make it an attractive reporter for gene expression studies ***(91)***. GFP can be visualized in living cells without prior fixation and/or permeabilization, which makes it an ideal passenger protein to study protein import into subcellular compartments or for labeling of the compartments themselves. To be able to detect GFP at low expression levels, random mutagenesis has been performed on the wild-type gene to obtain mutant forms of GFP, which show eightfold (Ser65Thr mutant) ***(92)*** or even 40-fold enhanced intensity ***(93)*** of the emitted light. Also shifts in the wavelength of the emitted light have been observed in the mutagenesis experiments, resulting in a blue- and a red-shifted fluorescent protein ***(92,94,95)***.

1.4.2. GFP in Peroxisomes of P. pastoris

Hybrid proteins have been synthesized in *P. pastoris* consisting of the GFP fused to the C-terminal PTS1, SKL ***(96)***, and to the N-terminal PTS2 sequence of *S. cerevisiae* thiolase ***(53)***, and the localization compared to wild-type GFP and endogenous *P. pastoris* proteins. The hybrid proteins containing a PTS are sorted to peroxisomes, whereas GFP itself remains in the cytosol. GFP-PTS1 is efficiently sorted to peroxisomes, since hardly any fluorescence is observed in the cytosol in methanol-grown cells synthesizing the protein under control of the alcohol oxidase promoter. In cells containing the PTS2-GFP, peroxisomal fluorescence is obvious, although part of the pool of hybrid protein remains in the cytosol. Synthesis of this protein, controlled by the constitutive glyceraldehyde 3-phosphate dehydrogenase (GAPDH) promoter, in either glucose-, methanol-, or oleate-grown cells revealed that the efficiency of import is affected by the growth conditions applied (Wenzel and Subramani, unpublished results), a phenomenon also observed for PTS2 proteins in the yeast *H. polymorpha* ***(97)***. This indicates that (at least part of) the import pathway involved in import of PTS2-containing proteins is induced during growth on oleate and repressed during growth on methanol.

GFP-PTS1 has also been used to label the peroxisomal compartment in mammalian cells ***(98)***. The use of GFP as a vital stain for peroxisomes provides a tool to analyze aspects of peroxisome biogenesis, degradation, movement, segregation, inheritance, and targeting of proteins in vivo in real time.

1.4.3. Use of GFP in pex *Mutant Screens*

The subcellular localization of GFP-PTS1 and/or PTS2-GFP has been used as a primary analytical screen to select strains affected in peroxisome biogenesis from a pool of mutagenized cells, which were unable to metabolize peroxisome-requiring substrates ***(103)***. The laborious biochemical and ultrastructural analysis of potential *pex* mutants was followed by fluorescence microscopy of living cells, selecting only those strains that showed a clear aberrant localization of the GFP reporter molecule(s). Putative *P. pastoris pex* strains were selected basically as described by Elgersma et al. ***(23)*** for baker's yeast, using engineered strains synthesizing two hybrid proteins, either bleomycin-PTS1 and GFP-PTS1 or PTS2-bleomycin and PTS2-GFP. This allowed the isolation of mutants showing either a PTS1-specific, a PTS2-specific, or a general import defect. Several new complementation groups have been obtained through this screen, which are currently under investigation.

1.5. Concluding Remarks and Prospects

In recent years, *P. pastoris* has emerged as an excellent model organism to study peroxisome biogenesis. A large collection of organelle-specific mutants has been generated, and many genes have been cloned and characterized. Several of these genes have led to the cloning and characterization of the human homologs, and reveal the molecular basis of a fatal human disease. The detailed analysis of the growing number of *P. pastoris* genes involved in peroxisome biogenesis will undoubtedly increase our knowledge about peroxisome biogenesis in general and serve as a source to identify new human genes involved in this process. In addition to the isolation of mutants hampered in peroxisome biogenesis, the main lines of current research are the identification of protein–

protein interactions among the *Pex* proteins and the determination of the effects of controlled expression of *PEX* genes.

2. Materials

2.1. Plasmid Rescue

1. SD medium: 0.67% (w/v) yeast nitrogen base, 2% (w/v) dextrose, supplemented with the appropriate amino acids (histidine and arginine at 20 mg/L).
2. L-buffer: 50 m*M* dextrose, 25 m*M* Tris-HCl, 10 m*M* EDTA, pH 8.0.
3. 1% (w/v) SDS, 0.2 *N* NaOH.
4. 3 *M* sodium acetate, pH 5.3.
5. 96% Ethanol.
6. 70% Ethanol.
7. Acid washed glass beads (425–600 μm) (Sigma, St. Louis, MO).

2.2. Generation of Temperature-Sensitive Alleles by PCR

1. SD medium: 0.67% (w/v) yeast nitrogen base, 2% (w/v) dextrose, supplemented with the appropriate amino acids, if needed.
2. 10 ng DNA template.
3. *Taq* polymerase (Life Technologies, Gaithersberg, MD).
4. 10X PCR buffer: 200 m*M* Tris-HCl, pH 8.4, 500 m*M* KCl.
5. 100 m*M* $MgCl_2$.
6. 25 m*M* of each dNTP.
7. 10 m*M* $MnCl_2$.

2.3. Immunofluorescence Microscopy

1. YPD medium: 1% (w/v) yeast extract, 2% (w/v) Bacto-peptone, 2% (w/v) dextrose. Add 2% (w/v) Bacto-agar to prepare plates.
2. YPM: 2% Bacto-peptone, 1% yeast extract, 0.5% methanol.
3. YPOL: 2% Bacto-peptone, 1% yeast extract, 0.2% oleic acid, 0.02% Tween 40.
4. 37% Formaldehyde.
5. 100 m*M* potassium phosphate buffer, pH 7.4.
6. SP buffer: 1.2 *M* sorbitol, 20 m*M* potassium phosphate buffer, pH 7.4.
7. Zymolyase 20-T (ICN Biochemicals, Inc., Aurora, OH 44202).
8. β-Mercaptoethanol: stock solution is 14.3 *M*.
9. Poly-L-lysine solution (0.1%) (Sigma).
10. 100% Methanol (–20°C).
11. Phosphate-buffered saline (PBS): 8 g NaCl, 0.2 g KCl, 1.44 g Na_2HPO_4, 0.24 g KH_2PO_4 in 1 L distilled water, pH 7.4.
12. Primary antibody.
13. Fluorochrome-linked secondary antibody (Jackson ImmunoResearch Laboratories, Inc., West Grove, PA) mounting medium ***(99)***.
14. Mowiol solution: 2.4 g Mowiol 4–88 (Hoechst, Sumerville, NJ) in 6 g glycerol, stir, add 6 mL water, and leave for several hours at room temperature. Add 12 mL

of 0.2 *M* Tris-HCl (pH 8.5) and heat to 50°C for 10 min with occasional mixing. After Mowiol is dissolved, clarify by centrifugation at 5000*g* for 15 min. For fluorescence detection, add 1,4-diazobicyclo-[2.2.2]-octane (DABCO) to 2.5% to reduce fading. Aliquot in air-tight containers, and store at –20°C. Stocks are stable at room temperature for several weeks after thawing.

2.4. EM/Photooxidation of GFP

1. PHEM/KC*N* buffer: 60 m*M* PIPES, 25 m*M* HEPES, 5 m*M* glycine, 10 m*M* EGTA, 2 m*M* $MgSO_4$, 10 m*M* KCN, pH 6.5.
2. Paraformaldehyde (freshly prepared from powder (3%): 1.5 g paraformaldehyde (Sigma) is dissolved in 25 mL water and heated for 10 min at 60°C. A few drops of 1 *N* NaOH are added to clear the solution. Solution is cooled, and 25 mL 2X PBS are added (*see* **Subheading 2.3.** for PBS recipe). Final pH should be 7.4.
3. EM-grade formaldehyde (Ted Pella Inc., Redding, CA).
4. EM-grade glutaraldehyde (Ted Pella Inc.).
5. 125 m*M* HEPES, pH 7.3, 1 mg/mL diaminobenzidine (DAB) and 10 m*M* KCN.
6. 0.2 *M* cacodylate buffer, pH 7.2.
7. DAB (PolySciences, Warrington, PA); *see* **item 5**.
8. 1% (v/v) Osmium tetroxide in 0.2 *M* cacodylate buffer, pH 7.2.
9. Ethanol (100%).
10. Propylene oxide (Ted Pella Inc.).
11. Epoxy resin Araldite-EMbed 812 (Electron Microscopy Sciences, Fort Washington, PA).
12. Uranyl acetate (2% in water).
13. Lead citrate (Reynold's solution) ***(101)***: Add 1.33 g lead nitrate and 1.76 g sodium citrate to 30 mL of distilled CO_2-free water. (Prepare CO_2-free water by boiling for 8 min.) Mix and then add 8.0 mL of 1 *N* NaOH, and dilute solution to 50 mL with distilled CO_2-free water. Mix gently by inversion. Solution is ready to use when clear. The pH should be 12.0. Reynold's solution stocks may be used for several weeks if kept tightly stoppered. If precipitate appears, discard solution.
14. Beem capsules (Electron Microscopy Sciences).

2.5. Isolation of Peroxisomes

1. YP medium: 1% (w/v) yeast extract, 2% (w/v) Bacto-peptone, supplemented with 2% dextrose to make YPD, 0.5% methanol to make YPM, or 0.2% oleate and 0.02% Tween 40 to make YPOT.
2. S medium: 0.67% (w/v) yeast nitrogen base, supplemented with 2% dextrose to make SD, 0.5% methanol to make SM, or 0.2% oleate and 0.02% Tween 40 to make SOT. S media are supplemented with the appropriate amino acids (histidine and arginine at 20 mg/L each) according to the requirements of the strain in use.
3. Zymolyase 20-T (ICN Biochemicals, Inc.).
4. Nycodenz: 5-(*N*-2,3-dihydroxypropylacet-amido)-2,4,6-triiodo-*N*,*N*'-bis(2,3-dihydroxypropyl)-isophthalamide ($C_{16}H_{26}I_3N_3O_3$; mol wt 821.1) (Sigma) (*see* **Note 1**).

5. 100 m*M* Tris-HCl, 50 m*M* EDTA, 200 m*M* β-mercaptoethanol, pH 7.5.
6. KPi/sorbitol: 10 m*M* potassium phosphate buffer, pH 7.4 1.2 *M* sorbitol. Buffers containing sorbitol are freshly made each time. Use 50X KPi (0.5 *M* K_2HPO_4, 0.5 *M* KH_2PO_4, pH 7.4) as a stock solution.
7. Dounce buffer: 5 m*M* morpholinoethanesulfonic acid (MES), pH 6.0, 0.5 m*M* EDTA-Na, 1 m*M* KCl, 0.1% ethanol, 0.8 *M* sorbitol, plus protease inhibitors. 10X Dounce buffer (50 m*M* MES, 5 m*M* EDTA, 10 m*M* KCl, 1% ethanol, pH 6.0), stored at 4°C, can be used as a stock solution.
8. 100X stock solutions for the protease inhibitors:
   a. 100 m*M* PMSF (in 100% ethanol; H_2O inactivates PMSF, so make sure to add it just prior to Dounce homogenization).
   b. 1.25 mg/mL leupeptin (in H_2O).
   c. 0.5 mg/mL aprotinin (in H_2O).
   d. 0.5 *M* NaF (in H_2O).
9. Dounce tissue grinder, Fisher (Pittsburgh, PA).
10. Vertical rotor, for example VTi50, Beckmann Instruments (Fullerton, CA).
11. 40-mL quick-seal tubes, Beckmann Instruments.

3. Methods

3.1. Plasmid Rescue

1. Inoculate 20 mL of liquid SD medium with a fresh *P. pastoris* colony.
2. Incubate with vigorous shaking until culture has reached A_{600} >1.0 (~16 h).
3. Centrifuge cells for 5 min at 2000*g* in a tabletop centrifuge.
4. Wash cells once by centrifugation with L buffer, resuspend in 400 μL L-buffer, and transfer to a Eppendorf tube filled with an equal volume of glassbeads.
5. Vortex vigorously for 2 min, and place tubes on ice for 1 min. Repeat this procedure once.
6. Centrifuge for 2 min at maximum speed in a minicentrifuge to pellet intact cells and debris. All following centrifugations are performed at room temperature in a minicentrifuge set at maximum speed (16,000*g*).
7. Transfer 100 μL of the supernatant to a fresh tube. (The remaining steps are essentially a plasmid isolation from *Escherichia coli* using the alkaline lysis method.)
8. Add 200 μL 1% (w/v) SDS and 0.2 *N* NaOH, and mix by inverting three times.
9. Add 150 μL 3 *M* sodium acetate, pH 5.3. Mix by inverting three times and spin for 5 min.
10. Pour the clear supernatant into a fresh tube filled with 500 μL ethanol, mix, and spin for 5 min.
11. Discard supernatant, and wash pellet once with 400 μL 70% (v/v) ethanol.
12. Dry pellet, and dissolve in 20 μL water.
13. Transform *E. coli* with the plasmid preparation, and analyze transformants for plasmid content according to standard techniques. Plasmid yield may be low. High-efficiency electroporation is recommended.

3.2. Generation of Temperature-Sensitive (ts) Alleles by PCR

A ts allele of a particular gene encodes a protein that can be inactivated by cultivation at restrictive temperature. Such alleles are useful for several types of experiments. For example, a ts allele can be used in screens designed to identify proteins that, when overexpressed, will compensate for the reduced activity of the ts protein and complement the defect at the restrictive temperature. Often, suppression of this type indicates that the ts protein and the overexpressed protein physically interact, suggesting that the proteins interact with each other in wild-type cells. Additionally, it is possible to gain clues about a protein's function by examination of cells following temperature-mediated inactivation (or activation) of the protein of interest.

We have modified the mutagenesis method of Muhlrad et al. ***(100)*** to generate ts alleles of a cloned *P. pastoris* gene. In this protocol, the DNA of interest is first amplified using mutagenic PCR. This product, together with a gapped plasmid whose ends contain homology to the ends of the PCR product, are then used to cotransform *P. pastoris*. The gapped plasmid must contain the elements required for propagation of the plasmid in *E. coli*, as well as a marker for selection of yeast transformants and an origin of replication to maintain the plasmid as an autonomous element in *P. pastoris*. Homologous recombination in vivo between the mutagenized DNA and the gapped plasmid recircularizes the plasmid. This allows selection for transformants, subsequent screening for desired phenotypes, and recovery of the plasmid DNA.

1. Amplify DNA region of interest using mutagenic PCR. We used the following conditions in 50-μL reactions: 10 ng DNA template,1 μL BRL Taq polymerase, 5 μL BRL 10X PCR buffer (10X buffer is 200 m*M* Tris-HCl, pH 8.4, 500 m*M* KCl), 2 m*M* $MgCl_2$, 0.25 m*M* of each dNTP, and 0.1 m*M* $MnCl_2$ ($MnCl_2$ lowers *Taq* polymerase fidelity). Amplify in PCR cycler for 27 cycles of 1 min at 94°C, 2 min at 50°C, and 3 min at 72°C.
2. Prepare a gapped plasmid by restriction digestion and isolation of desired fragment from gel. The DNA deleted from the plasmid will be replaced by the PCR product following homologous recombination between the ends of the PCR product and the ends of the gapped vector (*see* **Note 2**).
3. Use the PCR product and gapped vector to cotransform a *P. pastoris* strain and spread the cells on SD plates. This will allow for selection of transformants by the selectable marker in our experiment *HIS4* (*see* **Notes 3** and **4**). For transformation protocols, *see* Chapter 3.
4. Incubate the cells at 30°C for 3–6 d until colonies appear.
5. Patch transformants onto fresh SD plates, and grow for 2–3 d at 30°C. (Use the blunt end of a sterile toothpick to pick a transformed colony, and then make an approx 1 × 3 mm streak of cells on an SD plate.) Approximately 100 colonies can be patched onto a standard size (10-cm diameter) Petri plate.

6. Replica plate transformants onto three identical selective plates, and incubate the plates at 25, 30, and 34°C.
7. Identify strains that do not grow at 34 and/or 30, but do grow at 25°C. Strains unable to grow at 30 and/or 34°C can be considered ts for growth on S plates.
8. Restreak the putative ts strains on selective plates, and confirm their ts phenotype on agar plates and in liquid medium.
9. Rescue plasmids from transformed strains (*see* **Subheading 3.1.**) displaying the desired phenotype, and retransform appropriate strains to confirm that the phenotype is linked with the plasmid.

3.3. Immunofluorescence Microscopy

This method is essentially the same as that described by Zhang et al. ***(24)***.

1. Culture cells overnight in YPD medium.
2. Harvest cells by centrifugation, inoculate into desired medium, and culture for the appropriate period of time (*see* **Note 5**).
3. Fixation of the cells: Add a 1/10 vol of 37% formaldehyde to the culture, and incubate for 0.5 h (*see* **Note 6**).
4. Transfer enough of the cells to a 1.5-mL minicentrifuge tube so that the tube will contain approx 20–100 μL of cell pellet. Wash cells by centrifugation two times with 100 m*M* potassium phosphate buffer (pH 7.4) and once with SP buffer.
5. Convert the cells to spheroplasts: Centrifuge cells and resuspend in SP buffer with 100 μg/mL Zymolyase 20-T and 1 μL/mL β-mercaptoethanol, and incubate at room temperature for 20 min (*see* **Note 7**).
6. Wash the spheroplasts two times with SP buffer, and resuspend in 30 μL SP buffer.
7. Apply 1 μL of cell suspension to microscope slide coated with poly-L-lysine (*see* **Note 8**), and air-dry for 5 min.
8. Immerse the slide in –20°C methanol for 5 min to permeabilize the cells. Wash immobilized cells 10 times with 50 μL of 1% BSA in PBS.
9. Apply approx 15 μL of diluted primary antibody (diluted in PBS containing 1% BSA) to the immobilized cells. Incubate at room temperature for 2 h.
10. Wash the cells 10 times with 1% BSA in PBS (*see* **Note 9**).
11. Apply fluorochrome-linked secondary antibody, and incubate for 1 h.
12. Wash cells 10 times with 1% BSA in PBS.
13. Add mounting medium (Mowiol) containing an antiphotobleaching agent ***(99)***, and apply coverslip.
14. Allow Mowiol to set overnight in dark.
15. Examine the samples by fluorescence microscopy.

3.4. Photooxidation of GFP for Electron Microscopy

3.4.1. Photooxidation of Yeast Expressing GFP

The wavelengths for excitation (395 nm) and emission (509 nm) of GFP are well separated, and the latter wavelength can be used to oxidize DAB to induce the deposition of electron-dense oxidized DAB at intracellular locations of

GFP. These deposits of oxidized DAB are easily visualized by electron microscopy. Since the coupling of GFP photoemission to DAB oxidation is accomplished with little perturbation of cellular structures, this technique affords simultaneous visualization of the location of GFP (and GFP derivatives) and subcellular morphology ***(96)***.

1. Culture GFP-synthesizing cells under desired conditions. Collect ~50–200 µL of cell pellet cells by centrifugation (1 min at 3000*g*) in a minicentrifuge. Wash cells for 5 min at room temperature in PHEM/KCN buffer. The KCN is added to prevent oxidation of DAB by mitochondria.
2. Prefixation: Centrifuge cells in a minicentrifuge, resuspend in 1 mL PHEM/KC*N* buffer containing 3% (w/v) freshly prepared paraformaldehyde (*see* **Note 10**), and incubate for 20 min at room temperature.
3. Fixation: Centrifuge cells, resuspend in 1 mL of a mixture of 3% (w/v) formaldehyde and 0.2% (v/v) glutaraldehyde in PHEM/KCN, and incubate for 20 min at room temperature.
4. Wash fixed cells five times with 1 mL PHEM/KCN buffer, and follow with a 10-min incubation in PHEM/KCN prewarmed to 48–50°C. (This step abolishes DAB oxidation in the vacuoles).
5. DAB photooxidation: Centrifuge cells, resuspend in 0.5 mL 125 m*M* HEPES, pH 7.3, containing 1 mg/mL DAB and 10 m*M* KCN, and transfer to UV-transparent cuvets. Allow DAB to infuse into cells for 5 min in the dark (DAB will oxidize in visual light), and then irradiate for 10 min at 395 nm in a spectrophotometer. A plastic tubing attached to a peristaltic pump is used to "bubble" air slowly into the bottom of the cuvet and through the solution to provide adequate mixing of the cells during exposure to UV light.
6. Transfer cells to a 1.5-mL minicentrifuge tube, centrifuge, and wash twice in 1 mL 0.2 *M* cacodylate buffer, pH 7.2. Let tube with cells suspended in cacodylate buffer stand for 5 min between each wash.

3.4.2. Electron Microscopy

1. Postfixation: After the photooxidation procedure, centrifuge cells, resuspend them in 1% (v/v) osmium tetroxide in 0.2 *M* cacodylate buffer, pH 7.2, and incubate for 45 min at room temperature.
2. Centrifuge cells, and wash twice with distilled water.
3. Dehydration: Resuspend cells in 500 µL distilled water, and add 500 µL 100% ethanol. Mix by inversion, and pellet cells. Remove 500 µL, and replace with 500 µL 100% ethanol. Repeat this step five times. Infiltrate with propylene oxide: Add 500 µL propylene oxide to the cells already in 500 µL ethanol and mix. Centrifuge cells, remove 500 µL, and add 500 µL propylene oxide. Mix. Centrifuge and resuspend in 1 mL 100% propylene oxide. Embed in Araldite-EMbed 812 as follows: Centrifuge cells, remove 500 µL propylene oxide, add 500 µL Araldite-EMbed 812, and mix (repeat two times). Centrifuge cells, and resuspend in Araldite-EMbed 812. Transfer the cell suspension to Beem capsules, and polymerize overnight at 60°C.

4. Section polymerized blocks, and double-stain with aqueous solutions of uranyl acetate (2% in water) and lead citrate ***(101)***.

3.5. Isolation of Peroxisomes

3.5.1. Growth Conditions

1. Inoculate a culture of *P. pastoris* in 500 mL YPD medium, and grow cells overnight at 30°C.
2. Collect cells by centrifugation (10 min at 2000*g*), wash the pellet once with sterile water and inoculate cells into 2 L of S*M* or SOT medium at a starting OD_{600} of 0.5.
3. Grow cells for about 16 h to induce peroxisome proliferation. The OD_{600} will increase to approx 2.

3.5.2. Preparation of Spheroplasts

1. Measure the OD_{600} of the culture, and collect cells by centrifugation at 2000*g* for 10 min at room temperature (for example, in a Sorvall GS3 rotor at 4000 rpm).
2. Resuspend cells in 30 mL of 100 m*M* Tris-HCl, pH 7.5, 50 m*M* EDTA, and 200 m*M* β-mercaptoethanol. Transfer the suspension into a 50 mL centrifuge tube, and incubate at room temperature for 20 min.
3. Centrifuge cells at 2000*g* for 10 min at room temperature (for example, in a Sorvall HB4 rotor at 3500 rpm or in tabletop centrifuge at maximum rpm), wash once in 30 mL of KPi/Sorbitol buffer, and resuspend cells again in 30 mL KPi/Sorbitol buffer.
4. Add Zymolyase T-20 (6 mg per 1,000 OD units [= OD_{600} × volume in mL]). Incubate the suspension for 30–45 min at 30°C with gentle rotation (*see* **Note 11**).

3.5.3. Dounce Homogenization and Differential Centrifugation

1. Collect spheroplasts by centrifugation at 2000*g* at 4°C.
2. Discard the supernatant, and resuspend the pellet in 20 mL of ice-cold Dounce buffer.
3. Pour suspension into a Dounce homogenizer, and break spheroplasts by applying 10 firm strokes.
4. Pour the homogenate back into the centrifuge tube, and spin at 2000*g* for 10 min to pellet unbroken spheroplasts, cell debris, and nuclei.
5. Transfer the supernatant to a clean tube, and centrifuge at 27,000*g* for 20 min at 4°C (for example, in a Sorvall SS34 rotor at 15,000 rpm).
6. Save the supernatant for further analysis (*see* **Notes 12** and **13**), and resuspend the pellet consisting of organelles (mainly mitochondria and peroxisomes) in 5–7 mL of Dounce buffer.

3.5.4. Preparation of Nycodenz Density Gradients

1. Dissolve Nycodenz in Dounce buffer at concentrations of 60, 50, 35, and 28% (w/v).
2. Pour a step gradient by carefully layering Nycodenz solutions over each other into a 40-mL Quick-Seal centrifuge tube (Beckman Instruments, Inc.). The volumes are 4 mL of 60%, 6 mL of 50%, 12 mL of 35%, and 10 mL of 28% of Nycodenz solution (from bottom to top).

3. Freeze the step gradient at –20°C. Gradients can be prepared in advance and are stable for a long time, and can be thawed whenever they are needed.

3.5.5. Density Gradient Centrifugation

1. Load the resuspended organelle fraction on top of the (thawed) Nycodenz density gradient, and use Dounce buffer to fill the tubes completely. Prepare a second tube as the counterbalance, and seal the tubes using a tube sealer.
2. Centrifuge the gradients at 90,000*g* for 2 h in a vertical rotor (for example, at 33,000 rpm in the VTi 50 from Beckmann Instruments Inc.). Acceleration and deceleration should be slow (brake off).
3. After centrifugation, inspect the gradients. A prominent band of peroxisomes (pale-green-colored owing to catalase) should be apparent in the lower third of the gradient. A band of mitochondria should be localized in the upper half of the gradient. Gradient fractions can be collected from the top or from the bottom of the gradient by using a needle attached to a peristaltic pump (*see* **Note 12**). If only one band is to be collected, a needle can be inserted underneath the band, which can then be eluted with a syringe.

4. Notes

1. Sucrose gradient centrifugation can also be used to separate peroxisomes from other organelles (mainly mitochondria in this protocol). Nycodenz, though more expensive, gives better results in our hands.
2. We have performed this procedure with a PCR product and gapped vector that shared 1200 bp at one end and 230 bp at the other end. The PCR product bridged a 1070-bp gap. Studies with *S. cerevisiae* demonstrated that a PCR product and a gapped plasmid with overlapping ends of as little as 65 and 69 nucleotides will support efficient recombination ***(100)***. It is possible that *P. pastoris* will also support homologous recombination between gapped plasmids and PCR products that contain similarly short regions of homologous overlap.
3. Our gapped plasmid contained incompatible ends to prevent recirculization of the plasmid independent of a recombination event between the plasmid and the PCR product. The transformation efficiency of the gapped plasmid alone was <5% of that seen for cotransformations of the gapped plasmid plus the PCR product. This indicates that *P. pastoris* does not efficiently circularize plasmids with incompatible ends. We do not know whether this is the same for plasmids containing compatible ends.
4. Cotransformations contained 0.1 µg gapped vector and 0.5 µg PCR product. We have not tried to optimize these conditions.
5. Cells grown in minimal media are difficult to permeabilize. Thus, peroxisomes are induced in YPM or YPOL. We have found that 7 h of growth in either medium is sufficient for significant induction of peroxisomal enzymes and peroxisomes are clearly visualized by immunofluorescence microscopy.
6. Incubation for 0.5 h in 3.7% formaldehyde is sufficient to fix the cells. The cells can be left for an extended period (several days) in formaldehyde and the immunofluorescence performed at a later stage.

7. Cells grown in minimal media require more Zymolyase for spheroplasting than cells grown in YP-based media. Since spheroplasting is essential for making cells permeable to antibody, it is worth trying several Zymolyase concentrations when attempting to optimize spheroplasting conditions for cells grown in a particular medium.
8. Poly-L-lysine is used according to manufacturer's (Sigma) instructions to immobilize the yeast cells onto the slide.
9. Antibodies used for immunofluorescence are usually used at a concentration five times that determined to be optimal for Western blotting.
10. Freshly prepared paraformaldehyde is sometimes referred to as partially depolymerized paraformaldehyde.
11. The successful generation of spheroplasts can be confirmed by light microscopy. Spheroplasts lyse when diluted in water and this can be easily observed with a light microscope. Alternatively, 10 µL of spheroplasts can be diluted in 1 mL of 0.1% SDS. Although untreated cells will stay as a turbid suspension, spheroplasts will lyse instantly leading to a clear solution. This can be followed by eye or even be quantified by spectrophotometry at 600-nm wavelength.
12. The 27,000*g* supernatant can be further subfractionated into a microsomal subfraction and a cytosolic fraction by centrifugation at 100,000*g* for 1 h at 4°C.
13. Analysis of the gradient: We analyze the gradient fractions routinely for activity for catalase and cytochrome-c oxidase as marker enzymes for peroxisomes and mitochondria, respectively ***(102)***. The density of each fraction is determined by use of a refractometer.

References

1. Veenhuis, M. and Harder, W. (1989) Occurrence, proliferation and metabolic function of yeast microbodies. *Yeast* **5,** 517–524.
2. Veenhuis, M. and Harder, W. (1991) Microbodies, in *The Yeasts*, vol. 4, 2nd ed. (Rose, A. H. and Harrison, J. S., eds), Academic, London, UK, pp. 601–653.
3. Gould, S. J., McCollum, D., Spong, A. P., Heyman, J. A., and Subramani, S. (1992) Development of the yeast *Pichia pastoris* as a model organism for a genetic and molecular analysis of peroxisome assembly. *Yeast* **8,** 613–628.
4. Liu, H., Tan, X., Veenhuis, M., McCollum, D., and Cregg, J. M. (1992) An efficient screen for peroxisome-deficient mutants of *Pichia pastoris*. *J. Bacteriol.* **174,** 4943–4951.
5. Veenhuis, M., Mateblowski, M., Kunau, W., and Harder, W. (1987) Proliferation of microbodies in *Saccharomyces cerevisiae*. *Yeast* **3,** 77–84.
6. Veenhuis, M. (1992) Peroxisome biogenesis and function in *Hansenula polymorpha*. *Cell. Biochem. Funct.* **10,** 175–84.
7. Fukui, S. and Tanaka, A. (1979) *Yeast* peroxisomes. *Trends Biochem. Sci.* **4,** 246–249.
8. Couderc, R. and Barratti, J. (1980) Oxidation of methanol by the yeast *Pichia pastoris*. Purification and properties of alcohol oxidase. *Agric. Biol. Chem.* **44,** 2279–2289.
9. Rachubinski, R. A. and Subramani, S. (1995) How proteins penetrate peroxisomes. *Cell* **83,** 525–528.
10. Purdue, P. E. and Lazarow, P. B. (1994) Peroxisomal biogenesis: multiple pathways of protein import. *J. Biol. Chem.* **269,** 30,065–30,068.

11. Braverman, N., Dodt, G., Gould, S. J., and Valle, D. (1995) Disorders of peroxisome biogenesis. *Hum. Mol. Genet.* **4,** 1791–1798.
12. Subramani, S. (1993) Protein import into peroxisomes and biogenesis of the organelle. *Ann. Rev. Cell Biol.* **9,** 445–478.
13. Van den Bosch, H., Schutgens, R. B. H., Wanders, R. J. A., and Tager, J. M. (1992) *Biochemistry* of peroxisomes. *Ann. Rev. Biochem.* **61,** 157–197.
14. De Duve, C. and Baudhuin, P. (1966) Peroxisomes (microbodies and related particles). *Physiol. Rev.* **46,** 323–357.
15. Opperdoes, F. R. and Borst, P. (1977) Localization of nine glycolytic enzymes in a microbody-like organelle in *Trypanosoma brucei*: the glycosome. *FEBS Lett.* **80,** 360–364.
16. Breidenbach, R. W. and Beevers, H. (1967) Association of the glyoxylate cycle enzymes in a novel subcellular particle from castor bean endosperm. *Biochem. Biophys. Res. Commun.* **27,** 462–469.
17. Marvin-Sikkema, F. D., Kraak, M. N., Veenhuis, M., Gottschal, J. C., and Prins, R. A. (1993) The hydrogenosomal enzyme hydrogenase from the anaerobic fungus *Neocallimastix* sp. L2 is recognized by antibodies, directed against the C-terminal microbody protein targeting signal SKL. *Eur. J. Cell Biol.* **61,** 86–91.
18. Wiemer, E. A. and Subramani, S. (1994) Protein import deficiencies in human peroxisomal disorders. *Mol. Genet. Med.* **4,** 119–152.
19. Erdmann, R., Veenhuis, M., Mertens, D., and Kunau, W.-H. (1989) Isolation of peroxisome-deficient mutants of *Saccharomyces cerevisiae*. *Proc. Natl. Acad. Sci. USA* **86,** 5419–5423.
20. Cregg, J. M., Van der Klei, I. J., Sulter, G. J., Veenhuis, M., and Harder, W. (1990) Peroxisome-deficient mutants of *Hansenula polymorpha*. *Yeast* **6,** 87–97.
21. Nuttley, W. M., Brade, A. M., Gaillardin, C., Eitzen, G. A., Glover, J. R., Aitchison, J. D., and Rachubinski, R. A. (1993) Rapid identification and characterization of peroxisomal assembly mutants in *Yarrowia lipolytica*. *Yeast* **9,** 507–517.
22. Van der Leij, I., Van der Berg, M., Boot, R., Franse, M. M., Distel, B., and Tabak, H. F. (1992) Isolation of peroxisome assembly mutants from *Saccharomyces cerevisiae* with different morphologies using a novel positive selection procedure. *J. Cell Biol.* **119,** 153–162.
23. Elgersma, Y., Van den Berg, M., Tabak, H. F., and Distel, B. (1993) An efficient positive selection procedure for the isolation of peroxisomal import and peroxisome assembly mutants of *Saccharomyces cerevisiae*. *Genetics* **135,** 731–740.
24. Zhang, J. W., Han, Y., and Lazarow, P. B. (1993) Novel peroxisome clustering mutants and peroxisome biogenesis mutants of *Saccharomyces cerevisiae*. *J. Cell Biol.* **123,** 1133–1147.
25. Distel, B., Erdmann, R., Gould, S. J., Blobel, G., Crane, D. I., Cregg, J. M., Dodt, G., Fujiki, Y., Goodman, J. M., Just, W. W., Kiel, J. A. K. W., Kunau, W.-H., Lazarow, P. B., Mannaerts, G. P., Moser, H. W., Osumi, T., Rachubinski, R. A., Roscher, A., Subramani, S., Tabak, H. F., Tsukamoto, T., Valle, D., Van der Klei, I. J., Van Veldhoven, P. P., and Veenhuis, M. (1996) A unified nomenclature for peroxisome biogenesis factors. *J. Cell Biol.* **135,** 1–3.

26. Roa, M. and Blobel, G. (1983) Biosynthesis of peroxisomal enzymes in the methylotrophic yeast *Hansenula polymorpha. Proc. Natl. Acad. Sci. USA* **80,** 6872–6876.
27. Goodman, J. M., Scott, C. W., Donahue, P. N., and Atherton, J. P. (1984) Alcohol oxidase assembles post-translationally into the peroxisome of *Candida boidinii. J. Biol. Chem.* **259,** 8485–8493.
28. Fujiki, Y., Rachubinski, R. A., Zentella, D. A., and Lazarow, P. B. (1986) Induction, identification, and cell-free translation of mRNAs coding for peroxisomal proteins in *Candida tropicalis. J. Biol. Chem.* **261,** 15,787–15,793.
29. de Wet, J. R., Wood, K. V., DeLuca, M., Helinski, D. R., and Subramani, S. (1987) Firefly luciferase gene: structure and expression in mammalian cells. *Mol. Cell. Biol.* **7,** 725–737.
30. Keller, G., Gould, S. J., Deluca, M., and Subramani, S. (1987) Firefly luciferase is targeted to peroxisomes in mammalian cells. *Proc. Natl. Acad. Sci. USA* **84,** 3264–3268.
31. Gould, S. J., Keller, G. A., and Subramani, S. (1987) Identification of a peroxisomal targeting signal at the carboxy terminus of firefly luciferase. *J. Cell Biol.* **105,** 2923–2931.
32. Gould, S. J., Keller, G. A., Hosken, N., Wilkinson, J., and Subramani, S. (1989) A conserved tripeptide sorts proteins to peroxisomes. *J. Cell Biol.* **108,** 1657–1664.
33. Gould, S. J., Keller, G. A., Schneider, M., Howell, S. H., Garrard, L. J., Goodman, J. M., Distel, B., Tabak, H. F., and Subramani, S. (1990) Peroxisomal protein import is conserved between yeast, plants, insects and mammals. *EMBO J.* **9,** 85–90.
34. Keller, G. A., Krisans, S., Gould, S. J., Sommer, J. M., Wang, C. C., Schliebs, W., Kunau, W.-H., Brody, S., and Subramani, S. (1991) Evolutionary conservation of a microbody targeting signal that targets proteins to peroxisomes, glyoxysomes, and glycosomes. *J. Cell Biol.* **114,** 893–904.
35. Gould, S. J., Krisans, S., Keller, G. A., and Subramani, S. (1990) Antibodies directed against the peroxisomal targeting signal of firefly luciferase recognize multiple mammalian peroxisomal proteins. *J. Cell Biol.* **110,** 27–34.
36. Swinkels, B. W., Gould, S. J., Bodnar, A. G., Rachubinski, R. A., and Subramani, S. (1991) A novel, cleavable peroxisomal targeting signal at the amino-terminus of the rat 3-ketoacyl-CoA thiolase. *EMBO J.* **10,** 3255–3262.
37. Osumi, T., Tsukamoto, T., Hata, S., Yokota, S., Miura, S., Fujiki, Y., Hijikata, M., Miyazawa, S., and Hashimoto, T. (1991) Amino-terminal presequence of the precursor of peroxisomal 3-ketoacyl-CoA thiolase is a cleavable signal peptide for peroxisomal targeting. *Biochem. Biophys. Res. Commun.* **181,** 947–954.
38. Glover, J. R., Andrews, D. W., Subramani, S., and Rachubinski, R. A. (1994) Mutagenesis of the amino targeting signal of *Saccharomyces cerevisiae* 3-ketoacyl-CoA thiolase reveals conserved amino acids required for import into peroxisomes in vivo. *J. Biol. Chem.* **269,** 7558–7563.
39. Tsukamoto, T., Hata, S., Yokota, S., Miura, S., Fujiki, Y., Hijikata, M., Miyazawa, S., Hashimoto, T., and Osumi, T. (1994) Characterization of the signal peptide at the amino terminus of the rat peroxisomal 3-ketoacyl-CoA thiolase precursor. *J. Biol. Chem.* **269,** 6001–6010.

40. Gietl, C., Faber, K. N., Van der Klei, I. J., and Veenhuis, M. (1994) Mutational analysis of the N-terminal topogenic signal of watermelon glyoxysomal malate dehydrogenase using the heterologous host *Hansenula polymorpha. Proc. Natl. Acad. Sci. USA* **91,** 3151–3155.
41. Motley, A., Hettema, E., Distel, B., and Tabak, H. (1994) Differential protein import deficiencies in human peroxisome assembly disorders. *J. Cell Biol.* **125,** 755–767.
42. Van der Klei, I. J., Faber, K. N., Keizer-Gunnink, I., Gietl, C., Harder, W., and Veenhuis, M. (1993) Watermelon glyoxysomal malate dehydrogenase is sorted to peroxisomes of the methylotrophic yeast, *Hansenula polymorpha. FEBS Lett.* **334,** 128–132.
43. Faber, K. N., Keizer, G. I., Pluim, D., Harder, W., Ab, G., and Veenhuis, M. (1995) The N-terminus of amine oxidase of *Hansenula polymorpha* contains a peroxisomal targeting signal. *FEBS Lett.* **357,** 115–120.
44. Small, G. M., Szabo, L. J., and Lazarow, P. B. (1988) Acyl-CoA oxidase contains two targeting sequences each of which can mediate protein import into peroxisomes. *EMBO J.* **7,** 1167–1173.
45. Kamiryo, Y., Sakasegawa, Y., and Tan, H. (1989) Expression and transport of *Candida tropicalis* peroxisomal acyl-coenzyme A oxidase in the yeast *Candida maltosa. Agric. Biol. Chem.* **53,** 179–186.
46. Zhang, J. W. and Lazarow, P. B. (1996) Peb1p (Pas7p) is an intraperoxisomal receptor for the NH_2-terminal, type 2, peroxisomal targeting signal of thiolase: Peb1p itself is targeted to peroxisomes by an NH_2-terminal peptide. *J. Cell Biol.* **132,** 325–334.
47. Glover, J. R., Andrews, D. W., and Rachubinski, R. A. (1994) *Saccharomyces cerevisiae* peroxisomal thiolase is imported as a dimer. *Proc. Natl. Acad. Sci. USA* **91,** 10,541–10,545.
48. McNew, J. A. and Goodman, J. M. (1994) An oligomeric protein is imported into peroxisomes in vivo. *J. Cell Biol.* **127,** 1245–1257.
49. Walton, P. A., Hill, P. E., and Subramani, S. (1995) Import of stably folded proteins into peroxisomes. *Mol. Biol. Cell* **6,** 675–683.
50. McCammon, M. T., McNew, J. A., Willy, P. J., and Goodman, J. M. (1994) An internal region of the peroxisomal membrane protein PMP47 is essential for sorting to peroxisomes. *J. Cell Biol.* **124,** 915–925.
51. McNew, J. A. and Goodman, J. M. (1996) The targeting and assembly of peroxisomal proteins: some old rules do not apply. *Trends Biochem. Sci.* **21,** 54–58.
52. Dyer J. M., McNew, J. A., and Goodman, J. M. (1996) The sorting sequence of the peroxisomal integral membrane protein PMP47 is contained within a short hydrophilic loop. *J. Cell Biol.* **133,** 269–280.
53. Wiemer, E. A. C., Lüers, G., Faber, K. N., Wenzel, T., Veenhuis, M., and Subramani, S. (1996) Isolation and characterization of Pas2p, a peroxisomal membrane protein essential for peroxisome biogenesis in the methylotrophic yeast *Pichia pastoris. J. Biol. Chem.* **271,** 18,973–18,980.

54. Höhfeld, J., Veenhuis, M., and Kunau, W.-H. (1991) PAS3, a *Saccharomyces cerevisiae* gene encoding a peroxisomal integral membrane protein essential for peroxisome biogenesis. *J. Cell Biol.* **114,** 1167–1178.
55. Baerends, R. J. S., Rasmussen, S. W., Hilbrands, R. E., Van der Heide, M., Faber, K. N., Reuvekamp, P. T. W., Kiel, J. A. K. W., Cregg, J. M., Van der Klei, I. J., and Veenhuis, M. (1996) The *Hansenula polymorpha* PER9 gene encodes a peroxisomal membrane protein essential for peroxisome assembly and integrity. *J. Biol. Chem.* **271,** 8887–8894.
56. Bodnar, A. G. and Rachubinski, R. A. (1991) Characterization of the integral membrane polypeptides of rat liver peroxisomes isolated from untreated and clofibrate-treated rats. *Biochem. Cell Biol.* **69,** 499–508.
57. Waterham, H. R., Titorenko, V. I., Swaving, G. J., Harder, W., and Veenhuis, M. (1993) Peroxisomes in the methylotrophic yeast *Hansenula polymorpha* do not necessarily derive from pre-existing organelles. *EMBO J.* **12,** 4785–4794.
58. Heyman, J. A., Monosov, E., and Subramani, S. (1994) Role of the PAS1 gene of *Pichia pastoris* in peroxisome biogenesis. *J. Cell Biol.* **127,** 1259–1273.
59. Waterham, H. R., De Vries, Y., Russell, K. A., Xie, W., Veenhuis, M., and Cregg, J. M. (1996) The *Pichia pastoris* PER6 gene product is a peroxisomal integral membrane protein essential for peroxisome biogenesis and has sequence similarity to the Zellweger syndrome protein PAF-1. *Mol. Cell Biol.* **16,** 2527–2536.
60. Crane, D. I., Kalish, J. E., and Gould, S. J. (1994) The *Pichia pastoris* PAS4 gene encodes a ubiquitin-conjugating enzyme required for peroxisome assembly. *J. Biol. Chem.* **269,** 21,835–21,844.
61. McCollum, D., Monosov, E., and Subramani, S. (1993) The pas8 mutant of *Pichia pastoris* exhibits the peroxisomal protein import deficiencies of Zellweger syndrome cells—the PAS8 protein binds to the COOH-terminal tripeptide peroxisomal targeting signal, and is a member of the TPR protein family. *J. Cell Biol.* **121,** 761–774.
62. Spong, A. P. and Subramani, S. (1993) Cloning and characterization of *PAS5*: a gene required for peroxisome biogenesis in the methylotrophic yeast *Pichia pastoris*. *J. Cell Biol.* **123,** 535–548.
63. Liu, H., Tan, X., Russell, K. A., Veenhuis, M., and Cregg, J. M. (1995) PER3, a gene required for peroxisome biogenesis in *Pichia pastoris*, encodes a peroxisomal membrane protein involved in protein import. *J. Biol. Chem.* **270,** 10,940–10,951.
64. Kalish, J. E., Theda, C., Morrell, J. C., Berg, J. M., and Gould, S. J. (1995) Formation of the peroxisome lumen is abolished by loss of *Pichia pastoris* Pas7p, a zinc-binding integral membrane protein of the peroxisome. *Mol. Cell Biol.* **15,** 6406–6419.
65. Kalish, J. E., Keller, G.-A., Morrell, J. C., Mihalik, S. J., Smith, B., Cregg, J. M., and Gould, S. J. (1996) Characterizaion of a novel component of the peroxisomal protein import apparatus using fluorescent peroxisomal proteins. *EMBO J.* **15,** 3275–3285.
66. Gould, S. J., Kalish, J. E., Morrell, J. C., Bjorkman, J., Urquhart, A. J., and Crane, D. I. (1996) *Pex*13p is an SH3 protein of the peroxisomal membrane and a docking factor for the predominantly cytosolic PTS1 receptor. *J. Cell Biol.* **135,** 85–95.

67. Wilson, D. W., Wilcox, C. A., Flynn, G. C., Chen, E., Kuang, W. J., Henzel, W. J., Block, M. R., Ullrich, A., and Rothman, J. E. (1989) A fusion protein required for vesicle-mediated transport in both mammalian cells and yeast. *Nature* **339,** 355–359.
68. Rothman, J. E. (1994) Mechanisms of intracellular protein transport. *Nature* **372,** 55–63.
69. Eakle, K. A., Bernstein, M., and Emr, S. D. (1988) Characterization of a component of the yeast secretion machinery: identification of the SEC18 gene product. *Mol. Cell. Biol.* **8,** 4098–4109.
70. Terlecky, S. R., Nuttley, W. M., McCollum, D., Sock, E., and Subramani, S. (1995) The *Pichia pastoris* peroxisomal protein PAS8p is the receptor for the C-terminal tripeptide peroxisomal targeting signal. *EMBO J.* **14,** 3627–3634.
71. Kiebler, M., Becker, K., Pfanner, N., and Neupert, W. (1993) Mitochondrial protein import: specific recognition and membrane translocation of preproteins. *J. Membr. Biol.* **135,** 191–207.
72. Lithgow, T., Glick, B. S., and Schatz, G. (1995) The protein import receptor of mitochondria. *Trends Biochem. Sci.* **20,** 98–101.
73. Van der Leij, I., Franse, M. M., Elgersma, Y., Distel, B., and Tabak, H. F. (1993) PAS10 is a tetratricopeptide-repeat protein that is essential for the import of most matrix proteins into peroxisomes of *Saccharomyces cerevisiae*. *Proc. Natl. Acad. Sci. USA* **90,** 11,782–11,786.
74. Van der Klei, I. J., Hilbrands, R. E., Swaving, G. J., Waterham, H. R., Vrieling, E. G., Titorenko, V. I., Cregg, J. M., Harder, W., and Veenhuis, M. (1995) The *Hansenula polymorpha* PER3 gene is essential for the import of PTS1 proteins into the peroxisomal matrix. *J. Biol. Chem.* **270,** 17,229–17,236.
75. Nuttley, W. M., Szilard, R. K., Smith, J. J., Veenhuis, M., and Rachubinski, R. A. (1995) The *PAH2* gene is required for peroxisome assembly in the methylotrophic yeast *Hansenula polymorpha* and encodes a member of the tetratricopeptide repeat family of proteins. *Gene* **160,** 33–39.
76. Szilard, R. K., Titorenko, V. I., Veenhuis, M., and Rachubinski, R. A. (1995) Pay32p of the yeast *Yarrowia lipolytica* is an intraperoxisomal component of the matrix protein translocation machinery. *J. Cell Biol.* **131,** 1453–1469.
77. Marzioch, M., Erdmann, R., Veenhuis, M., and Kunau, W. H. (1994) PAS7 encodes a novel yeast member of the WD-40 protein family essential for import of 3-oxoacyl-CoA thiolase, a PTS2-containing protein, into peroxisomes. *EMBO J.* **13,** 4908–4918.
78. Zhang, J. W. and Lazarow, P. B. (1995) PEB1 (PAS7) in *Saccharomyces cerevisiae* encodes a hydrophilic, intra-peroxisomal protein that is a member of the WD repeat family and is essential for the import of thiolase into peroxisomes. *J. Cell Biol.* **129,** 65–80.
79. Rehling P., Marzioch, M., Niesen, F., Wittke, E., Veenhuis, M., and Kunau, W.-H. (1996) The import receptor for the peroxisomal signal 2 (PTS2) in *Saccharomyces cerevisiae* is encoded by the *PAS7* gene. *EMBO J.* **15,** 2901–2913.
80. Dodt, G., Braverman, N., Wong, C., Moser, A., Moser, H. W., Watkins, P., Valle, D., and Gould, S. J. (1995) Mutations in the PTS1 receptor gene, *PXR1*, define complementation group 2 of the peroxisome biogenesis disorders. *Nature Genet.* **9,** 115–125.

81. Fransen, M., Brees, C., Baumgart, E., Vanhooren, J. C. T., Baes, M., Mannaerts, G. P., and Van Veldhoven, P. P. (1995) Identification and characterization of the putative human peroxisomal C-terminal targeting signal import receptor. *J. Biol. Chem.* **270,** 7731–7736.
82. Wiemer, E. A. C., Nuttley, W. M., Bertolaet, B. L., Li, X., Francke, U., Wheelock, M. J., Anne, U. K., Johnson, K. R., and Subramani, S. (1995) Human peroxisomal targeting signal-1 receptor restores peroxisomal protein import in cells from patients with fatal peroxisomal disorders. *J. Cell Biol.* **130,** 51–65.
83. Tsukamoto, T., Miura, S., and Fujiki, Y. (1991) Restoration by a 35K membrane protein of peroxisome assembly in a peroxisome-deficient mammalian cell mutant. *Nature* **350,** 77–81.
84. Yahraus, T., Braverman, N., Dod, G., Kalish, J. E., Morrell, J. C., Moser, H. W., Valle, D., and Gould, S. J. (1996) The peroxisome biogenesis disorder group 4 gene, *PXAAA1*, encodes a cytoplasmic ATPase required for stability of the PTS1 receptor. *EMBO J.* **15,** 2914–2923.
85. Erdmann, R., Wiebel, F. F., Flessau, A., Rytka, J., Beyer, A., Frohlich, K. U., and Kunau, W.-H. (1991) PAS1, a yeast gene required for peroxisome biogenesis, encodes a member of a novel family of putative ATPases. *Cell* **64,** 499–510.
86. Voorn-Brouwer, T., Van der Leij, I., Hemrika, W., Distel, B., and Tabak, H. F. (1993) Sequence of the PAS8 gene, the product of which is essential for biogenesis of peroxisomes in *Saccharomyces cerevisiae. Biochim. Biophys. Acta* **1216,** 325–328.
87. Nuttley, W. M., Brade, A. M., Eitzen, G. A., Veenhuis, M., Aitchison, J. D., Szilard, R. K., Glover, J. D., and Rachubinski, R. R. (1994) *PAY4*, a gene required for peroxisome assembly in the yeast *Yarrowia lipolytica*, encodes a novel member of a putative ATPase. *J. Biol. Chem.* **269,** 556–566.
88. Tsukamoto, T., Miura, S., Nakai, T., Yokota, S., Shimozawa, N., Suzuki, Y., Orii, T., Fujiki, Y., Sakai, F., and Bogaki, A. (1995) Peroxisome assembly factor-2, a putative ATPase cloned by functional complementation on a peroxisome-deficient mammalian cell mutant. *Nature Genet.* **11,** 395–401.
89. Subramani, S. (1997) *PEX* genes on the rise. *Nature Genet.* **15,** 331–333.
90. Inouye, S. and Tsuji, F. I. (1994) Evidence for redox forms of the Aequorea green fluorescent protein. *FEBS Lett.* **351,** 211–214.
91. Chalfie, M., Tu, Y., Euskirchen, G., Ward, W. W., and Prasher, D. C. (1994) Green fluorescent protein as a marker for gene expression. *Science* **263,** 802–805.
92. Cubitt, A. B., Heim, R., Adams, S. R., Boyd, A. E., Gross, L. A., and Tsien, R. Y. (1995) Understanding, improving and using green fluorescent proteins. *Trends Biochem. Sci.* **20,** 448–455.
93. Crameri, A., Whitehorn, E. A., Tate, E., and Stemmer, W. P. C. (1996) Improved Green Fluorescent Protein by molecular evolution using DNA shuffling. *Nature Biotechnol.* **14,** 315–319.
94. Heim, R., Prasher, D. C., and Tsien, R. Y. (1994) Wavelength mutations and post-transcriptional autooxidation of green fluorescent protein. *Proc. Natl. Acad. Sci. USA* **91,** 12,501–12,504.

95. Delagrave, S., Hawtin, R. E., Silva, C. M., Yang, M. M., and Youvan, D. C. (1995) Red-shifted excitation mutants of the green fluorescent protein. *Bio/Technology* **13,** 151–154.
96. Monosov, E. Z., Wenzel, T. J., Lüers, G. H., Heyman, J. A., and Subramani, S. (1996) Labeling of peroxisomes with green fluorescent protein in living *P. pastoris* cells. *J. Histochem. Cytochem.* **44,** 581–589.
97. Faber, K. N., Haima, P., Gietl, C., Harder, W., Ab, G., and Veenhuis, M. (1994) The methylotrophic yeast *Hansenula polymorpha* contains an inducible import pathway for peroxisomal matrix proteins with an N-terminal targeting signal (PTS2 proteins). *Proc. Natl. Acad. Sci. USA* **91,** 12,985–12,989.
98. Wiemer, E. A. C., Wenzel, T. J., Deerinck, T. J., Ellisman, M. H., and Subramani, S. (1997) Visualization of the peroxisomal compartment in living mammalian cells: dynamic behavior and associationwith microtubules. *J. Cell Biol.* **136,** 71–80.
99. Harlow, E. and Lane, D. (1988) *Antibodies: A Laboratory Manual.* Cold Spring Harbor Laboratory, Cold Spring Harbor, NY.
100. Muhlrad, D., Hunter, R., and Parker R. (1992) A rapid method for localized mutagenesis of yeast genes *Yeast* **8,** 79–82.
101. Reynolds, E. S. (1963) The use of lead citrate at high pH as an electronopaque stain in electron microscopy. *J. Cell Biol.* **17,** 208–212.
102. Storrie, B. and Madden, E. A. (1990) Isolation of subcellular organelles, in *Guide to Protein Purification* (Deutscher, M. P., ed.), Academic, San Diego, CA, pp. 203–225.
103. Elgersma, Y., Elgersma-Hoorsma, M., Wenzel, T., McCaffrey, M. J., Farquhar, M. G., and Subramani, S. (1998) A mobile receptor for peroximal protein import in *Pichia pastoris. J. Cell Biol.*, in press.
104. Faber, K. N., Heyman, J. A., and Subramani, S. (1998) The AAA-family peroxins, PpPex1p and PpPex6p, interact with each other in an ATP-dependent manner and are associated with different subcellular membranous structures distinct from peroxisomes. *Mol. Cell Biol.*, in press.

11

Secretion of Recombinant Human Insulin-Like Growth Factor I (IGF-I)

Russell A. Brierley

1. Introduction

The development of efficient recombinant protein production processes can be a critical factor in whether or not a pharmaceutical therapeutic protein can enter human clinical trials and ultimately the marketplace. This is especially true for therapeutic proteins that need to be administered on a daily basis for prolonged periods or if dosage requirements are very high. The use of *Pichia pastoris* as a recombinant expression host strain can be an excellent choice for such situations. *P. pastoris* has the potential for high expression levels ***(1,2)***, efficient secretion, and proper protein folding ***(3–5)***, and is a robust fermentation organism capable of high cell density on inexpensive simple basal salts medium ***(6)***. The development of an insulin-like growth factor I (IGF-I) production process in which *P. pastoris* is used as the recombinant host strain is discussed in this chapter.

IGF-I consists of 70 amino acids with a mol wt of 7648 Dalton. This single-chain protein has three intrachain disulfide bridges. IGF-I belongs to a heterogeneous family of peptides that share some of the biological and chemical properties of insulin. IGF-I promotes growth by mediating the effects of growth hormone. Thus, such processes as skeletal growth, cell replication, and other growth-related processes are affected by IGF-I levels. Physiological concentrations of IGF-I have been shown to be influenced by such conditions as thyroid disease, diabetes, and malnutrition ***(7)***. IGF-I has also been shown to act synergistically with other growth factors, such as accelerating the healing of soft and mesenchymal tissue wounds ***(8)*** and enhancing the growth of mammalian cells in serum-free tissue-culture medium ***(9)***. IGF-I has been indicated as a possible treatment for renal failure, dwarfism, insulin-resistant diabetes, and

From: *Methods in Molecular Biology, Vol. 103:* Pichia *Protocols*
Edited by: D. R. Higgins and J. M. Cregg © Humana Press Inc., Totowa, NJ

a variety of anabolic disorders ***(10,11)***. Additionally, IGF-I has potential in the treatment of a variety of neuronal disorders and injuries, such as amyotrophic lateral sclerosis (ALS), peripheral neuropathies caused by chemotherapy, diabetes of genetic origin, and rescue of injured CNS neurons caused by trauma or stroke ***(12,13)***.

IGF-I has been expressed in *Escherichia coli* and *Saccharomyces cerevisiae*. However, reported expression yields are quite low from *S. cerevisiae*, and IGF-I is produced intercellularly and in an insoluble state in *E. coli* ***(14–17)***. With expression in *P. pastoris*, IGF-I is secreted into the medium in a soluble form and at relatively high expression levels. Many critical factors that can affect development of a successful recombinant protein production process in *P. pastoris* are discussed in this chapter, such as gene copy number, controlling proteolytic degradation of product by both genetic manipulation and fermentation growth conditions, methanol-utilization phenotype, secretion and folding of the recombinant protein, and good analytical methodologies for characterization of the protein product.

2. Materials

2.1. Host Strains, Vectors, and Cloning Reagents

1. *P. pastoris* host strain GS115, NRRL Y-15851 (Invitrogen, San Diego, CA).
2. *E. coli* MC1061.
3. *P. pastoris* expression vector pAO815 (Invitrogen).
4. *P. pastoris* PEP4 disruption vector pDR421.
5. Synthetic human IGF-I DNA (Beckman, catalog #267421, Fullerton, CA).
6. *S. cerevisiae* prepro α-mating factor secretion signal.
7. *Eco*RI, *Bgl*II, *Bam*HI, *Stu*I, restriction enzymes.
8. 5-fluororoctic acid (5-FOA) plates: 0.67% yeast nitrogen base without amino acids, 2% agar, 2% glucose, 750 mg/L 5-FOA, 48 mg/L uracil.

2.2. Culture Media

1. YPD medium: yeast extract (10 g/L), peptone (20 g/L), dextrose (20 g/L).
2. YNB medium: yeast nitrogen base without amino acids (6.7 g/L), monobasic potassium phosphate (11.6 g/L), dibasic potassium phosphate (2.7 g/L).
3. Fermentation basal salts medium: phosphoric acid, 85% (46 g/L), calcium sulfate·$2H_2O$ (0.96 g/L), potassium sulfate (18.2 g/L), magnesium sulfate·$7H_2O$ (14.9 g/L), potassium hydroxide (4.1 g/L), glycerol (40.0 g/L).
4. PTM1 trace salts: biotin (0.2 g/L), boric acid (0.02 g/L), cobalt chloride·$6H_2O$ (0.5 g/L), sodium iodide (0.08 g/L), sodium molybdate·$2H_2O$ (0.2 g/L), cupric sulfate·$5H_2O$ (6.0 g/L), ferrous sulfate·$7H_2O$ (65.0 g/L), manganese sulfate·H_2O (3.0 g/L), zinc chloride (20.0 g/L), sulfuric acid (5.4 g/L).
5. Ammonium hydroxide, 28%.
6. KFO 673 antifoam (Kabo, Jackson, WY).

7. Glycerol.
8. Methanol.

2.3. Analytical Materials

2.3.1. IGF-I RIA Based on Incstar Antihuman Antisera

1. Rabbit anti-IGF-I antisera raised against last 17 amino acid carboxy-terminus (Incstar catalog #22275, Stillwater, MN).
2. ^{125}I-IGF-I (Incstar catalog #22303).
3. Pansorbin.
4. RIA buffer: 50 m*M* sodium phosphate, 0.1% BSA, 0.1% NaN_3, and 0.1% Triton X-100, pH 7.4.

2.3.2. Nichols Antihuman RIA Kit

This is available from Nichols Institute Diagnostic (San Juan Capistrano, CA).

2.3.3. Tricine SDS-PAGE

1. 13% Acrylamide separating gels in 1 *M* Tris-HCl, pH 8.5, 0.1% SDS, and 3% crosslinker.
2. 4% Acrylamide stacking gels in 1 *M* Tris-HCl, pH 8.5, 0.1% SDS, and 3% crosslinker.
3. Cathode running buffer: 0.1 *M* Tris-HCl, 0.1 *M* tricine, 0.1% SDS, pH 8.25.
4. Anode running buffer: 0.2 *M* Tris-HCl, pH 8.9.
5. 2X Sample buffer: 4% SDS, 12% glycerol, 0.1 *M* Tris-HCl, pH 6.8, 0.004% Coomassie brilliant blue G, 0.002% pyronin Y, and 0.1 *M* dithiothreitol (if samples are to be reduced).
6. Ethanol, acetic acid, trichloroacetic acid (TCA), glutaraldehyde.
7. 0.1% Silver nitrate.
8. Developer buffer: 3% sodium carbonate, 0.05% formaldehyde (37%).
9. Citric acid (2.3 *M*).

2.3.4. Western Blot

1. 0.1 μm Nitrocellulose.
2. Towbin buffer: 25 m*M* Tris, pH 8.3, 190 m*M* glycine, 20% methanol.
3. Blocking buffer: 0.25% gelatin, phosphate-buffered saline, 0.05% Tween 20, 0.02% sodium azide.
4. Rabbit anti-IGF-I antisera raised against last 14 amino acid carboxy-terminus of IGF-I conjugated to α-globulin.
5. ^{125}I-Protein A (0.02 μCi/mL).
6. IGF-I standard (Amgen, Thousand Oaks, CA).

2.3.5. Reverse-Phase (RP) HPLC

1. Vydac C4 RP column, 4.6 × 50 mm, (Vydac Catalog #214TP5405, Hesperia, CA).
2. Vydac C8 RP column, 4.6 × 150 mm (Vydac Catalog #280TP5415).

3. Acetonitrile, HPLC-grade.
4. Trifluoroacetic acid, HPLC-grade.
5. SP Sperodex M resin (BioSepra, Marlborough, MA).
6. Acetic acid.
7. Sodium chloride.
8. Sodium hydroxide.
9. Methanol.

2.3.6. Protease Overlay Assay Screen

1. YPD plates (YPD + 2% agarose).
2. Overlay medium: 0.6% agarose, 40% dimethylformamide (DMF), 1.2 mg/mL *N*-acetyl-DL-phenylalanine-β-naphthyl ester (APNE).
3. Fast garnet salt, 5 mg/mL.

3. Methods

3.1. Vector Construction and Transformation

3.1.1. Construction of Single-Copy Vector in pAO815

In order to direct the secretion of recombinant human IGF-I in *P. pastoris*, the synthetic IGF-I gene is fused to the DNA sequence encoding the prepro region of the *S. cerevisiae* mating hormone, α-mating factor (MF). The MF prepro sequence is an 89 amino acid polypeptide, which can direct peptides fused to it through the secretory pathway. The native MF prepro protein sequence contains three processing sites, one KEX2 cleavage site (Lys–Arg) and two dipeptidase cleavage sites $(\text{Glu–Ala})_2$, which are susceptible to the proteolytic action of two specific proteases localized in the secretion pathway. Proteolytic cleavage of an MF–IGF-I gene fusion protein at the processing site junctions allows the mature peptide to exit the cell. However, only one proteolytic processing site (Lys–Arg) is used in the construction of the MF–IGF-I gene fusion, since one processing site simplifies the maturation mechanism for the mature peptide to exit the cell. Additional signals for the secretion of IGF-I in *P. pastoris* have also been examined, but are not presented ***(18)***. The specific method used for fusing the IGF-I gene to the MF secretion signal and insertion into the pAO815 expression vector is given below.

1. A *Hin*dIII–*Bam*HI fragment containing the IGF-I gene is inserted into the *Hin*dIII–*Bam*HI site of a pUC18-based vector directly downstream of the DNA encoding the MF prepro region, including the three proteolytic processing sites.
2. M-13 mutagenesis is performed in order to prepare the MF–IGF-I gene fusion for insertion into an expression vector. The first mutagenesis is performed in order to remove the codon for the initial methionine on the IGF-I gene, the two MF dipeptidase processing sites (the $[\text{Glu–Ala}]_2$ residues), the *Hin*dIII cloning site, and the polylinker attached to the IGF-I synthetic gene. The mutagenized DNA is trans-

formed into *E. coli* JM103 cells and screened for isolates containing the desired sequence using an oligonucleotide probe.

3. A second mutagenesis is performed in order to insert an *Eco*RI site immediately following the translation termination codon of the IGF-I gene. The mutagenized DNA is transformed back into *E. coli* JM103 cells, screened for isolates containing the desired sequence using an oligonucleotide probe, and sequenced to confirm the changes.
4. Next, the MF–IGF-I fusion is inserted into the pAO815 expression vector. The MF–IGF-I fusion is isolated on a 478-bp *Eco*RI fragment and inserted into the *Eco*RI site of pAO815. The *P. pastoris* expression vector, pAO815, is shown in **Fig. 1**. This vector contains the ampicillin resistance gene and the origin of replication for *E. coli* in order to shuttle the vector into *E. coli* hosts to facilitate construction steps of the expression vector. The *P. pastoris AOX1* promoter and regulatory regions (5'), as well as the *AOX1* transcription termination and polyadenylation signals (3') are present on *Bgl*II–*Eco*RI and *Eco*RI–*Bam*HI fragments, respectively. In addition, pAO815 contains the *P. pastoris HIS4* gene used for selection in *his4 P. pastoris* hosts and *AOX1* 3' structural sequences. The *HIS4* and *AOX1* genes can be used to direct integration of the vector at either the *HIS4* or *AOX1* locus in the host genome. One of the unique features of the pAO815 vector is the *Bam*HI–*Bgl*II sites, which excise the entire *AOX1* promoter (5') and termination (3') regions, which allow isolation of an entire expression cassette from this vector in order to build in vitro multicassette expression vectors, as discussed below.
5. The resulting 8187-bp vector, pIGF201 (**Fig. 1**), contains an MF–IGF-I expression cassette (1755 bp) starting at the *Bgl*II site at position 6768 to the *Bam*HI site at position 336. The vector is transformed into *E. coli* MC1061 cells and screened by restriction enzyme digestion. After selection of one transformant containing the correct size insert, the entire MF–IGF-I fusion gene and ~50 nucleotides each of the promoter and termination regions of pIGF201 are then sequenced to verify the construction. The DNA sequence and corresponding amino acid sequence are given in **Fig. 2**.

3.1.2. Construction of Multicopy Expression Cassette Vector

1. The IGF-I expression cassette is isolated from pIGF201 as a 1755-bp *Bgl*II–*Bam*HI fragment. This expression cassette, encompassing the *AOX1* 5'-promoter, the MF–IGF-I fusion, and the *AOX1* 3'-terminator regions, is then used to generate additional copies of itself for multicassette vectors.
2. After the IGF-I expression cassette is excised from pIGF201 by *Bgl*II–*Bam*HI digestion, the correct DNA fragment containing the expression cassette is isolated and purified on an agarose gel.
3. The 1755-bp expression cassette is then inserted back into the unique *Bam*HI site of pIGF201 (*Bam*HI-digested and calf alkaline phosphatase-treated pIGF201). MC1061 *E. coli* cells are transformed with the ligation reaction.
4. Ampicillin-resistant colonies are screened by restriction digestion. Analysis of restriction enzyme digests of the resulting two-cassette vector, pIGF202, is then

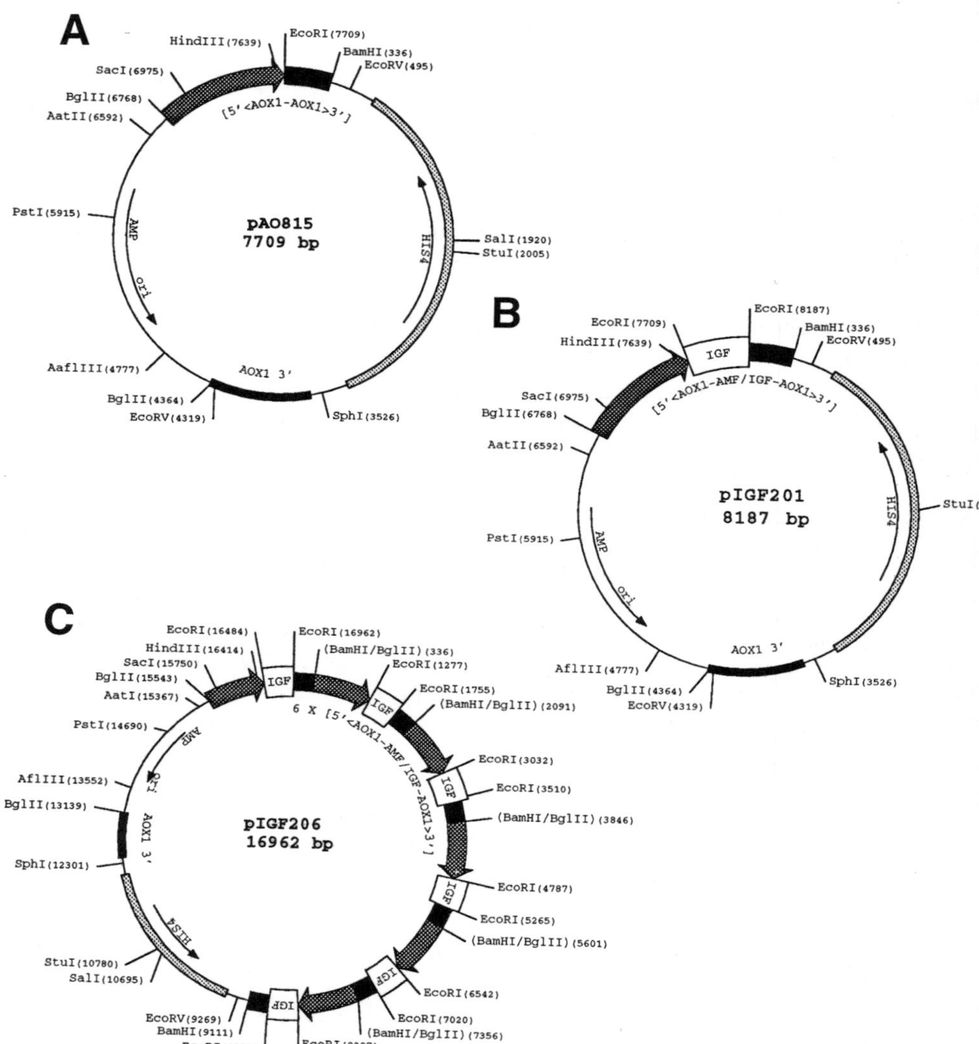

Fig. 1. Construction of single- and multicopy IGF-I expression vectors. The IGF-I gene fused to the *S. cerevisiae* α-MF secretion signal was inserted into the *Eco*RI site of pAO815 **(A)** to create the one-copy vector, pIGF201 **(B)**. The entire IGF-I expression cassette, containing the 5'-*AOX1* promoter—MF secretion signal fused to the IGF-I gene—3'-*AOX1* terminator, was isolated on a 1755-bp *Bam*HI–*Bgl*II fragment and inserted back into the *Bam*HI site of pIGF201 to create a two-copy vector, pIGF202. To construct a four-copy vector, pIGF204, the 3510-bp two-copy cassette from pIGF202 was isolated on a *Bgl*II–*Bam*HI fragment and inserted into the *Bam*HI site of pIGF202. To construct the six-copy vector pIGF206 **(C)**, the two-copy cassette from pIGF202 was cloned into the *Bam*HI site of pIGF204.

```
  1  AATTGCGACT GGTTCCAATT GACAAGCTTT TGATTTTAAC GACTTTTAAC
     AOX1 5' →                HindIII

 51  GACAACTTGA GAAGATCAAA AAACAACTAA TTATTCGAAA CGAGGAATTC
                                                 EcoRI

101  GatgAGATTT CCTTCAATTT TTACTGCAGT TTTATTCGCA GCATCCTCCG
      M  R  F  P  S  I  F  T  A  V  L  F  A  A  S  S  A

151  CATTAGCTGC TCCAGTCAAC ACTACAACAG AAGATGAAAC GGCACAAATT
       L  A  A  P  V  N  T  T  T  E  D  E  T  A  Q  I

201  CCGGCTGAAG CTGTCATCGG TTACTCAGAT TTAGAAGGGG ATTTCGATGT
     P  A  E  A  V  I  G  Y  S  D  L  E  G  D  F  D  V

251  TGCTGTTTTG CCATTTTCCA ACAGCACAAA TAACGGGTTA TTGTTTATAA
      A  V  L  P  F  S  N  S  T  N  N  G  L  L  F  I  N

301  ATACTACTAT TGCCAGCATT GCTGCTAAAG AAGAAGGGGT ATCTTTGGAT
       T  T  I  A  S  I  A  A  K  E  E  G  V  S  L  D

351  AAAAGAggaC CGGAGACGCT CTGCGGGGCT GAGCTCGTGG ATGCTCTGCA
     K  R  G  P  E  T  L  C  G  A  E  L  V  D  A  L  Q

401  GTTCGTGTGT GGAGACAGGG GCTTTTATTT CAACAAGCCC ACAGGGTATG
      F  V  C  G  D  R  G  F  Y  F  N  K  P  T  G  Y  G

451  GCTCCAGCAG TCGACGGGCG CCTCAGACAG GCATCGTGGA TGAGTGCTGC
       S  S  S  R  R  A  P  Q  T  G  I  V  D  E  C  C

501  TTCCGGAGCT GTGATCTAAG GAGGCTCGAG ATGTATTGCG CACCCCTCAA
     F  R  S  C  D  L  R  R  L  E  M  Y  C  A  P  L  K

551  GCCTGCCAAG TCAGCTTGAt aaGAATTCGC CTTAGACATG ACTGTTCCTC
      P  A  K  S  A  *  *    EcoRI    → AOX1 3' →

601  AGTTCAAGTT GGGCACTTAC GAGAAGACCG GTCTTGCTAG ATTCTAATCA

651  AGAG
```

Fig. 2. The pIGF201 sequence analysis containing 5'-*AOX1* (1–100), MF secretion signal (101–356), IGF-I (357–572), and 3'-*AOX1* (573–654).

performed to verify that the two expression cassettes are joined as tandem repeat units rather than inverted repeat units. Since the internal *Bgl*II site is destroyed if the two cassettes are ligated in tandem, *Bgl*II–*Bam*HI digestion liberates a 3510-bp fragment containing two expression cassettes.

5. To construct a four-cassette vector, the 3510-bp two-copy expression cassette is isolated on a *Bgl*II–*Bam*HI fragment from pIGF202 and gel-purified.
6. The fragment is then inserted into the unique *Bam*HI site in calf alkaline phosphatase-treated pIGF202 to create a four-cassette vector, pIGF204.
7. The ligation mixture is then transformed into MC1061 *E. coli* cells, and the ampicillin-resistant colonies are screened by restriction enzyme analyses.
8. To construct expression vector pIGF206, which contains six copies of the IGF-I expression cassette, the *Bgl*II–*Bam*HI fragment from pIGF202 (containing the two-copy IGF-I expression cassette) is cloned into the *Bam*HI site in calf alkaline phosphatase-treated pIGF204.

9. MC1061 *E. coli* cells are then transformed, and ampicillin-resistant colonies are screened by restriction enzyme analyses. The restriction digests of the vector DNA are examined to verify the number of expression cassettes and that they were joined as tandem-repeat units. The six-cassette vector, pIGF206, is shown in **Fig. 1**.

3.1.3. Transformation of Expression Vectors into P. pastoris

An *his4* mutant of *P. pastoris*, GS115 (NRRL Y-15851), is used as the host for the transformations. Histidine auxotrophy was created in GS115 by mutating a wild-type strain of *P. pastoris*, NRRL Y-11430 ***(19)***. For methanol-utilization positive strains (Mut^+), each vector is linearized with the restriction enzyme *Stu*I, which is located within the *HIS4* gene on the vector, prior to transformation. This directs integration of the entire *Stu*I-linearized expression vector into the *HIS4* locus of the GS115 host genome by an additive homologous recombination event. For methanol-utilization slow strains (Mut^s), the pIGF206 vector is linearized with the restriction enzyme *Bgl*II. Sites for this enzyme are located within the *AOX1* structural gene on the vector. In Mut^s strains, the *AOX1* chromosomal gene is disrupted by integration of the vector at the *AOX1* locus. Mut^s strains are still able to utilize methanol owing to the presence of the *AOX2* gene, but at a much slower rate than the Mut^+ strains.

3.2. Disruption of Protease PEP4 *Gene*

A protease-deficient derivative of an expression strain can be created by disrupting the *PEP4* gene, which encodes proteins that directly or indirectly affect protease activities of the cell. Disruption of this gene results in a reduction of a portion of the protease activity of the cell, including proteinase A and carboxypeptidase Y (CPY) activities. For IGF expression, a strain containing the six-copy expression vector, pIGF206, was used for transformation with a *PEP4* gene disruption vector, pDR421 ***(20)***. Vector pDR421 contains a 450-bp DNA fragment from the internal portion of the *P. pastoris PEP4* gene, which integrates into the host genome at the *PEP4* locus to generate two incomplete and nonfunctional copies of the *PEP4* gene. Also present in pDR421 is the 2000-bp *P. pastoris URA3* gene. Both these genes were inserted into the *E. coli* vector, pUC19, to create pDR421. More detail on protease-defective host strains can be found in Chapter 7. Details for creating a *pep4* production strain are given below.

1. In order to facilitate identification of transformants that incorporate the *PEP4* disruption vector pDR421, a *ura3* mutation is established in the strain prior to transformation. This is accomplished by isolation of a colony that, through spontaneous mutation, becomes auxotrophic for uracil (Ura^-). This isolation is accomplished by selecting for cells that are able to grow in the presence of 5-

FOA and uracil on YNB agar plates. 5-FOA is toxic for cells that contain a functional *URA3* gene. However, the compound is not toxic to *ura3* cells. Cells are plated at ~5 × 10^7 cells onto a 5-FOA-containing medium (0.67% yeast nitrogen base, 2% agarose, 2% glucose, 750 mg/L 5-FOA, and 45 mg/L uracil). After a 1–2 wk incubation at 30°C, colonies can be isolated that are Ura^-.

2. The *ura3* strain is transformed with the *PEP4* disruption vector, pDR421, and screened for its ability to grow on uracil-lacking medium.
3. Ura^+ transformants are then analyzed for CPY activity using a colony-overlay colorimetric screening procedure ***(21)***. The following is a basic overview of the method:
   a. Transformants are grown on YPD plates at 30°C.
   b. Plates are then overlaid with overlay medium (0.6% agarose, 40% DMF, 1.2 mg/mL APNE).
   c. After the overlay medium has hardened, the plates are soaked in a solution of 5 mg/mL fast garnet salt.

   APNE is cleaved by the esterolytic activity of CPY. The products of this reaction bind the fast garnet salt to produce a red color in the colony. Colonies lacking CPY activity do not bind the salt and, therefore, stain less intensely than the colonies with higher protease activity.
4. Colonies showing low CPY activity are isolated, subcultured, and rescreened using the overlay assay.
5. Confirmation of *pep4* disruption is then performed by Southern blot analysis of the final selected clone.

3.3. IGF-I Expression

3.3.1. Shake-Flask Analysis

Transformants containing the integrated vectors are first tested in a shake flask to evaluate IGF-I expression levels with respect to cassette copy number and Mut phenotype. The procedure is as follows:

1. Inoculate cultures in 250-mL triple-baffled flasks containing 50 mL of phosphate-buffered YNB + 2% glycerol (*see* **Note 1**).
2. Incubate flasks in a rotary shaker incubator at 250 rpm at 30°C.
3. After 24 h, harvest cells by centrifugation, wash with water, and resuspend at a seeding density of 0.1–0.2 OD (absorbance at 600 nm) in phosphate-buffered YNB + 1% methanol.
4. Incubate flasks at 250 rpm at 30°C.
5. Mut^+ strains reach an OD of 5–6 after ~2 d and are analyzed for IGF-I expression levels at that time. Mut^s strains reach an OD of 3–4 after ~4 d.

Results of the shake-flask studies are compared in **Table 1**. Supernatants from the shake-flask cultures comparing copy number in Mut^+ strains were analyzed for IGF-I levels by Western blot using an antisera raised against the last 14 amino acid carboxy-terminus. In this method, samples were first run on

Table 1
Expression of IGF-I in *Pichia pastoris*: Shake-Flask Studies

Cassette copy #	Mut phenotype	Western analysis	Incstar RIA
1	Mut^+	1.4 mg/L	nd[a]
2	Mut^+	2.7 mg/L	nd
4	Mut^+	4.3 mg/L	nd
6	Mut^+	8.1 mg/L	18 mg/L
6	Mut^s	nd	14 mg/L

nd = not determined.

a reduced SDS-PAGE gel and then transferred onto nitrocellulose for Western blot (*see* **Subheading 3.5.**). IGF-I levels for shake-flask cultures comparing Mut phenotype in strains containing the six-copy expression vector were measured by RIA using an antisera raised against the last 17 amino acid carboxy terminus (*see* **Subheading 3.5.**). As the data indicate, there is clearly a copy number effect on IGF-I expression in *P. pastoris* (*see* **Note 2**) ***(22)***. In addition, expression in Mut^+ and Mut^s phenotypes is comparable. Although there is some difference in the IGF-I levels determined by Western blot and RIA analysis (8.1 vs 18 mg/L for six-copy Mut^+), given the sensitivity of these assays, they are in fairly good agreement.

3.3.2. Fermentation Analysis

After shake-flask analysis indicated that IGF-I expression was optimal with the Mut^+ six-copy strain, this strain was tested under fermentation conditions. During the course of developing an IGF-I production process, the fermentation process was standardized for optimal production conditions. Although some of the early fermentations varied slightly from this protocol, all the fermentation followed the same basic protocol.

The fermentation process for production of IGF-I from *P. pastoris* includes three distinct stages. First, cells are grown in a batch mode using excess glycerol as the carbon source. The excess glycerol allows a rapid production of cell mass, although no IGF-I is produced owing to repression of the *AOX1* promoter by excess glycerol. Following exhaustion of the glycerol, a limited glycerol feed is initiated in a fed-batch mode, such that glycerol does not accumulate. During this second stage, cell mass continues to accumulate rapidly, but since the glycerol is limited in the medium, the *AOX1* promoter becomes derepressed causing small amounts of methanol utilization pathway enzymes to be produced, which prepares the culture for growth on methanol. The third stage of the fermentation is initiated by replacing the glycerol with methanol as the carbon source fed into the fermenter. Cell mass accumulates more slowly in this phase, and IGF-I is secreted into the fermentation medium at a constant rate over a 3-d period of methanol feeding. Since this stage is carried out in a fed-batch mode, volume in the fermenter will increase steadily, nearly doubling the initial volume. The detailed fermentation protocol is as follows.

3.3.2.1. Glycerol Batch Phase

1. Fermenters are sterilized with an initial volume of fermentation basal salts medium. It is best to use a starting volume of ~50% of the maximum working volume of the fermentation vessel.
2. After sterilization and cooling, set temperature to 30°C, and set agitation and aeration to operating conditions (usually mid to maximum rpm and 1–2 vvm [volume gas/volume liquid/min] air). Adjust the pH of the basal salts medium to 5.0 with 28% ammonium hydroxide. Take care not to overshoot pH or excessive medium precipitation will occur. Some medium precipitation is expected at pH 5.0.
3. Add 2 mL of PTM1 trace salts/L of basal salts medium (*see* **Note 3**).
4. The fermenter is inoculated with ~5–10% initial fermentation volume from seed culture generated in inoculum shake-flasks or seed fermenters.
5. The batch culture continues until the glycerol is completely consumed (12–24 h, depending on the size of the inoculum used). This is indicated by a sharp rise in dissolved oxygen levels. A cellular yield of 90–150 g/L wet cells is expected from 4% glycerol utilized for this stage, and no product is produced.
6. Temperature is controlled at 30°C, pH at 5.0, with the addition of 28% ammonium hydroxide and excess foam with the addition of 5–100% Kabo antifoam. Maintain the dissolved oxygen level at >20% saturation by increasing agitation, fermenter back pressure, air flow rate, or by initiating oxygen feeding. If oxygen feeding is used, balance the air and oxygen feeds to give a constant flow rate into the fermenter, preferably at 1 vvm total air flow and up to a maximum of 2 vvm total air flow (*see* **Note 4**).
7. Manual operator measurements are carried out at regular intervals to confirm proper control, and to measure the consumption rates of ammonium hydroxide, antifoam, glycerol, and methanol. The culture is also measured for carbon limitation during glycerol and methanol fed-batch stages by performing dissolved oxygen (DO) spikes.

3.3.2.2. Glycerol Fed-Batch Phase

1. Initiate by starting a 50% w/v glycerol feed at a feed rate of ~20 mL/h/L initial fermentation volume. The glycerol feed should not be initiated until 15 min or more after the batch phase is completed. This is indicated by a sudden rise in DO and a leveling off of the DO at a high level, typically above 70% (*see* **Note 5**).
2. Glycerol feeding is carried out for 4 h or until ~80 mL/L (initial fermentation volume) 50% glycerol has been fed into fermenter. A cellular yield of 180–220 g/L wet cells should be achieved at the end of this stage, but no appreciable product is produced.

3. DO spikes should be performed 15–30 min after initiating the glycerol feed and at 1–2 h intervals to ensure that the culture is limited for glycerol. DO spikes are performed by terminating the carbon feed and timing how long it takes for the DO to rise 10%, after which the carbon feed is resumed.

3.3.2.3. Methanol Fed-Batch Phase

1. Initiate by terminating the glycerol feed and starting a 100% methanol feed at a feed rate of approx 3.5 mL/h/L initial volume. The glycerol feed is left off for at least 5 min before proceeding with the methanol feed.
2. For low-pH fermentations, the pH of the fermentation medium can be decreased to 3.0 by changing the control set point to 3.0 and allowing the metabolic activity of the culture to lower the pH slowly to 3.0 over 4–5 h (*see* **Note 6**).
3. During the first 2–3 h, methanol will accumulate in the fermenter, and the dissolved oxygen will steadily decrease. DO spikes are not performed during this time. Increase agitation, aeration, pressure, or oxygen feeding during this phase to maintain the DO above 20%. If the DO cannot be maintained above 20%, stop the methanol feed, wait for the DO to spike, and continue on with the current methanol feed rate.
4. After the culture is fully adapted to methanol utilization (2–4 h) and the culture is limited on methanol, the feed rate is doubled to ~7 mL/h/L initial fermentation volume. This will be indicated by a steady DO reading and a fast DO spike time (generally under 1 min). It is recommended to maintain the lower methanol feed rate under limited conditions for 1 h before doubling the feed.
5. After increasing the feed rate to 7 mL/h/L, DO spikes should be performed after ~15 min and as often as necessary to ensure that the culture is limited on methanol (*see* **Note 7**).
6. After 2 h at the 7 mL/h/L feed rate, the methanol feed rate is again increased to approx 11 mL/h/L initial fermentation volume. This feed rate is maintained for a total of 24 h on methanol when the methanol feed can be further increased to 13 mL/h/L in order to ensure full methanol addition by harvest time.
7. The entire methanol fed-batch phase lasts ~70 h with a total of ~740 mL methanol fed/L of initial volume. For many recombinant proteins, a direct correlation between amount of methanol consumed and the amount of product produced has been observed. Therefore, some attention needs to be given to the total amount of methanol fed during the fermentation. The cell density continues to increase during the methanol fed-batch phase to a final level of ~450 ± 100 g/L wet cells. Because most of the fermentation is carried out in a fed-batch mode, the final fermentation volume will be approximately double the initial fermentation volume (*see* **Notes 8** and **9**).

3.3.2.4. Fermentation Results with Six-Copy Mut$^+$ Strain

The results of the fermentation studies are summarized in **Table 2**. These fermentations were carried out in 2-L vessels starting out at a 1-L volume. As stated previously, the Mut$^+$ six-copy strain was first tested under standard fer-

Table 2
Expression of IGF-I in *Pichia pastoris*: Fermentation Studies

Fermentation	Cassette copy #	Phenotype	pH	Western analysis, mg/L	Incstar RIA, mg/L	Nichols RIA, mg/L	C4 RP-HPLC, mg/L[a]	Cell density, wet g/L
A	6	Mut^+	5	117	306	15	3	325
B	6	Mut^+	3	555	1550	140	121	385
C	4	Mut^+	3	nd[b]	1400	174	103	350
D	2	Mut^+	3	nd	740	65	39	430
E	1	Mut^+	3	21	167	35	14	415
F	6	Mut^s	3	nd	745	nd	nd	370

[a]HPLC values reported as authentic IGF-I values that represent ~20% of the total IGF-I forms present in HPLC.
[b]nd = not determined.

mentation conditions (**Table 2**, line A) at pH 5.0 throughout. Cell density increased as expected throughout the fermentation. Although IGF-I was secreted into the medium during the fermentation with the six-copy Mut^+ strain, the levels of IGF-I produced were not as high as expected, based on the shake-flask studies. Also, it was noted that on studying the time-course of the fermentation, the levels of IGF-I stopped increasing early in the fermentation. It was hypothesized that proteolytic degradation of IGF-I might be occurring during the fermentation. In order to test this hypothesis, IGF-I standard was added into the final fermentation medium, which had been adjusted to several different pH ranges (pH 3.0–5.0) using phosphoric acid (*see* **Note 6**). It was discovered that all of the IGF-I would ultimately be completely degraded after several days at pH 5.0; at pH 3.0, most of the IGF-I would remain intact with time (data not shown). As a result, the fermentation was repeated at pH 3.0 during the methanol fed-batch phase with the six-copy Mut^+ strain (**Table 2**, line B). Results demonstrate that IGF-I levels were improved significantly (555 vs 117 mg/L by Western, 1550 vs 306 mg/L by Incstar RIA), and a steady increase in IGF-I levels was observed throughout the fermentation time-course.

Several additional procedures were employed to evaluate analytically the results of the fermentations. One of these procedures was a second RIA from Nichols Institute, which is based on antisera raised against the entire IGF-I protein instead of the last 17 amino acids of the carboxy-terminus with the Incstar antisera. A C4 RP HPLC procedure was also developed in order to characterize better the IGF-I being secreted into the medium by *P. pastoris* (*see* **Subheading 3.5.**). As **Table 2** indicates, the Nichols RIA and RP-HPLC values are lower than the Western and Incstar RIA values, in particular for the pH 5.0 fermentation at 15 and 3 mg/L, respectively. However, the pH 3.0 fermentations showed a similar improvement in IGF-I assay values for the Nichols RIA (140 vs 15 mg/L) and RP-HPLC (121 vs 3 mg/L) as observed using the Western and Incstar RIA assays.

Despite the improvement in assay results in the pH 3.0 fermentations, the levels of IGF-I measured by Nichols RIA and RP-HPLC are consistently lower than those determined by the Western blot and Incstar RIA. The reason for these lower values was elucidated during further development and characterization work (*see* **Subheading 3.4.**). The discrepancy in IGF-I values for each assay is owing to the fact that there are a variety of forms of IGF-I secreted into the fermentation broth by *P. pastoris*, but each assay measures only certain forms. The major IGF-I forms include correctly folded monomeric IGF-I (authentic), misfolded monomeric IGF-I, degraded or nicked IGF-I, and multimeric (dimeric, trimeric) IGF-I. The majority of these forms are readily separated and quantified by C4 RP-HPLC (**Fig. 3**), with the exception of the

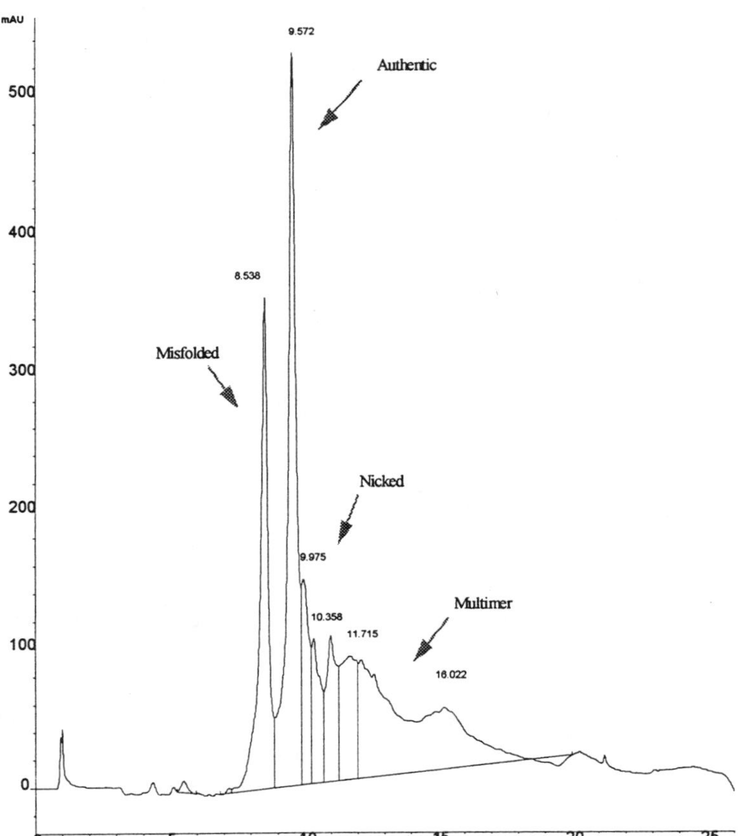

Fig. 3. RP-HPLC chromatogram of cation-pretreated fermentation broth generated from *P. pastoris* six-copy, Mut$^+$, *pep4* IGF-I production strain. Separation accomplished on a C4 Vydac column using a 1%/min gradient from 25–42% mobile phase B (95% acetonitrile, 0.1% TFA)/58% mobile phase A (water, 0.1% TFA).

trimer and higher oligomeric forms of IGF-I. With all of the *P. pastoris* IGF-I-expressing strains, authentic IGF-I generally represents about 20–30% of the total IGF-I identified by RP-HPLC, irrespective of the length of the fermentation. Therefore, the total IGF-I produced by fermentation with the six-copy strain is estimated at ~600 mg/L as measured by RP-HPLC. This value agrees well with the Western analysis results (**Table 2**). The data indicate that the Nichols RIA primarily recognizes the authentic IGF-I form with some additional crossreactivity to some of the other IGF-I forms, whereas the Western blot and Incstar RIA recognize the majority of IGF-I forms secreted into the medium.

Table 3
Expression of IGF-I in *pep4 Pichia pastoris*: Fermentation Studies

Fermentation	Phenotype	pH	C4 RP-HPLC, authentic mg/L	C4 RP-HPLC, nicked mg/L	Cell density, wet g/L
A	Mut⁺, *pep4*	3	139	43	450
B	Mut⁺, *pep4*	5	167	40	363
C[a]	Mut⁺, *pep4*	5	54	74	383
D[a]	Mut⁺, *pep4*	3	138	48	387
E[a]	Mut⁺, *pep4*	3	135	35	361
F[a]	Mut⁺, *PEP4*	3	103	68	385
G[a]	Mut⁺, *PEP4*	3	77	71	350
H[a]	Mut⁺, *PEP4*	5	0	0	330

[a]Fermentations carried out at 10-L scale. All other fermentations are at 2-L scale.

3.3.2.5. Fermentation Results Comparing Copy Number and Mut Phenotype

Copy number in the Mut$^+$ phenotype was evaluated in pH 3.0 fermentations. As the cassette copy number decreased (compare **Table 2**, lines B–E), expression level decreased in each assay used to measure IGF-I levels. Likewise, the six-copy Mut^s strain appears to produce lower levels than the six-copy Mut$^+$ strain (**Table 2**, line B vs F, 1550 vs 745 mg/L by the Incstar RIA). One observation pertaining to copy number is that there is little difference between four and six copies. Also, there appears to be slightly more degradation of IGF-I in fermentations of the six-copy vs the four-copy strains as determined by RP-HPLC (data not shown).

3.3.2.6. Fermentation with Six-Copy Mut$^+$ *PEP4* Strain

The six-copy Mut$^+$ strain was disrupted in its *pep4* protease gene to decrease the amount of protease present in the culture medium. Although *PEP4* activates vacuolar proteases, cell lysis during fermentation causes these proteases to be released into the medium. The *pep4* fermentation results are shown in **Table 3**, lines A–E. Since the RP-HPLC method gives much more information about IGF-I secreted into the fermentation medium, it was the only method chosen for analysis. Lines A and B compare 2-L fermentations at pH 3.0 and 5.0. The results show that the *pep4* strain is able to produce similar levels of IGF-I at pH 5.0 and 3.0. However, on scale-up to a 10-L fermenter, authentic IGF-I levels dropped significantly (**Table 3**, line C). The 10-L fermentation results for *pep4* strains at pH 3.0 are shown in **Table 3**, lines D and E. Comparable 10-L fermentation data are also given for the *PEP4* strains. The combination of the *pep4* strain and low-pH fermentation also showed a significant decrease in the ratio of authentic IGF-I to degraded or nicked IGF-I when compared to the *PEP4* strains. A combination of six copies of the IGF-I expression cassette, *pep4* phenotype, and low pH was optimal for IGF-I production and controlling protease degradation of IGF-I during fermentation. This strain and fermentation process has been successfully scaled up to a 1500-L scale with a linear increase in fermentation yields with volume and producing 140 ± 40 mg/L of authentic IGF-I in each of more than 100 fermentations at this scale.

3.4. Characterization and Purification of IGF-I

Having detailed characterization of the product produced after fermentation is critical for determining which strain is optimal for recombinant protein production as well as providing important information on how to purify the product from the fermentation medium. The characterization and purification of proteins, such as IGF-I from fermentation medium, are specific to each protein. However, some general approaches taken with IGF-I should be applicable to other proteins expressed in *P. pastoris*. Therefore, a general overview of some characterization and purification strategies is given below. More detailed discussions can be found elsewhere ***(23–25)***.

3.4.1. IGF-I Forms Found in Culture Broth

Figure 3 shows a representative HPLC chromatogram of cation-exchange pretreated fermentation broth. The major forms of IGF-I present in the fermentation broth are labeled. Proportions of the major IGF-I forms are: ~20–30% authentic, 10–20% misfolded, 7–15% nicked, and 30–50% multimer forms. These forms have been identified by several analytical techniques (including SDS-PAGE), reactivity to an antibody directed against IGF-I in Western blot analysis, size-exclusion chromatography, N-terminal sequence analysis, and biological activity. The peak that elutes by HPLC at ~9.5 min corresponds to the authentic form of IGF-I. The identity of this form was confirmed initially on the basis of elution time by RP-HPLC, which is identical to that of an IGF-I standard. Furthermore, this peak corresponds to the protein that was purified and subjected to a variety of physical-chemical characterization methods. These methods include reducing and nonreducing SDS-PAGE, isoelectric focusing, size-exclusion chromatography, N-terminal sequence analysis, amino acid sequence of the entire molecule, and peptide mapping. This molecule was also fully active in an in vitro biological assay.

The protein that elutes from the HPLC column at ~8.5 min (**Fig. 3**) has been identified as a misfolded form of IGF-I. This protein was isolated by HPLC and hydrophobic interaction chromatography, and characterized by SDS-PAGE, Western blot, and protein sequence analysis. SDS-PAGE analysis of reduced and nonreduced samples of this protein demonstrated that this form comigrates with authentic IGF-I. Western blot analysis of this protein using an antibody directed against the C-terminus of IGF-I showed that it is immunoreactive. Amino-terminal protein sequencing of this protein showed that it is identical to IGF-I. Also, when the disulfide bonds of this molecule are reduced, it elutes at the same position as reduced authentic IGF-I on RP-HPLC. Similar results are also reported for a misfolded form of IGF-I produced from a recombinant *S. cerevisiae* strain expressing IGF-I ***(15)***. Refolding experiments by oxidizing the reduced form of IGF-I (generated from either the misfolded or authentic forms) show generation of two predominant peaks on HPLC corresponding to the misfolded and authentic forms. The generation and HPLC analysis of misfolded forms of IGF-I generated by refolding of *E. coli*-produced IGF-I has also been carried out yielding similar results ***(16,17)***.

The proteins that elute from the HPLC column at 10–11 min (**Fig. 3**) have been identified as nicked or degraded forms of IGF-I (i.e., IGF-I molecules containing two or more peptide fragments, generated by cleavage of one or more peptide bonds, and held together by disulfide bonds). There appear to be at least two peaks by HPLC analysis of pretreated broth that correspond to the nicked IGF-I. The protein represented by the major peak was isolated by SP cation-exchange chromatography. SDS-PAGE analysis of nonreduced samples of this isolated species reveal that it is a single band that comigrates with authentic IGF-I. Gels of reduced samples of this protein, however, exhibited a doublet representing two peptides of approx 3–4 kDa each (approximately half the size of intact IGF-I). Amino-terminal protein sequence analysis of the protein confirms that the molecule is nicked prior to residue 40 of IGF-I, since both residues 1–5 and 40–44 are identified in the first five cycles of sequencing. Western blot analysis of reduced and nonreduced samples of this isolated nicked IGF-I molecule shows that it is reactive with the IGF-I antibody, but less reactive than intact IGF-I.

The last set of proteins detected in HPLC analysis of cell-free broth, which elute from the HPLC column after 11–18 min, have been identified as disulfide-bonded multimeric forms of IGF-I. The presence of disulfide-bonded IGF-I multimers in *P. pastoris* broth is indicated in SDS-PAGE gels, Western blots, N-terminal sequencing, and size-exclusion chromatography. The putative multimers migrate as IGF-I dimers and trimers on nonreduced SDS-PAGE gels, and are reactive with antibodies directed against the C-terminus of IGF-I. When these multimers are reduced, they comigrate with authentic IGF-I on SDS-PAGE gels, which indicates that these are disulfide-bonded IGF-I monomers (**Fig. 4**, lanes 3 and 9). N-terminal sequence analysis of purified IGF multimers shows the sequence to be identical with IGF-I. Furthermore, multimeric IGF-I (apparent dimer and trimer species) was isolated on a gel-filtration column, and analyzed by HPLC and SDS-PAGE. The isolated multimers elute from the HPLC column at the expected times.

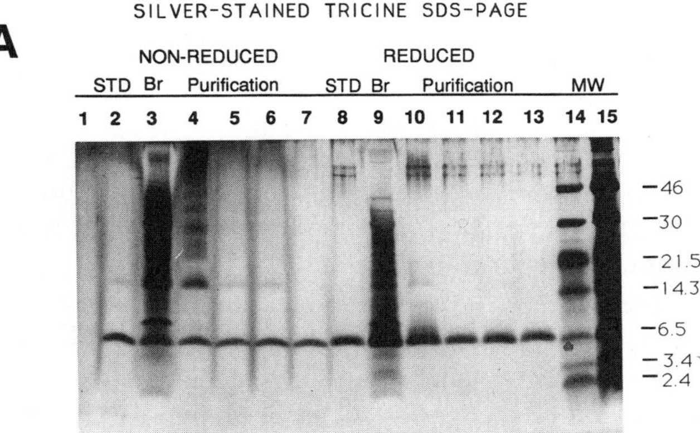

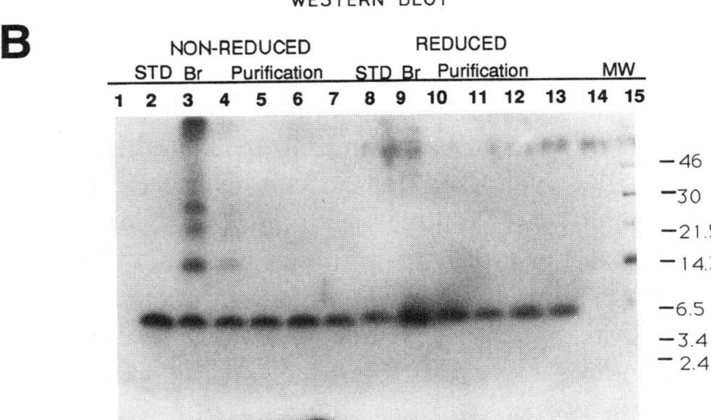

Fig. 4. Recombinant IGF-I expression and purification from fermentation broth generated from *P. pastoris* six-copy, Mut$^+$, *pep4* IGF-I production strain, silver-stained SDS-PAGE **(A)** and Western blot **(B)** loaded identically with the following samples: Lane 1, blank; lane 2, nonreduced (NR) IGF-I reference standard; lane 3, NR fermentation broth; lane 4, NR SP-cation recovery; lane 5, NR HIC chromatography; lane 6, NR SP2-chromatography; lane 7, NR gel-filtration chromatography; lane 8, reduced (R) IGF-I reference standard; lane 9, R fermentation broth; lane 10, R SP-cation recovery; lane 11, R HIC chromatography; lane 12, R SP2-chromatography; lane 13, R gel-filtration chromatography; lanes 14 and 15, mol wt markers.

Finally, there are several minor forms of IGF-I that are not readily resolved by C4 RP-HPLC. These forms include O-linked glycosylated, oxidized, acetylated, and amino-terminal and carboxy-terminal clipped species (data not shown). These forms typically amount to <10% of total IGF-I forms in the fermentation medium. The presence of O-linked glycosylated IGF-I forms with *S. cerevisiae* expression has also been reported ***(15)***.

3.4.2. IGF-I Purification

Two approaches have been developed to purify authentic IGF-I from *P. pastoris* fermentation broth. A process flow diagram outlining the two processes is shown in **Fig. 5**. The first approach involves direct purification of only the authentic IGF-I form. Although there is a heterogeneous mixture of IGF-I forms present in the fermentation medium, the use of traditional low-pressure chromatography using three basic modes of separation was successful in producing IGF-I of acceptable quality and yield for clinical trials. These three modes of separation are based on charge (cation-exchange chromatography, SP550C, TosoHaas), hydrophobicity (hydrophobic interaction chromatography, Butyl 650M, TosoHaas), and size (gel filtration, HW 50F, TosoHaas). SDS-PAGE and Western blot analysis illustrating the purity of samples taken from the fermentation broth and at selected points during purification is presented in **Fig. 4**.

The second approach for purification of IGF-I includes a refolding step to convert many of the misfolded and multimer forms to authentic IGF-I. Because all of the IGF-I forms are recovered in the cation-exchange step, the refold step was employed after this point. With the refold purification scheme, an RP step (low-pressure chromatography, Amberchrom CG100sd, TosoHaas) was used in place of the gel-filtration step, but the other steps remained essentially the same. One of the advantages of a refolding step for a yeast-secreted protein is that the product is soluble, unlike many products from *E. coli*, which are mostly insoluble.

The stages of the production process immediately following fermentation and preceding the purification are considered recovery stages. There are some unique aspects of *P. pastoris* fermentations that impact the development of the recovery process. The recovery operations typically involve two steps for a secreted protein. First, cell removal must be accomplished; second, product must be recovered or captured from the fermentation medium. Owing to the high cell densities achieved with *P. pastoris* fermentations, cell removal is not always an easy step to accomplish, since the final fermentation volume typically contains in excess of 40% wet cell mass. Microfiltration and centrifugation are two basic options that have worked equally well for the IGF-I production process. However, each step has its own advantages and disadvantages. Centrifugation is generally a faster process than microfiltration. However, larger facility space needs to be dedicated to this operation owing to the large process tanks that are necessary. A secondary cell removal step, such as

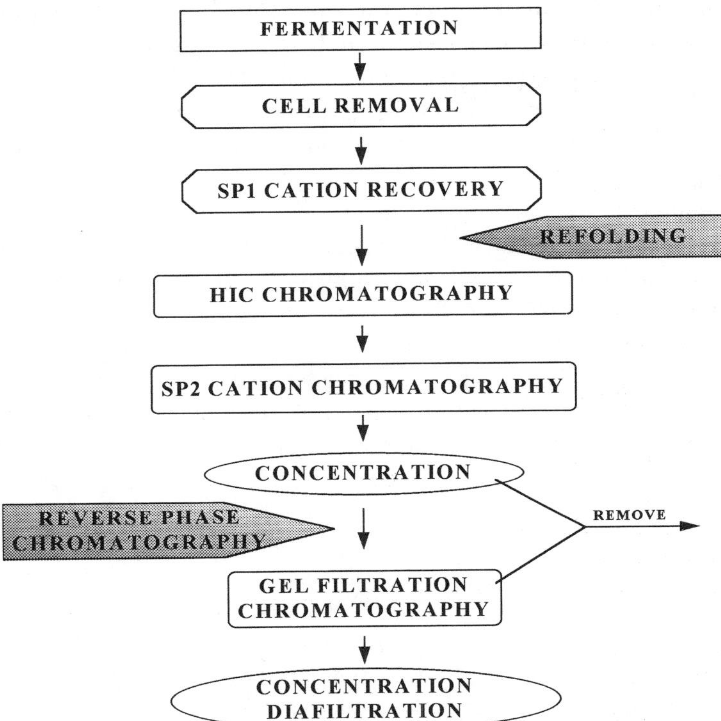

Fig. 5. Process flow diagram illustarting the two IGF-I production process strategies used to produce purified IGF-I from *P. pastoris* fermentations. The first strategy (nonrefold) is shown in the open block steps. In this strategy, the authentic form of IGF-I is purified from all the other IGF-I forms present in the fermentation broth. The differences in the second refold strategy are depicted in the shaded block steps. With this strategy, many of the other forms of IGF-I are converted to authentic IGF-I by inclusion of a refold step.

depth filtration, is needed for centrifugation, but is not a necessity with microfiltration. Microfiltration requires more development work, since protein compatibility with filters can be critical. Because the cell density starts off so high for recovery, care must be taken not to overconcentrate the cell phase during microfiltration. Similarly, either a predilution or postcell washing step is needed for centrifugation, since a large amount of a secreted product can be trapped with the cell-containing phase.

For the capture step, a strong cation-exchange resin, such as SP550C, has worked very well for IGF-I. If a protein's isoelectric point is above 7, often very little dilution of the *P. pastoris* fermentation broth is needed in order for the recombinant protein to bind to the cation-exchange matrix, especially if a lower pH is maintained (pH 3.0 for IGF-I). After cation-exchange recovery, nearly all of the IGF-I forms detected by RP-HPLC are recovered from the *P. pastoris* fermentation broth, and >90% of other yeast proteins are removed. Therefore, cation-exchange chromatography is an excellent capture step.

The remaining steps in each of the purification process strategies are unique to IGF-I, as stated above. Both purification schemes efficiently remove *P. pastoris* yeast proteins to <150 ng/mg and DNA to <20 pg/mg product. Although there is very little difference in purity of the final product from these two purification strategies, the final product yield was threefold greater with the refolding scheme owing to the conversion of most of the multimeric and misfolded forms to authentic IGF-I. Both of these purification strategies have worked well when scaled up for purification at the 1500–4500 L fermentation scales. C8 RP-HPLC chromatograms for final products produced from both processes are shown in **Fig. 6**.

3.5. Analytical Methods

3.5.1. IGF-I RIA Based on Incstar Antihuman Antisera

1. A 1:5000 final dilution of rabbit antihuman IGF-I antisera (Incstar), 10,000–12,000 cpm of ^{125}I-IGF-I (Incstar), and various dilutions of IGF-I standard and unknown samples are incubated overnight at 4°C in a final volume of 0.5 mL in 12 × 75-mm polystyrene tubes.
2. After incubation, 100 µL of Pansorbin are added to the tubes and incubated for 15 min at room temperature.
3. Add 2 mL RIA buffer to each tube before centrifugation at 3200 rpm for approx 70 min at 4°C.
4. Following centrifugation, the supernatant is decanted, and the radioactivity associated with the pellet is determined with a γ-counter.

3.5.2. Nichols Antihuman IGF-I RIA

Follow instructions supplied by the manufacturer (Nichols Institute Diagnostic).

3.5.3. Tricine SDS-PAGE

The Tricine SDS-PAGE system is that developed by Schagger and von Jagow ***(26)*** for the separation of proteins ranging from 5–20 kDa.

1. Separating gels are 13% acrylamide and 3% crosslinker; the stacking gels are 4% acrylamide and 3% crosslinker.
2. Electrophoresis samples are denatured by placing them in a boiling water bath for 2–3 min after adding 2X sample buffer. When samples are to be reduced, 100 m*M* dithiothreitol (DTT) is added to the sample buffer.

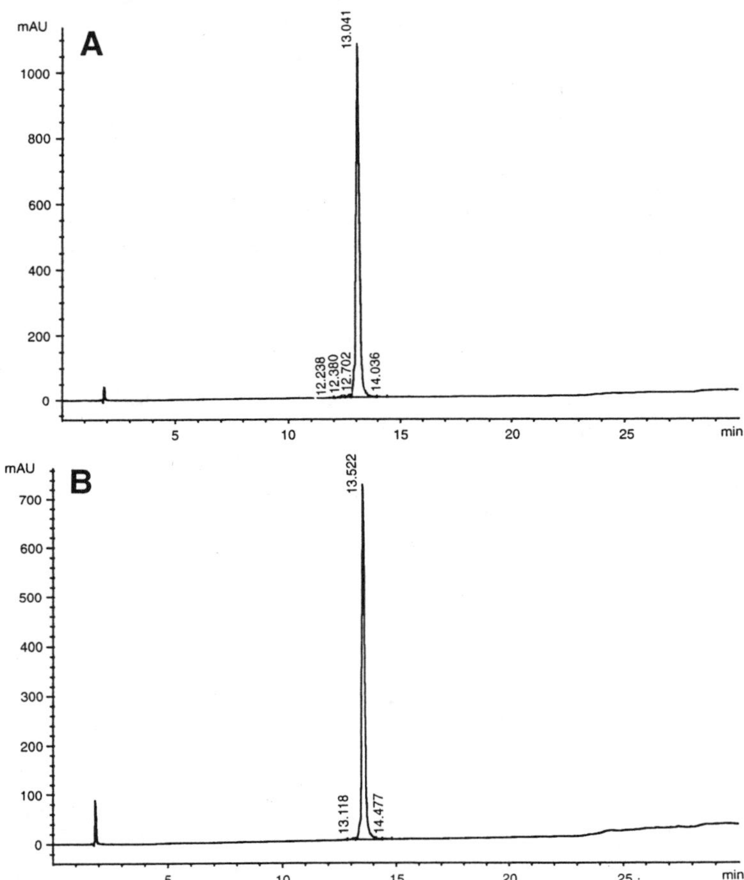

Fig. 6. RP-HPLC chromatogram of final purified products generated from *P. pastoris* six-copy, Mut⁺, *pep4* IGF-I production strain: **(A)** nonrefold purification process; **(B)** refold purification process. Separation accomplished on a C8 Vydac column using a 1%/min gradient from 20–40% mobile phase B (acetonitrile, 0.1% TFA)/60% mobile phase A (water, 0.1% TFA).

3. A constant sample volume is added to each well. Electrophoresis is conducted at a maximum of 100 V until the tracking dye migrates to the bottom of gels.
4. Gels that are silver-stained are treated with a first wash of 50% ethanol, 10% acetic acid, and 5% TCA, and a second wash of 10% ethanol, 7% acetic acid, and 1% TCA for 30 min.
5. Gels are fixed in 10% glutaraldehyde for 30 min. The fixed gels are then soaked in 5 μg/mL DTT for 30 min, and then equilibrated in 0.1% silver nitrate for another 30 min.

6. The color is developed with 3% sodium carbonate, and 0.05% formaldehyde (37%). The developing reaction is stopped by addition of 5% 2.3 *M* citric acid. After incubating for 10 min, the gel is transferred to water.

3.5.4. Western Blotting

1. After separation by tricine SDS-PAGE, gels are electroblotted onto 0.1 µm nitrocellulose in Towbin buffer for at least 90 min at 20 V/cm.
2. The filter is incubated for 1 h in blocking buffer.
3. The filter is then incubated with rabbit anti-IGF-I antisera (raised against the last 14 amino acids of the carboxy-terminus of IGF-I) diluted 1:2000 with blocking buffer for a minimum of 2 h.
4. The filter is washed with blocking buffer for 1 h and then incubated with ^{125}I-protein A (0.02°Ci/mL) for 45 min.
5. After washing the filter for 1 h with blocking buffer, the filter is air-dried and exposed to X-ray film with intensifying screen at –75°C.
6. IGF-I concentrations in samples are determined by comparing intensities of samples relative to known amounts of standard IGF-I loaded on the same gel.

3.5.5. RP-HPLC

3.5.5.1. Pretreatment of Fermentation Broth

P. pastoris-produced IGF-I exists in several forms in the fermentation broth. HPLC analysis of crude cell-free broth from fermentations of IGF-I-expressing *P. pastoris* does not adequately resolve the various IGF-I species. In order to distinguish these IGF-I species by HPLC, the broth must be pretreated on a small-scale cation-exchange column.

1. Equilibrate ~0.3 mL sulfylpropyl (SP) cation-exchange resin, SP-spherodex, with 2 mL 0.2 *M* acetic acid followed by 2 mL 0.02 *M* acetic acid. A 10-mL plastic disposable column (Bio-Rad, Hercules, CA) works well for the pretreatment procedure.
2. Typically, 1–4 mL of cell-free fermentation broth is diluted in half with 0.02 *M* acetic acid and bound to the resin.
3. After loading the sample, the resin is washed with 1–2 mL 0.02 *M* acetic acid.
4. The IGF-I is then eluted in 1 mL 0.05 *M* sodium acetate, pH 5.5, containing 1 *M* NaCl. The eluate can be directly analyzed by RP-HPLC.

3.5.5.2. C4 RP-HPLC

1. A Vydac C4 column (0.46 × 5 cm) with a guard column is used to resolve components of the *P. pastoris*-produced IGF-I preparations in pretreated fermentation broth.
2. Fermentation broth samples that have been pretreated as described above are loaded onto the column at a flow rate of 1 mL/min and are eluted in a trifluoroacetic acid (TFA)/acetonitrile-TFA gradient.
3. The eluant is prepared by using a mobile phase A (0.1% TFA) to dilute mobile phase B (95% acetonitrile, 5% water, 0.1% TFA).

4. A 1%/min gradient from 25–42% mobile phase B is passed through the column over 17 min at a flow rate of 1 mL/min to elute the IGF-I from the column.
5. The column is regenerated with 100% mobile phase B at a flow rate of 2 mL/min for 4 min, followed by 25% mobile phase B for 4 min at 2 mL/min. The flow rate is then reduced to 1 mL/min, and the column is equilibrated for 2 min before injection of the next sample. The detector is set at a wavelength of 215 nm for maximum sensitivity.

3.5.5.3. C8 RP-HPLC

1. A Vydac C8 column (0.46 × 15 cm) is used to resolve components of the *P. pastoris*-produced IGF-I preparations in purification samples.
2. The eluant is prepared by using a mobile phase A (water, 0.1% TFA) to dilute mobile phase B (100% acetonitrile, 0.1% TFA).
3. A 1%/min gradient from 20–40% mobile phase B is passed through the column for over 20 min at a flow rate of 1 mL/min to elute the IGF-I from the column.
4. The column is regenerated with 85% mobile phase B at a flow rate of 1 mL/min for 5 min. The column is equilibrated for 5 min before injection of the next sample. The detector is set at a wavelength of 215 nm for maximum sensitivity.

4. Notes

1. An alternative method for shake-flask analysis that gives equivalent or better results for Mut^+ strains is to reduce the glycerol from 2% to 0.05–0.1%, so that cultures will reach a maximum OD of approx 1.0 on glycerol. After ~24 h of culturing on glycerol, methanol is then added directly to the flask at a final concentration of 1%. Methanol addition is then repeated at ~48 and 72 h, and the flasks are harvested at 96 h, typically at an OD of >8.0.
2. Although increasing copy number usually increases expression levels of recombinant proteins in *P. pastoris*, this is not always the case. In some cases, increasing copy number can decrease expression levels ***(22)***. Often, this effect must be empirically determined for each recombinant protein. Therefore, copy number effect should always be compared to a one copy number strain. Also, choice of secretion signal can sometimes have an impact on whether or not a copy number effect will be observed.
3. During the early development stage, much higher levels of PTM1 trace salts were added during the fermentation. Originally, approx 4 mL/L of basal salts medium were added directly to the fermenter, and 12 mL PTM1/L of glycerol and methanol feeds were added as well. During the development of the fermentation process, it was found that the PTM1 trace salts could be reduced to a single addition of 2 mL/L basal salts medium at the start of the fermentation.
4. Automatic DO control schemes are not recommended during the fed-batch phases. These automatic control schemes should not be necessary, since as long as glycerol and methanol feed rates are at fixed rates, the rate of oxygen utilization will also be fixed within a certain range. One exception to this strategy may be to add automatic control of the methanol and glycerol feed rates, if for some uncontrolled reason the DO level falls below 20%.

5. The level of cell mass produced on glycerol can have an impact on how much recombinant product is produced. A range of 50–300 g/L wet cells is recommended by varying the amount of glycerol in the batch and fed-batch phases. A maximum level of 6% glycerol is recommended in the batch phase.
6. Low-pH fermentations have worked well for several proteins that appear to be sensitive to proteases in the fermentation medium, in particular for IGF-I. When operating at the low-pH ranges, samples should be measured off-line to ensure that fermenter pH readings remain accurate. A pH of below 2.8 will affect cellular metabolism. It is best to wait until the methanol feed to reduce pH to minimize flocculation.

   The exact pH that might be optimal for a given protein is empirically determined. For some proteins, other pH values (e.g., 3.0–7.0) have been found to be optimal for product stability. If standards of the recombinant protein are available in early stages of strain and fermentation development, a spiking experiment can be done with spent fermentation medium from a control of pH 5.0 fermentation in which pH is varied. This may aid in determining what pH ranges to test during fermentation development.
7. If DO falls below 20% after the methanol feed rate is increased to 7 mL/h/L, the methanol feed should be stopped, but nothing further should be done to increase oxygen (agitation, pressure, aeration, or oxygen feeding) until the DO level increases sharply (spikes). At this point, adjustments can be made to agitation, aeration, pressure, or oxygen feeding. This is primarily the reason why DO control is not recommended at stated in **Note 4**. Although in many cases not following these general guidelines concerning oxygen control will only cause subtle differences in expression levels, in other cases, the entire culture will stop growing because of overproduction of formaldehyde, the product of methanol oxidation by alcohol oxidase.
8. Maintaining the DO above 20% may be difficult depending on the oxygen transfer rate (OTR) of the fermenter. Pressure can be used to increase OTR up to 15–30 psi. Also, the 1–2 vvm of fermentation air feed can be supplemented with oxygen, generally at 0.1–0.5 vvm to achieve adequate oxygen levels. One 200-L T-size tank of compressed oxygen gas should be enough for at least 10 1-L fermentations. Liquid oxygen can also serve as a good source for oxygen gas. If the necessary level of oxygen cannot be supplied to a fermentation culture, then the methanol feed rate should be reduced accordingly. This is one of the major advantages of the Mut$^+$ strains in that since they are always fed methanol at a growth-limiting rate, the amounts of cooling and oxygen required are directly proportional to this feed rate. If Mut^s strains are grown in fermentations with excess methanol, as typically applied, the amount of heating and cooling cannot be controlled with feed rate. For Mut$^+$ fermentations, the time can be increased to deliver similar levels of methanol at the lower feed rate if it desired to reach maximum product levels. Conversely, higher methanol feed rates are possible to decrease the fermentation time to 48 h while still feeding the 740 mL methanol/L, but high oxygen and cooling requirements tend to make this difficult. The amount

of heat produced by *P. pastoris* has been measured at 2.6 kcal/g glycerol consumed/h and 4.5 kcal/g methanol consumed/h. The amount of oxygen required by *P. pastoris* has been measured at 13 m*M* oxygen/g glycerol consumed/h and 35 m*M* oxygen/g methanol consumed/h.

9. A similar fermentation protocol was used for Mut^s strains. Methanol feeds were started at 1 mL/h/L and then increased to ~3–6 mL/h/L. Another alternative strategy for both Mut^+ and Mut^s strains is the use of mixed feeds. Instead of feeding pure methanol into the fermenter, a mixture of methanol and glycerol can be used. DO spikes are still used to ensure a limited carbon feed. The ratio of methanol to 50% glycerol will have to be determined empirically. Good ratios to start with are 4:1, 2:1, 1:1, 1:2, and 1:4 methanol:glycerol at the feed rate specified for the Mut^+ strains. This technique has been found to be very useful for proteins that appear toxic to the strain or for improving the productivity of Mut^s fermentations. Mixed feeds will also decrease the overall cooling and oxygen demand of the fermentation ***(6)***.

Acknowledgments

IGF-I expression vectors and stains were constructed by Geneva Davis. The *PEP4* gene disruption was performed by Martin Gleeson and Bradley Howard. Purification and characterization was contributed by Gregory Holtz and Tom Vedvick. RIA development work was contributed by Sally Provow. Fermentation work was contributed by Bob Siegel. More recent process development was accomplished by Joan Abrams, Anthony Attipoe, Timothy Garner, John Hanson, Sanjay Jain, Khalil Kerdahi, Francis Maslanka, Steve Pick, Harold Ross, and Judy Wynn.

References

1. Cregg, J. M., Vedvick, T. S., and Raschke, W. C. (1993) Recent advances in the expression of foreign genes in *Pichia pastoris*. *Bio/Technology* **11,** 905–909.
2. Clare, J. J., Rayment, F. B., Ballantine, S. P., Sreekrishna, K., and Ramanos, M. A. (1991) High-level expression tetanus toxin fragment C in *Pichia pastoris* strains containing multiple tandem integrations of the gene. *Bio/Technology* **9,** 455–460.
3. Tschopp, J. F., Sverlow, G., Kosson, R., Craig, W., and Grinna, L. (1987) High-level secretion of glycosylated invertase in the methylotrophic yeast, *Pichia pastoris*. *Bio/Technology* **5,** 1305–1308.
4. Digan, M. E., Lair S. V., Brierley, R. A., Siegel, R. S., Williams, M. E., Ellis, S. B., Kellaris, P. A., Provow, S. A., Craig, W. S., Velicelebi, G., Harpold, M. M., and Thill, G. P. (1989) Continuous production of a novel lysozyme via secretion from the yeast, *Pichia pastoris*. *Bio/Technology* **7,** 160–164.
5. Vedvick, T., Buckholz, R. G., Engel, M., Urcan, M., Kinney, J., Provow, S., Siegel, R. S., and Thill, G. P. (1991) High level secretion of biologically active aprotinin from the yeast, *Pichia pastoris*. *J. Ind. Microbiol.* **7,** 197–202.

6. Brierley, R. A., Bussineau, C., Kosson, R., Melton, A., and Siegel, R. S. (1990) Fermentation development of recombinant *Pichia pastoris* expressing the heterologous gene: bovine lysozyme. *Ann. NY Acad. Sci.* **589,** 350–362.
7. Preece, M. A. (1983) The somatomedins, in *Hormones in Blood* (Gray, C. H. and James, V. H. T., eds.) Academic, London, pp. 87–108.
8. Lynch, S. E., Nixon, J. C., Colvin, R. B., and Antoniades, H. N. (1987) Role of platelet-derived growth factor in wound healing: synergistic effects with other growth factors. *Proc. Natl. Acad. Sci. USA* **84,** 7696–7700.
9. Burleigh, B. D. and Meng, H. (1986) Development of biosynthetic somatomedin-C/IGF-I as a product for cell culture. *Am. Biotechnol. Lab.* **4,** 48–53.
10. Quin, J. D. (1992) The insulin-like growth factors. *Q. J. Med.* **82,** 81–90.
11. Cotterill, A. M. (1992) The therapeutic potential of recombinant human insulin-like growth factor-I. *Clin. Endocrinol. Oxford* **37,** 11–15.
12. Lewis M. E., Neff, N. T., Contreras, P. C., Stong, D. B., Oppenheim, R. W., Grebow, P. E., and Vaught, J. L. (1993) Insulin-like growth factor-I: potential for treatment of motor neuronal disorders. *Exp. Neurol.* **124,** 73–88.
13. Gluckman, P., Klempt, N., Guan, J., Mallard, C, Sirimanne, E., Dragunow, M., Klempt, M., Singh, K., Williams, C., and Nikolics, K. (1992) The role for IGF-I in the rescue of CNS neurons following hypoxic-ischemic injury. *Biochem. Biophys. Res. Commun.* **182,** 593–599.
14. Bayne, M. L., Applebaum, J., Chicchi, G. G., Hayes, N. S., Green, B. G., and Cascieri, M. A. (1988) Expression, purification and characterization of recombinant human insulin-like growth factor I in yeast. *Gene* **66,** 235–244.
15. Elliott, S., Fagin, K. D., Nahri, L. O., Miller, J. A., Jones, M., Koski, R., Peters, M., Hsieh, P., Sachdev, R., Rosenfeld, R. D., Rohde, M. F., and Arakawa, T. (1990) *Yeast*-derived recombinant human insulin-like growth factor I: production, purification, and structural characterization. *J. Protein Chem.* **9,** 95–104.
16. Meng, H., Burleigh, B. D., and Kelly, G. M. (1988) Reduction studies on bacterial recombinant human somatomedin C/insulin-like growth factor I. *J. Chromatogr.* **443,** 183–192.
17. Chang, J. Y. and Swartz, J. R. (1993) Single step solubilization and folding of IGF-I aggregates from *Escherichia coli*, in *Protein Folding In Vivo and In Vitro* (Cleland, J. L., ed.), ACS Symp. Ser. **526,** American Chemical Society, Washington, DC, pp. 178–188.
18. Scott, R. W., Brierley, R. A., and Howland, D. S. (1996) Secretion sequence for the production of a heterologous protein in yeast. US Patent 5,521,086.
19. Stroman, D. W., Cregg, J. M., Harpold, M. M., and Sperl, G. T. (1989) Transformation of yeasts of the genus Pichia. US Patent 4,879,23.
20. Gleeson, M. A. and Howard, B. D. (1994) US Patent 5,324,660.
21. Jones, E. (1977) Proteinase mutants of *Saccharomyces cerevisiae*. *Genetics* **85,** 23–33.
22. Thill, G. P., Davis, G. R., Stillman, C., Holtz, G., Brierley, R. A., Engel, M., Buckholz, R., Kinney, J., Provow, S., Vedvick, T., and Siegel, R. S. (1990) Positive and negative effects of multi-copy integrated expression vectors on protein expression in

Pichia pastoris, in *Proceedings of 6th International Symposium on Genetics of Industrial Microorganisms*, vol. 2 (Heslot, H., Davies, J., Bobichon, L., Durand, G., and Penasse, L., eds.), Societé Française de Microbiologie, Paris, pp. 477–490.

23. Brierley, R. A., Davis, G. R., and Holz, G. C. (1994) Production of insulin-like growth factor-1 in methylotrophic yeast cells. US Patent 5,234,639.
24. Holz, G. C. and Brierley, R. A. (1993) Method for the purification of intact, correctly folded insulin-like growth factor-1. US Patent 5,231,178.
25. Brierley, R. A., Abrams, J. A., Hanson, J. M., and Maslanka, F. C. (1996) IGF-I purification process. International Patent Application, Publ. No. WO 96/32407.
26. Schagger, H. and von Jagow, G. (1987) Tricine-sodium dodecyl sulfate-polyacrylamide gel electrophoresis for the separation of proteins in the range from 1 to 100 kDa. *Anal. Biochem.* **166,** 368–379.

12

Secretion of scFv Antibody Fragments

Hermann Gram, Rita Schmitz, and Rüdiger Ridder

1. Introduction

Antibodies consist of two functional parts, the antigen binding site, a heterodimer composed of the V_H and V_L domains, and the Fc part, which facilitates in vivo the effector functions and stability of the antibody. Recombinant antibodies were first successfully expressed by secretion from mammalian cells, whereas attempts to express whole antibodies in *Escherichia coli* remained unsuccessful. Later, the relatively small antigen binding portion of an antibody were functionally expressed as Fab or single-chain Fv (scFv) fragments in bacteria ***(1,2)***. Although intracellular expression of scFv fragments as inclusion bodies has been reported, most scFv antibodies are expressed in *E. coli* by secretion into the periplasmic space. By secretion across the cytoplasmic membrane, the heavy- and light-chain domains can fold and assemble properly, and intra- and interdomain disulfide bonds are formed ***(3,4)***. Single-chain antibodies purified from the periplasmic space of *E. coli* are primarily monomeric and, in most cases, functional. Although respectable yields of more than 200 mg/L of scFv fragment have been reported from large-scale *E. coli* fermentations ***(5)***, the yields obtained from small-scale expression cultures utilizing shake flasks are typically below 1 mg/L ***(4)***. Fermentation approaches gave rise to much higher yields, but these require thorough optimization and expensive equipment.

The concept of combinatorial immunoglobulin libraries and the display of scFv antibody fragments on the surface of filamentous bacteriophage has led to the isolation of an increasing number of scFv antibodies ***(6,7)***. These successes have created a demand for a suitable expression system for rapid production of scFv fragments. One of the major limitations for the practical application of the phage display technology as an alternative to the well-established hybridoma generation is the low yield of functional antigen binding pro-

From: *Methods in Molecular Biology, Vol. 103:* Pichia *Protocols*
Edited by: D. R. Higgins and J. M. Cregg © Humana Press Inc., Totowa, NJ

teins produced in bacterial hosts. Even though methods for affinity purification of scFv fragments from *E. coli* cultures have been described, the achieved yield is often not sufficient for extended binding studies, the development of immunoassays, or in vivo experiments. Therefore, *E. coli* is not the best host for scFv production in our opinion, and we have explored alternative expression systems.

The methylotrophic yeast *Pichia pastoris* has been shown to be a suitable host for high-level expression of various heterologous proteins. Large amounts of soluble and active protein showing posttranslational modifications similar to those of higher eukaryotes can be secreted into the culture medium ***(8)***. *P. pastoris* combines the features of eukaryotic secretion machinery with a capacity for fast growth in noncomplex bacterial growth media. These features make this yeast an attractive host for scFv expression, particularly since the high expression level of heterologous proteins requires only small-scale production in shake flasks, which can be performed with the standard equipment of a molecular biology laboratory. Furthermore, engineered *P. pastoris* strains can be fermented to high cell density, yielding product titers in the gram/liter range ***(9–11)***. We demonstrate here that functional scFv antibody fragments tagged with the immunoglobulin light-chain κ domain or peptide tags for affinity purification and detection can be produced in this system with a high yield. Also, scFv fragments expressed from *P. pastoris* can be engineered to dimerize, which increases their apparent affinity toward the antigen.

2. Materials

1. *P. pastoris* strain GS115 (*his4*) and the expression vectors pHIL-S1 and pPIC9 were obtained from Invitrogen (La Jolla, CA).
2. YPD: 1% yeast extract, 2% peptone, 2% glucose.
3. MD: 1.34% yeast nitrogen base, 1% glucose, 1.6 μ*M* biotin.
4. MM: 1.34% yeast nitrogen base, 0.5% methanol, 1.6 μ*M* biotin.
5. BMGY: 1% yeast extract, 2% peptone, 1.34% yeast nitrogen base, 1% glycerol, 1.6 μ*M* biotin, 100 m*M* K_2HPO_4, pH 6.0.
6. BMMY: 1% yeast extract, 2% peptone, 1.34% yeast nitrogen base, 0.5% methanol, 1.6 μ*M* biotin, 100 m*M* K_2HPO_4, pH 6.0.
7. Klenow reaction mix: 20 m*M* Tris-HCl, 10 m*M* $MgCl_2$, 40 m*M* NaCl, 0.4 m*M* DTT, and 150 μ*M* of each of the dNTPs at pH 7.5.
8. Recombinant human leukemia inhibitory factor (hLIF) was produced in *E. coli* by expression in inclusion bodies, denaturing, and subsequent refolding.
9. Nickel-chelate resin (Ni-NTA) was obtained from Qiagen (Hilden, Germany).
10. Ni-NTA column loading buffer: 300 m*M* NaCl and 50 m*M* sodium phosphate, pH 8.0.
11. Ni-NTA column washing buffer: 300 m*M* NaCl and 50 m*M* sodium phosphate, pH 6.5.
12. Ni-NTA column elution buffer: 300 m*M* NaCl and 50 m*M* sodium phosphate, pH 5.0.
13. Ni-NTA column elution buffer: 50 m*M* sodium phosphate, pH 7.5.

14. Ni-NTA column elution buffer plus imidazole: 300 m*M* NaCl, 50 m*M* sodium phosphate, and 500 m*M* imidazole, pH 5.0.
15. Hybridoma secreting 9E10 anti-*myc* antibody was obtained from the American Type Culture Collection (Rockville, MD). Purified monoclonal antibody (MAb) was coupled to a cyanogen bromide-activated Sepharose, yielding a resin with a capacity of 10 mg antibody/mL. The murine antihuman IL-3 MAb F15-216 recognizing the epitope Leu-Pro-Leu-Leu was generated by Lokker et al. ***(12)***.
16. PBS buffer: 137 m*M* NaCl, 2.7 m*M* KCl, 4.3 m*M* Na_2HPO_4, 1.4 m*M* KH_2PO_4, pH 7.2.
17. Affinity-column elution buffer: 100 m*M* glycine-HCl, pH 2.7.
18. 0.1 *N* NaOH.
19. Streptavidin conjugated to horseradish peroxidase for ELISA detection or Western blotting is available from Jackson Immunology Laboratories (catalog #016-030-084; Bar Harbor, ME).
20. OPD solution: dissolve 10 mg of OPD (Sigma [St. Louis, MO] P8287) in solution A (12.15 mL of 0.1 *M* citric acid) and solution B (12.85 mL of 0.2 *M* Na_2HPO_4). 10 µL H_2O_2 are added immediately before use.

3. Methods

3.1. Cloning of the scFv Fragments

The scFv3-3 antibody fragment binding to hLIF was selected by phage display from a combinatorial library prepared from rabbit spleen ***(13)***. Below is described the construction of four different expression vectors for *P. pastoris*, which directed the production of monomeric scFv3-3, a scFv3-3::IgCκ fusion protein, and a scFv3-3 dimer (**Fig. 1**). We used two different leader sequences, the α-mating factor leader and the *PHO1* leader present in the expression vectors pPIC9 and pHIL-S1, respectively. For detection of binding to hLIF in ELISA, the tetrapeptide sequence Leu-Pro-Leu-Leu derived from human interleukin 3 and against which the MAb F15-216 is available ***(12)***, was fused to the carboxyl-terminus of the scFv antibody. This tetrapeptide tag (IL3 tag) was followed by an His_5 sequence used for affinity purification on a nickel-chelate resin (**Figs. 1** and **2**). By using the murine IgCκ constant domain as a tag, affinity purification, detection and quantification of scFv fragments by commercially available reagents are possible. We have previously shown the expression of a scFv3-3::IgCκ fusion protein in COS cells ***(14)***. We further demonstrate here the successful expression of the dimerized scFv3-3 antibody fragment, which has a 40-fold higher apparent affinity for hLIF as determined by surface plasmon resonance analysis. The scFv3-3 antibody was dimerized by virtue of a helix-turn-helix motif previously used for the dimerization of an scFv fragment expressed from bacteria ***(3,5)***. Here, the helix-turn-helix dimerization motif is separated from the scFv antibody by a flexible hinge region

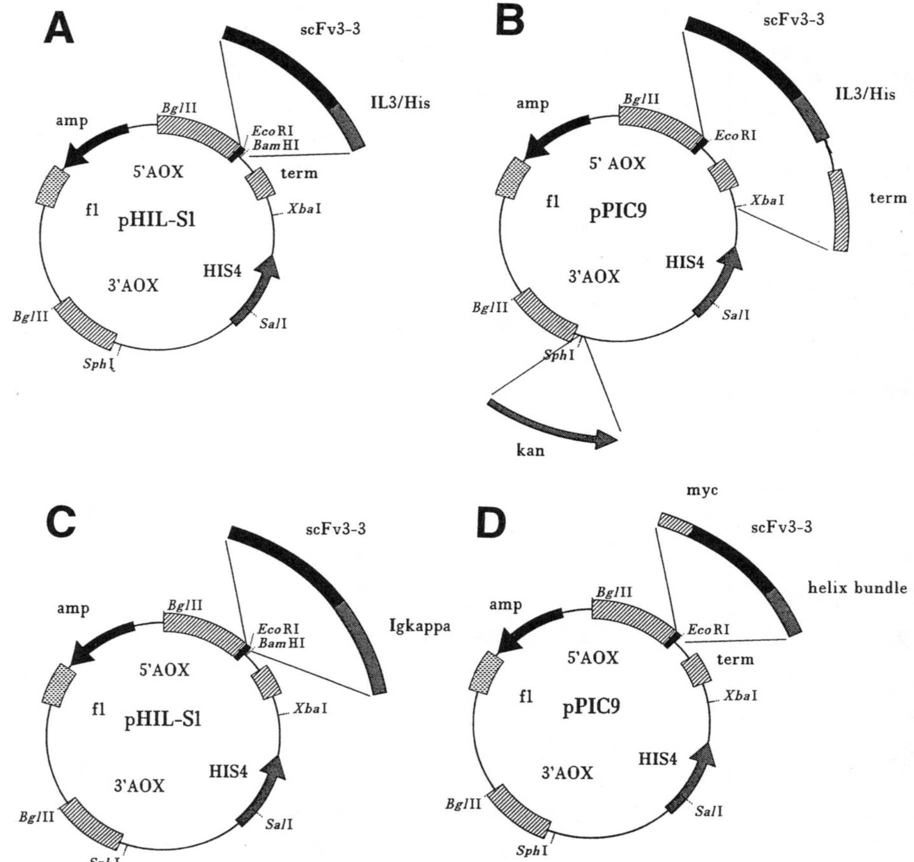

Fig. 1. Expression constructs for the rabbit scFv3-3 antibody. **(A)** The coding region for scFv3-3 containing the IL3 tag and a His_5 tag was cloned into pHIL-S1. **(B)** The same expression cassette was moved into the pPIC9 vector. In another experiment, this vector was modified to contain the bacterial *kan*^{r} gene for selection by G418. **(C)** The coding region for scFv3-3 was fused to the murine IgCκ gene. **(D)** For dimerization, the scFv3-3 was modified to contain a helix-turn-helix motif and a *myc* tag. Details of the constructions are given in the text and in **Fig. 2**. Abbreviations: amp, ampicillin resistance gene; f1, origin of replication of phage f1; *HIS4*, histidinol dehydrogenase gene; term, transcriptional terminator from *AOX1* gene; 3'AOX, sequence downstream of the *P. pastoris AOX1* open reading frame (ORF); 5'AOX, sequence upstream of the *AOX1* ORF containing the *AOX1* promoter.

derived from the murine IgG3 heavy-chain constant domain. Since the addition of the dimerization motif at the carboxyl-terminus did not permit the fusion of a further peptide tag that could facilitate purification, a decapeptide tag

	Leader	secreted scFv3-3	Tag
pPIC9 α-mat. leader IL3/His-tag	K R ... AAA AGA	scFv3-3 E A E A Y V E F E L D gag gct gaa gct tac gta gaa ttc GAG CTC GAT P S V A S Q P P ... ... CCA TCA GTC gct agc cag cca cct	IL-3 tag His L P L L D H H TTG CCT TTG CTG GAC CAT CAC tag H H H STOP CAT CAC CAC TAA TA GGATCC
pHIL-S1 pho1-leader IL3/His-tag	F A ... TTC GCT	E F E L D P S cga gaa ttc GAG CTC GAT ... ... CCA TCA V A S Q P P GTC gct agc cag cca cct	L P L L D H H TTG CCT TTG CTG GAC CAT CAC H H H STOP CAT CAC CAC TAA TA GGATCC
pHIL-S1 pho1-leader Igκ tag	F A ... TTC GCT	E F E L D P S cga gaa ttc GAG CTC GAT ... ... CCA TCA V A S GTC gct agt	Ig kappa R A D A A CGG GCT GAT GCT GCA ... R N E C STOP AGG AAT GAG TGT TAA GGATCC
		secreted myc-scFv3-3	Dimerization motif
pPIC9 myc-tag, hinge helix-turn-helix motif	K R ... AAA AGA	E A E A Y V E F P E Q gag gct gaa gct tac gta gaa ttc cca GAA CAA myc tag K L I S E E D L P E L AAG TTG ATT TCC GAA GAA GAC TTA cca GAA CTC hinge P S V A S *P K P* ... ... CCA TCA GTC gct agc *CCA AAG CCT* *S T P P G S S* *TCT ACT CCA CCT GGT TCT TCC*	G E L E E L L GGT GAA TTG GAA GAG TTA TTG K H L K E L L AAG CAC TTA AAA GAA CTC TTG K G P R K G E AAG ggt cca aga AAG GGT GAG L E E L L K H TTA GAA GAA TTG TTA AAA CAT L K E L L K G TTG AAG GAA TTG CTA AAA GGT STOP TAG GGATCC

Fig. 2. Construction of the expression vectors for scFv3-3. Shown is part of the leader sequence, the boundaries of the scFv3-3 DNA sequence, and the relevant tag or dimerization motif. The coding sequence for scFv3-3 is shown in capital letters; DNA sequences preceding and following the coding frame for the scFv3-3 antibody are shown in small letters; cloning sites are underlined. For the *myc*–scFv3-3 construct, the DNA sequence for the *myc* tag is written in capital letters and double underlined. The hinge region connecting the scFv3-3 to the dimerization motif is shown in capital italicized letters.

derived from the c-*myc* protooncogene and recognized by the MAb 9E10 ***(15)*** was fused to the amino-terminus of the scFv3-3.

1. For the construction of a suitable expression vector for *P. pastoris*, a bacterial pET3a-based expression vector containing the scFv3-3 coding sequence appended by codons for the IL-3 and the His_5 tag is used as a template in a PCR reaction. The scFv3-3 sequence is thereby modified using the oligonucleotides 5'-CGA GAATTCGAGCTCGATCT GACCCAGACT-3' (adding an *Eco*RI site, underlined, directly upstream of the scFv sequence) and 5'-GCTAGTTATT GCTCAGCGGT-3' (hybridizing within the terminator region of pet3A). The resulting PCR fragment is ligated after digestion with *Eco*RI and *Bam*HI into the equivalent cloning sites of *P. pastoris* expression vector pHIL-S1, thus generating vector pHIL-S1–scFv3-3–IL3 (**Figs. 1A** and **2**).
2. Create a derivative of the above vector with the leader sequence of the α-mating factor upstream of the scFv encoding sequence by replacing the *Eco*RI–*Xba*I DNA fragment of the *P. pastoris* vector pPIC9 with the *Eco*RI–*Xba*I-fragment of pHIL-S1–scFv3-3–IL3, resulting in pPIC9–scFv3-3–IL3 (**Figs. 1B** and **2**).

3. To generate a vector for the expression of a scFv3-3::IgCκ fusion protein in COS cells, fuse the scFv3-3 to the murine IgCκ domain ***(14)***. Recover the coding region for the scFv3-3::IgCκ fusion from the COS cell expression vector by PCR using the primers 5'-CGAGAATTCGAGCTCGAT CTGACCCAGACT-3' (adding an *Eco*RI site, underlined, directly upstream of the scFv sequence) and 5'-GGATGGATCCTTAA CACTCATTCCTGTTGAAG-3' (adding a *Bam*HI site, underlined, beyond the coding sequence of the murine IgCκ cDNA). Digest the resulting PCR fragment with *Eco*RI and *Bam*HI, and ligate into the same sites in pHIL-S1 (**Figs. 1C** and **2**).
4. To create a dimerization vector, extend the scFv3-3 coding region 3' by PCR reactions using the overlapping primers 5'-CAATTCACCGGAAGAACCAGG TGGAGTAGAAGGCTTTGGCTAGCGACTGATGG AGCCTT-3' (complementary to the 3' portion of the scFv3-3), 5'-CTTTCTTGGACCCTTC AATAGTTCTTTTAAGTGCTTCAATAACTCTTCCAATTCACCGGAAGA-3', 5'- TTTTAGCAATTCCTTCAAATGTTTTAACAATTCTTCTAACTCACCCTT TCTTGGACCCTT-3', and 5'-CTCGAATTCGGATCCCTAACCTTTTAGCAA TTC CTTC-3' (introducing a *Bam*HI site, underlined, at the end of the coding region for the helix-turn-helix motif). The extension encodes the hinge region of murine IgG3 and the helix-turn-helix dimerization motif (**Fig. 2**). Extend the amino-terminus of scFv3-3 by adding sequences encoding the decapeptide *myc* tag using the overlapping PCR primers 5'-CAGAACAA AAGTTGATTT CCGAAGAAGACTTACCAGAACTCGATCTGACCCAGAC-3' and 5'-TCTT GAATTCC CAGAACAAAAGTTGATTTC-3' (introducing an *Eco*RI site upstream of the coding region for the decapeptide *myc* tag). After digestion with *Eco*RI and *Bam*HI, clone the resulting PCR fragment into pHIL-S1, and subsequently transfer as an *Eco*RI–*Xba*I fragment into pPIC9 (**Figs. 1D** and **2**).
5. Prepare plasmid DNA of each vector by CsCl gradient centrifugation, and cleave each with *Bgl*II prior to transformation. Inactivate the restriction enzyme by extracting the DNA with phenol. Precipitate a cleaved DNA, and dissolve at a concentration of 1 μg/μL.

3.2. Construction of Expression Vectors for Multiple Integrations

Production of heterologous proteins in *P. pastoris* can be enhanced by selection of clones that bear multiple copies of the expression cassette ***(9,16)***. Since the integration of multiple copies is relatively rare, a selection scheme based on the copy number-dependent expression of the bacterial kana*myc*in resistance gene (*kan*^{r}) was developed by Scorer et al. ***(17)***. For this selection procedure, the *kan*^{r} gene is cloned downstream of the *HIS4* gene such that its transcription initiated by a cryptic promoter is counterclockwise and opposite to the *AOX1* promoter. It is important to control the orientation, since the opposite orientation of the *kan*^{r} gene will lead to a much higher transcription and antibiotic resistance. Subsequently, His$^+$ transformants are plated on different concentrations of the aminoglycoside antibiotic G418. Resistant colonies are then selected for Mut[s] phenotype as described below or screened directly for expression.

For construction of vectors conferring G418 resistance, isolate the *kan*[r] gene as a 1250-bp *Hinc*II fragment from plasmid pUC4K ***(18)***. Cut the pPIC9 expression vector containing the scFv3-3–IL3 with *Sph*I, which cleaves pPIC9 uniquely between the *HIS4* gene and the 3'-*AOX* segment. Treat the 3'-overhangs generated by *Sph*I with Klenow fragment of DNA polymerase I to generate blunt ends by incubating 1 μg of linearized plasmid with 1 U of Klenow in buffer containing 20 m*M* Tris-HCl, 10 m*M* $MgCl_2$, 40 m*M* NaCl, 0.4 m*M* DTT, and 150 μ*M* of each of the dNTPs at pH 7.5. Select for bacterial colonies resistant to both kanamycin (35 μg/mL) and ampicillin (100 μg/mL), and screen for proper orientation of the bacterial *kan*[r] gene by using the oligonucleotides 5'-GGGGCGATTCAGGCCTGGTATGAG-3' and 5'-CCCTA GCGCCTGGG ATCATCC-3', which are complementary to DNA sequences located in the *kan*[r] gene and the expression vector, respectively. Recombinant plasmids bearing the *kan*[r] gene in the desired orientation give rise to a PCR fragment of 650 bp in length. Prepare plasmid DNA by CsCl gradient centrifugation, and cleave with *Bgl*II prior to transformation. Inactivate the restriction enzyme by extracting the DNA with phenol. Precipitate the cleaved DNA, and dissolve at a concentration of 1 μg/μL.

3.3. Transformation into P. pastoris and Mut[s] Phenotype Selection

We used the spheroplast transformation method essentially as described by Cregg et al. (***19***; *see* Chapter 3) and in the manufacturer's manual supplied by Invitrogen.

1. After 6 d of incubation at 30°C, pick His^+ transformed colonies using a sterile disposable loop.
2. Streak the colonies on a 10-cm MD plate using a 52-square grid, and allow the clones to grow for 2 d.
3. Replica plate the colonies onto MD and M*M* plates using a velvet stamp (Lederberg stamp).

Mut[s] clones are found at a frequency of ~5–20% of total His^+ transformants and can easily be identified by their much slower growth on MM plates (*see* **Notes 1** and **2**).

3.4. Selection for G418-Resistant Colonies

1. His^+ transformants carrying the *kan*[r] gene are further selected for multiple integrated copies by replating on G418 antibiotic. To recover the transformants from the regeneration plate, scrape the soft agarose containing ~10^3 yeast colonies, place in a 50-mL conical tube, and resuspend in 10 mL of water by vortexing.

2. To remove agarose, allow agarose to settle to the bottom of the tube. Remove the supernatant containing the yeast cells, and allow residual agarose to settle again. Repeat this procedure until no visible traces of agarose are present in the cell preparation.
3. Titer the yeast cells using a Neubauer counting chamber.
4. Plate 10^5 cells on YPD plates containing selected concentrations of G418 (0, 0.25, 0.5, 1, and 2 mg/mL). With increasing concentrations of the antibiotic, decreasing numbers of colonies will be observed. At 2 mg/mL of G418, we usually obtain 10–150 colonies.
5. Examine clones from the plates with the highest concentrations of G418 for an Mut^s phenotype as described above. In contrast to clones selected only for the His$^+$ phenotype, we typically observe a high frequency of Mut^s colonies among His$^+$ G418-resistant clones (*see* **Notes 3** and **4**).

3.5. Small-Scale Expression of the Selected scFv Constructs

From each of the transformations, 10–15 recombinant *P. pastoris* clones with an Mut^s phenotype were randomly chosen for expression studies. The clones were first grown in BMGY medium with glycerol as the sole carbon source. For induction of the *AOX1* promoter and protein production, the growth medium was changed to the same medium except with 0.5% methanol in place of glycerol as the carbon source. Most of the clones examined expressed the recombinant scFvs in high yields, as judged by SDS-PAGE analysis of the culture supernatant (*see* **Notes 4** and **5**). In general, we observed little difference between individual clones of a given set, and even different scFv3-3 fusion proteins expressed at about the same level (**Fig. 3**). A three-fold higher expression, however, was seen with one clone derived from the pPIC9–scFv3-3–IL3 construct (**Fig. 3**, lane 1). We further compared pPIC9–scFv3-3–IL3 His$^+$ Mut^s clones with those derived with the same construct modified with the *kan*^{r} gene and selected for resistance to high levels of G418. Though the average level of secretion seemed a little higher from the G418-resistant series, we did not observe a large enhancement. The best producer of scFv3-3–IL3 was selected from the series without G418 selection and is shown in **Fig. 3**. Time-course studies with this clone indicated that maximum production under the conditions described below was reached after 40 h (**Fig. 3B**). At later time-points, the content of scFv in the culture did not increase further. We also examined Mut$^+$ clones, but in general, observed little or no expression of the scFv3-3 antibody. Culture supernatant from clones secreting high levels of scFv could be used directly in ELISA (**Fig. 4**) or surface plasmon resonance, a technique that measures the ability of scFv3-3 to bind to hLIF. The procedure for expression of svFv in shake-flask cultures is as follows:

1. Grow randomly chosen clones in 25 mL of BMGY medium at 30°C in a 100-mL Erlenmeyer flask with glycerol as the sole carbon source for 2 d until an OD_{660} of 20–40 is reached.

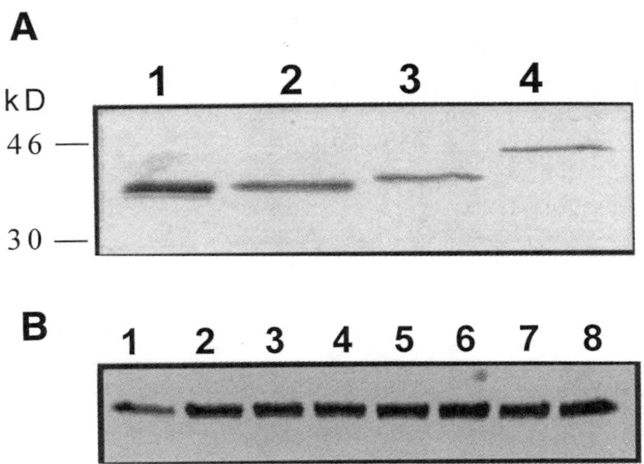

Fig. 3. Expression of scFv3-3 in culture supernatant of recombinant *P. pastoris*. **(A)** Culture supernatant (15 µL) from pPIC9 strain (lane 1), culture supernatant (30 µL) from pHIL-S1 strain (lane 2), culture supernatant (30 µL) from the scFv3-3-dimer-secreting strain (lane 3), and culture supernatant (30 µL) from scFv3-3::IgCκ secreting-strain (lane 4) were analyzed by SDS-PAGE on a 12% gel. Samples were taken 48 h after induction with methanol, and protein bands were visualized by staining with Coomassie brilliant blue. **(B)** Time-course of expression of scFv3-3–IL3. Samples were taken 16 h (lane 1), 24 h (lane 2), 40 h (lane 3), 48 h (lane 4), 64 h (lane 5), 72 h (lane 6), 88 h (lane 7), and 96 h (lane 8) after induction with methanol. Culture supernatant (20 µL) from each time-point was analyzed by SDS-PAGE as described in (A).

2. Dilute cells 1:10 in 25 mL of BMGY medium containing 1% casamino acids.
3. Shake cultures at 30°C for 6–8 h.
4. Induce protein production by shifting cultures by centrifugation to 25 mL of BMMY medium containing 0.5% methanol.
5. Allow cultures to grow at 30°C in a 300-mL Erlenmeyer flask under vigorous shaking.
6. Examine expression by harvesting a 20-µL aliquot of supernatant at 48 h after induction by SDS–PAGE. Visualize recombinant scFv3-3 antibodies by staining with Coomassie brilliant blue.

3.6. Affinity Purification of the His5- and myc-Tagged scFv3-3

To obtain purified scFv fragment for binding studies, 150 mL of culture supernatant from pPIC9–scFv3-3–IL3 were produced in a 2-L Erlenmeyer flask under the conditions described above. This 150-mL culture generated 17.5 mg of His_5-tagged scFv3-3–IL3 after purification for a calculated yield of 116 mg/L. Characterization of this material by surface plasmon resonance revealed an

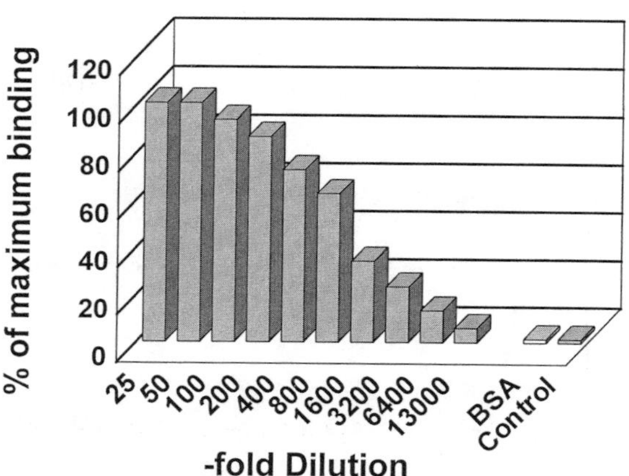

Fig. 4. ELISA of supernatant from recombinant secreting pPIC9–scFv3-3–IL3 *P. pastoris* strain. Dilutions of the supernatant sample shown in **Fig. 3**, lane 1, were analyzed for binding to hLIF as described in **Subheading 3.8**. Binding of a 50-fold dilution of the supernatant to BSA and binding of supernatant from a nonrecombinant *P. pastoris* strain to hLIF (control) are also shown.

equilibrium dissociation constant of 2.3×10^{-8} *M,* which is in good agreement with the value of 2.8×10^{-8} obtained for the scFv3-3 produced in *E. coli* ***(13)***. N-terminal sequence analysis of the purified scFv3-3 revealed that about 70% of the material starts with the sequence Y-V-E- and 30% starts with E-A-Y-V-E. The removal of the E-A-E-A repeats originally present in the sequence (**Fig. 2**) was probably owing to the activity of the *STE13* gene product. Since these deletions occur only within the artificially created amino-terminal extension and not within the structure of the scFv, its biological activity was not affected. Further characterization of the purified scFv3-3 by MALDI-TOF mass spectrometry surprisingly revealed a molecular mass that is about 900–1100 mass units greater than the calculated molecular weight. We also observed a somewhat broader mass distribution than one would expect for a protein of this size. Since there are no sites for N-linked glycosylation (N-X-T/S) in the scFv3-3–IL3, we speculate that the apparent multiple species and their higher mass might be owing to O-linked glycosylation or other posttranslational modification.

From 600 mL of culture supernatant containing the *myc*-tagged dimerized scFv, we could purify only 9.9 mg of material by affinity chromatography on a 9E10 affinity column corresponding to a calculated yield of 16.5 mg/L. Surface plasmon resonance analysis of the culture supernatant and the flow-through of the affinity column revealed that more than 65% of the binding activity was not absorbed to the column and was instead present in the flowthrough. This observation is consistent with SDS-PAGE analysis of the culture supernatant which suggested an expression level about threefold lower than that of the monomeric scFv3-3 (**Fig. 3**, lane 3). We estimate that the *myc*–scFv3-3 dimer was produced at about 50 mg/L. We reason that our failure to purify this modified scFv from the culture supernatant quantitatively was owing to the partial accessibility of the *myc* tag to the 9E10 antibody or to partial degradation of the *myc* tag by proteases secreted into the culture supernatant. We demonstrated dimerization of the purified *myc*–scFv3-3 by gel filtration and SDS-PAGE under nonreducing conditions (data not shown). By surface plasmon resonance analysis, the purified *myc*–scFv-3-3 dimer had a 40-fold higher equilibrium dissociation constant than the monomeric scFv3-3, which can be largely attributed to a 10-fold lower dissociation rate. The procedure for purification of His_5-tagged svFv is as follows:

1. Bring culture supernatant to 300 m*M* NaCl and 50 m*M* sodium phosphate, pH 8.0, and load at a flow rate of 22 cm/h onto a Ni-NTA resin column previously equilibrated with 300 m*M* NaCl and 50 m*M* sodium phosphate, pH 8.0.
2. Wash the column with 300 m*M* NaCl and 50 m*M* sodium phosphate, pH 8.0, and 300 m*M* NaCl and 50 m*M* sodium phosphate, pH 6.5, until no protein is detected in the effluent.
3. Elute the column with 300 m*M* NaCl and 50 m*M* sodium phosphate, pH 5.0, and dialyze fractions containing protein against 50 m*M* sodium phosphate buffer, pH 7.5. If the protein does not elute at pH 5.0, repeat the elution step with 300 m*M* NaCl, 50 m*M* sodium phosphate, and 500 m*M* imidazole, pH 5.0.

All steps are performed at 4°C.

The following procedure is used to purify *myc*-tagged svFvs:

1. Adjust pH of the culture supernatant to 7.4 with sodium phosphate buffer.
2. Load pH-adjusted supernatant onto an anti-*myc* 9E10 Sepharose column previously equilibrated with PBS (137 m*M* NaCl, 2.7 m*M* KCl, 4.3 m*M* Na_2HPO_4, 1.4 m*M* KH_2PO_4, pH 7.2) at a flow rate of 12 cm/h.
3. Wash the column PBS until no protein is found in the effluent.
4. Elute the recombinant protein from the column with 100 m*M* glycine-HCl, pH 2.7 (2 column volumes), and quickly neutralize the eluate by drop-wise addition of 0.1 *N* NaOH.
5. Dialyze against PBS.

3.7. ELISA

The secreted scFv3-3 antibody fragment could be used directly in binding studies by ELISA or for surface plasmon resonance studies. As an example, we show the binding of scFv3-3–IL3 to immobilized hLIF. As demonstrated in **Fig. 4**, supernatant from recombinant *P. pastoris* expression cultures could be used in ELISA assays, and because of the usually high concentrations of scFv fragments, even highly diluted supernatants provide a strong signal in this assay. The ELISA procedure is as follows:

1. Coat microtiter wells at 4°C for 16 h using 100 µL of a 10 µg/mL solution of recombinant hLIF/well.
2. Wash with PBS, and incubate for 2 h with 300 µL of 2% bovine serum albumin (BSA) in PBS to block nonspecific binding of the immunochemical reagents to the microtiter plate.
3. Wash five times with 300 µL of PBS + 0.05% Tween 20.
4. Add to each well 200 µL of culture supernatant containing the recombinant scFv3-3–IL3 that has been diluted with PBS containing 1% BSA and 0.05% Tween 20, and incubate for 2 h at ambient temperature.
5. Remove the nonspecifically bound scFv antibody by repeated washing (5–10 times) with 300 µL of PBS + 0.05% Tween 20.
6. Incubate with the biotinylated MAb F15-216 (1 µg/mL) for 2 h.
7. Incubate with streptavidin conjugated to horseradish peroxidase (0.2 µg/mL) for 30 min.
8. Remove unbound immunoconjugates by washing six times with PBS + 0.05% Tween 20 and then two times with water.
9. Detect bound horseradish peroxidase by incubating for 5–10 min at ambient temperature with 200 µL of OPD solution.
10. Stop the reaction by addition of 50 µL of 2 *M* H_2SO_4, and read absorbances at 490 and 650 nm.
11. Calculate the A_{650} was subtracted from A_{490}.

4. Notes

1. For transformation of *P. pastoris*, we also found electroporation to be a convenient method that requires fewer steps.
2. The selection for the Mut^s phenotype is done best by the replica plating technique using the Lederberg stamp rather than by streaking onto MD and MM plates directly from the regeneration plate. The latter approach saves 1–2 d, but the results are sometimes difficult to interpret.
3. To increase chances of finding clones with multiple inserts, colonies from several regeneration plates can be combined and plated on G418 antibiotic plates.
4. Although we did observe a considerable improvement in the production of many other heterologous proteins when using the selection protocol for multiple inserts, we did not *see* this with the expression of the scFv3-3–IL3 antibody.
5. We had more consistent results by inducing the cultures at relatively low cell density for the initial expression studies. Concentrating cells to higher densities prior to methanol induction should only be performed as an option during studies on optimizing expression and not for the initial screening.
6. It is crucial to perform the expression studies under good aeration. We generally achieve this by using <10% of the maximum flask volume (e.g., 25 mL of culture in a 300-mL shake flask) and by shaking at a high speed. We found this to be more convenient than using baffled flasks, which sometimes result in extensive foam-

ing. The initial screening of recombinant clones for expression can also be performed in 100-mL Erlenmeyer flasks with 10-mL cultures. For scFv fragments, supernatants from expression cultures also can be screened by ELISA rather than by SDS-PAGE.

References

1. Skerra, A. and Plückthun, A. (1988) Assembly of a functional immunoglobulin Fv fragment in *Escherichia coli*. *Science* **240,** 1038–1041.
2. Better, M., Chang, C. P., Robinson, R. R., and Horwitz, A. H. (1988) *Escherichia coli* secretion of an active chimeric antibody fragment. *Science* **240,** 1041–1043.
3. Plückthun, A. (1992) Mono- and bivalent antibody fragments produced in *Escherichia coli*: engineering, folding and antigen binding. *Immunol. Rev.* **130,** 151–188.
4. Skerra, A. (1993) Bacterial expression of immunoglobulin fragments. *Curr. Opinion Immunol.* **5,** 256–262.
5. Pack, P., Kujau, M., Schroeckh, V., Knuepfer, U., Wenderoth, R., Riesenberg, D., and Pluckthun, A. (1993) Improved bivalent miniantibodies, with identical avidity as whole antibodies, produced by high cell density fermentation of *Escherichia coli*. *BioTechnology* **11,** 1271–1277.
6. Nissim, A., Hoogenboom, H. R., Tomlinson, I. M., Flynn, G., Midgley, C., Lane, D., and Winter, G. (1994) Antibody fragments from a "single pot" phage display library as immunochemical reagents. *EMBO J.* **13,** 692–698.
7. Williamson, R. A., Burioni, R., Sanna, P. P., Partridge, L. J., Barbas, C. F., and Burton, D. R. (1993) Human monoclonal antibodies against a plethora of viral pathogens from single combinatorial libraries. *Proc. Natl. Acad. Sci. USA* **90,** 4141–4145.
8. Cregg, J. M., Vedvick, T. S., and Raschke, W. C. (1993) Recent advances in the expression of foreign genes in *Pichia pastoris*. *BioTechnology* **11,** 905–910.
9. Clare, J. J., Rayment, F. B., Ballantine, S. P., Sreekrishna, K., and Romanos, M. A. (1991) High-level expression of tetanus toxin fragment C in *Pichia pastoris* strains containing multiple tandem integrations of the gene. *BioTechnology* **9,** 455–460.
10. Laroche, Y., Storme, V., De Meutter, J., Messens, J., and Lauwereys, M. (1994) High-level secretion and very efficient isotopic labeling of tick anticoagulant peptide (TAP) expressed in the methylotrophic yeast *Pichia pastoris*. *BioTechnology* **12,** 1119–1124.
11. Romanos, M. A., Clare, J. J., Beesley, K. M., Rayment, F. B., Ballantine, S. P., Makoff, A. J., Dougan, G., Fairweather, N. F., and Charles, I. G. (1991) Recombinant *Bordetella pertussis* pertactin (P69) from the yeast *Pichia pastoris*: high-level production and immunological properties. *Vaccine* **9,** 901–906.
12. Lokker, N. A., Strittmatter, U., Steiner, C., Fagg, B., Graff, P., Kocher, H. P., and Zenke, G. (1991) Mapping the epitopes of neutralizing anti-human IL-3 monoclonal antibodies. *J. Immunol.* **146,** 893–898.
13. Ridder, R., Schmitz, R., Legay, F., and Gram, H. (1995) Generation of rabbit monoclonal antibody fragments from a combinatorial phage display library and their production in the yeast *Pichia pastoris*. *BioTechnology* **13,** 255–260.

14. Ridder, R., Geisse, S., Kleuser, B., Kawalleck, P., and Gram, H. (1995) A Cos-cell-based system for rapid production and quantification of scFv::IgCkappa antibody fragments. *Gene* **166,** 273–276.
15. Evan, G. I., Lewis, G. K., Ramsay, G., and Bishop, J. M. (1985) Isolation of monoclonal antibodies specific for human c-myc proto-oncogene product. *Mol. Cell. Biol.* **5,** 3610–3616.
16. Clare, J. J., Romanos, M. A., Rayment, F. B., Rowedder, J. E., Smith, M. A., Payne, M. M., Sreekrishna, K., and Henwood, C. A. (1991) Production of mouse epidermal growth factor in yeast: high-level secretion using *Pichia pastoris* strains containing multiple gene copies. *Gene* **105,** 205–212.
17. Scorer, C. A., Clare, J. J., McCombie, W. R., Romanos, M. A., and Sreekrishna, K. (1994) Rapid selection using G418 of high copy number transformants of *Pichia pastoris* for high-level foreign gene expression. *BioTechnology* **12,** 181–184.
18. Taylor, L. and Rose, R. E. (1988) A correction in the nucleotide sequence of the TN903 kanamycin resistance determinant in pUC4K. *Nucleic Acids Res.* **16,** 358,359.
19. Cregg, J. M., Barringer, K. J., Hessler, A. Y., and Madden, K. R. (1985) *Pichia pastoris* as a host system for transformations. *Mol. Cell Biol.* **5,** 3376–3385.

13

Expression of Tetanus Toxin Fragment C

Jeff Clare, Koti Sreekrishna, and Mike Romanos

1. Introduction

The *Pichia pastoris* system has now been used successfully to express a large number of different intracellular and secreted proteins. In some cases, extremely high levels have been obtained (e.g., tumor necrosis factor, TNF *[1]*; human serum albumin, HSA *[2]*). The aim of this chapter is to illustrate some of the important practical considerations in obtaining efficient expression, to highlight the problems that can occur, and to describe the different approaches that have been used to solve them.

The example presented here is the intracellularly expressed bacterial protein, tetanus toxin fragment C. This was one of the earliest examples of very high level expression in *P. pastoris (3)*, and in direct comparisons, the yield was significantly better than three other expression systems (*Esherichia coli*, *Saccharomyces cerevisiae*, and baculovirus; *see* **ref.** *4*). Fragment C exemplified the extreme clonal variation in expression level that can occur with *P. pastoris* and showed the importance of gene dosage in maximizing yields. This work led to the characterization of multicopy transplacement and the mechanism underlying this phenomenon. Fragment C was also used in a thorough comparison of various classes of transformants in an attempt to determine the optimum expression strain. Consideration of this example can therefore reveal several points of general relevance to the *P. pastoris* expression system.

2. Materials

2.1. P. pastoris *Strains and Expression Vectors*

The strains used were GS115 (*his4*) and KM71 (*his4 aox1*::*ARG4*). The vectors used were pPIC3, pPIC3K, pPIC9, and pPIC9K (or closely related derivatives), which are described in Chapter 5.

From: *Methods in Molecular Biology, Vol. 103:* Pichia *Protocols*
Edited by: D. R. Higgins and J. M. Cregg © Humana Press Inc., Totowa, NJ

2.2. Materials for Small-Scale Inductions

1. Yeast nitrogen base (YNB, Difco, Detroit, MI) medium containing 4 ng/mL biotin and 2% (w/v) glycerol or 1% (v/v) methanol.
2. Sterile 25-mL Universal screw-cap tubes (Nunc, Denmark) and 100-mL conical culture flasks.

2.3. Apparatus and Reagents for Large-Scale (Fermenter) Inductions

1. YNB medium containing 4 ng/mL biotin and 2% (w/v) glycerol).
2. 5X basal salts: 42 mL/L phosphoric acid, 1.8 g/L calcium sulfate. $2H_2O$, 28.6 g/L potassium sulfate, 23.4 g/L magnesium sulfate. $7H_2O$, 6.5 g/L potassium hydroxide. Autoclave and store at room temperature.
3. PTM1 salts solution: 6 g/L cupric sulfate; $5H_2O$, 0.08 g/L potassium iodide, 3 g/L manganese sulfate; H_2O, 0.2 g/L sodium molybdate, 0.02 g/L boric acid, 0.5 g/L cobalt chloride, 20 g/L zinc chloride, 65 g/L ferrous sulfate; $7H_2O$, 0.2 g/L biotin, 5 mL/L sulphuric acid. Autoclave and store at room temperature.
4. Ammonium hydroxide.
5. Methanol.
6. Fermenter with monitors and controls for for pH, temperature, dissolved O_2, air-flow, and stirring speed.

2.4. Materials for Chromosomal DNA Preparation

1. YPD medium.
2. 250-mL conical flasks.
3. SCE buffer: 1 *M* sorbitol, 10 m*M* sodium citrate, 10 m*M* EDTA. Autoclave and store at room temperature.
4. 1 *M* dithiothreitol: store at –20°C.
5. Zymolyase-100T (ICN, Costa Mesa, CA), 3 mg/mL suspension in distilled water. Make fresh.
6. 1% Sodium dodecyl sulfate (SDS).
7. 5 *M* potassium acetate, pH 8.9.
8. TE buffer: 10 m*M* Tris-HCl, pH 8.0, 0.1 m*M* EDTA.
9. Ribonuclease A (DNase-free), 10 mg/mL. Store at –20°C.
10. Isopropanol.

2.5. Apparatus and Reagents for Quantitative DNA Dot Blots

1. 96-Well dot-blot manifold (e.g., "minifold," Schleicher and Schuell, Keene, NH).
2. Vacuum pump.
3. Nitrocellulose filters.
4. 20X SSC stock solution: 3 *M* sodium chloride, 0.3 *M* sodium citrate.

3. Methods

Fragment C is a polypeptide derived from the C-terminus of the toxin produced by *Clostridium tetani*. The toxin consists of two subunits, light chain

Table 1
Fragment C Expression Levels in Different Types of Transformant

Clone	Type			Copy no.	Shake flask[a]	Fermenter[a]
4F	*HIS4*	Integrant[b]	Mut^s	1	0.44 ± 0.22	ND
10F	*HIS4*	Integrant[b]	Mut^s	2	1.16 ± 0.16	ND
97A	*AOX1*	Integrant[b]	Mut^s	1	0.30 ± 0.02	6.4 ± 0.1
98I	*AOX1*	Integrant	Mut^+	1	0.54 ± 0.05	8.3 ± 0.4
98B	*AOX1*	Integrant[c]	Mut^+	1	0.30 ± 0.05	ND
10C	*AOX1*	Integrant[c]	Mut^+	9	6.5 ± 1.5	19.4 ± 1.0
11C	*AOX1*	Transplacement	Mut^s	1	0.33 ± 0.06	4.5 ± 0.8
2E	*AOX1*	Transplacement	Mut^s	3	3.4 ± 0.5	ND
5C	*AOX1*	Transplacement	Mut^s	6	5.4 ± 0.5	ND
12D	*AOX1*	Transplacement	Mut^s	7	7.0 ± 0.6	ND
3H	*AOX1*	Transplacement	Mut^s	9	8.4 ± 0.3	24.6 ± 3.7
1F	*AOX1*	Transplacement	Mut^s	14	10.5 ± 1.4	27.4 ± 1.0

[a]Fragment C levels were determined after induction for 6 d in shake flasks or 4 d in the fermenter. Levels are given as a percentage of total cell protein and were determined by ELISA.
[b]Single-crossover integrants into strain KM71 (which carries a disrupted copy of *AOX1*).
[c]In these clones, the fragment C expression cassette has inserted into *AOX1* without disrupting it.

and heavy chain, which are derived from a precursor polypeptide by a single proteolytic cleavage. Fragment C can be generated from this precursor by cleavage with papain and is found to be completely nontoxic, but yet highly immunogenic. Fragment C is thus a potential subunit vaccine against tetanus. Expression studies in *S. cerevisiae* with fragment C led to the finding that foreign genes often contain AT-rich sequences that fortuitously cause premature termination of transcription in yeast ***(5)***. This problem may be a common reason for failure to express foreign genes and is more fully discussed in Chapter 14 on HIV-1 ENV expression. Therefore, the *P. pastoris* expression studies described here used a synthetic fragment C gene with increased GC content.

3.1. Fragment C Expression in Different Kinds of Transformants

Fragment C was used in a detailed study of some of the many factors that could affect the final yield of foreign proteins in *P. pastoris* ***(3)***. Expression levels in single-copy integrants were compared in order to determine the influence of Mut phenotype, site of vector integration (*HIS4* vs *AOX1*), and type of integration (nondisruptive insertion vs transplacement). Each of these variables were found to have only modest effects on the final yield of fragment C (**Table 1**). This is perhaps surprising, since there are reasons for expecting significant effects. Host cells that contain an undisrupted *AOX1* gene (Mut^+) might be expected to yield less foreign product than equivalent *AOX1* disrupted cells (Mut[s]) as a consequence of high-level expression of alcohol oxidase. However, even in high-efficiency fermenter inductions (*see* **Subheading 3.2.**), equivalent Mut[+] and Mut[s] transformants produced similar amounts of fragment C. Possibly the higher growth rate of Mut[+] cells during induction can compensate for the metabolic burden of high-level alcohol oxidase production. In other cases, for example, *Bordetella pertussis* pertactin ***(6)*** and hepatitis surface antigen (HBsAg; *7*), differences between Mut[+] and Mut[s] in product quality and yield were observed. However, this may reflect the use of nonoptimized induction conditions.

The relative stability of foreign DNA inserted by transplacement, as opposed to nondisruptive integration, might be expected to result in higher yields especially in large-scale, high-density cultures. This effect should be exacerbated in *HIS4* integrants using histidine prototrophy for selection, since excision of vector DNA between duplicated *HIS4* sequences can occur leaving an unmutated copy of the gene. However, this effect was not observed at the scale used for fragment C inductions.

3.1.1. Small-Scale Fragment C Inductions

1. Inoculate 10 mL of YNB containing 2% glycerol with cells from a single colony of transformant and incubate overnight with shaking in a universal container at 30°C. (A_{600} should be about 5.)
2. Dilute the starter culture into 10 mL of YNB containing 2% glycerol to an A_{600} of 0.25. Incubate at 30°C with vigorous shaking for 6–8 h to obtain an exponentially growing culture (*see* **Note 1**).
3. Harvest the cells by centrifugation (4°C, 2000*g*, 5 min). Resuspend the cell pellet in 10 mL sterile water. Harvest the cells by centrifugation, and resuspend in the same volume of YNB containing 1% methanol. Incubate in a 100 mL conical flask at 30°C with vigorous shaking for 1–5 d (*see* **Notes 2** and **3**). (A_{600} should be 5–10.)
4. Harvest the cells for analysis by centrifugation (4°C, 2000*g*, 5 min).

3.2. Optimization of Fragment C Expression

The yield of fragment C in shake-flask inductions of single-copy transformants was surprisingly low—only 0.3–0.5% of total protein was produced (**Table 1**)—compared to 2–3% obtained in *S. cerevisiae* with a 2-μ based vector. However, during the analysis of the transplacement transformants, considerable variation was observed in the level of fragment C produced (**Fig. 1**). Clonal variation was first observed by K. Sreekrishna, who found large differences in the level of TNF produced by different transplacement transformants (from 1 to 30% of total protein). This was shown to be owing to variation in gene copy number even though transplacement by double crossover would be expected to give only single-copy integrants. Although it was unclear how multicopy

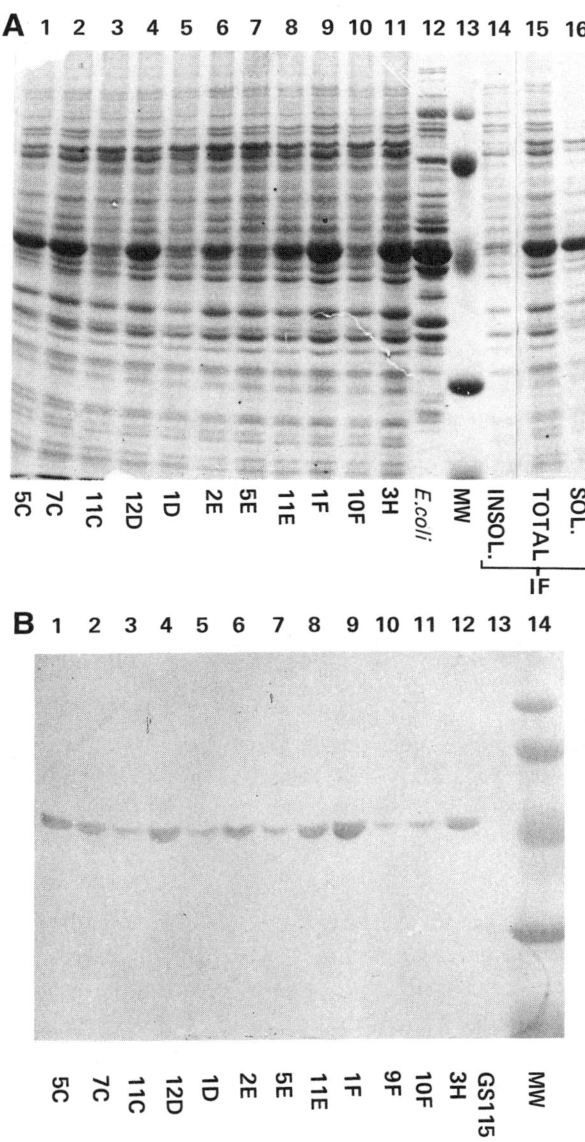

Fig. 1. Clonal variation in fragment C expression levels in different Mut[s] transplacement transformants. **(A)** Coomassie blue-stained SDS-polyacrylamide gel showing total cell extracts. Lanes 14 and 16 are the pellet and supernatant fractions, respectively (after centrifuging at 10,000*g* for 15 min) from an aliquot of the same extract as in lane 15. Lane 12 shows an extract from a fragment C-expressing *E. coli* strain as a control. **(B)** Immunoblot of a similar gel using fragment C-specific antiserum. A control extract from untransformed cells is shown in lane 13.

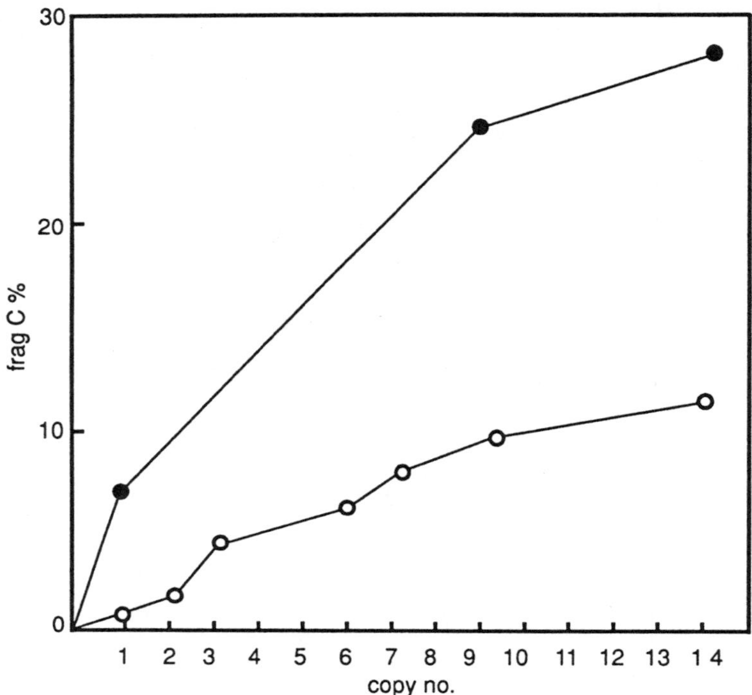

Fig. 2. Correlation between gene dosage and the level of expression of tetanus toxin fragment C (% of total protein) in Mut^s transformants. Closed circles represent high-density fermenter inductions, and open circles represent shake-flask inductions. Data are taken from **Table 1**.

transplacement could occur, we nevertheless examined DNA copy number in the fragment C transplacement strains. Transformants with a range of different copy numbers were identified, and a clear correlation between expression level and gene dosage was demonstrated (**Fig. 2**). These strains were authentic "multicopy transplacement" transformants since they contained a disrupted *AOX1* gene (*see* **Subheading 3.3.**) and were Mut^s. The transformant with the highest yield (10% of total protein in shake-flask inductions) had 14 copies of the fragment C gene. The data in **Fig. 2** predict that transfomants with even higher gene dosage should give higher yields still, suggesting that optimal expression levels can be obtained by maximizing DNA copy number.

3.2.1. Preparation of Chromosomal DNA

1. Grow an overnight culture of each transformant in 50 mL of YPD, incubating at 30°C.

2. Harvest cells by centrifugation at 2000*g* for 5 min, resuspend the cell pellet in distilled water, then repellet the cells, and resuspend in 2 mL of SCE buffer containing 10 m*M* dithiothreitol.
3. Add 100 mL of zymolyase (3 mg/mL), and incubate at 37°C for 1 h or until spheroplast formation approaches 100% (*see* **Note 4**).
4. Add 2 mL of 1% (w/v) SDS, mix gently by inversion, and incubate on ice for 5 min.
5. Add 1.5 mL of 5 *M* potassium acetate (pH 8.9), mix gently, and remove the precipitate by centrifuging at 10,000*g* for 10 min.
6. Add 2 vol of ethanol to the supernatant leave at room temperature for 15 min and then centrifuge at 10,000*g* for 15 min.
7. Allow the pellet to dry, add 3.0 mL of TE containing 0.5 mg/mL RNase A. Agitate gently at 30°C for 1–2 h to resuspend the pellet and digest RNA. If the sample still contains any insoluble material, it can be removed by centrifuging at 10,000*g* for 15 min.
8. Slowly add an equal volume of isopropanol to form a separate layer. Mix gently by swirling so the DNA forms a fibrous precipitate at the interface.
9. Remove the DNA by "spooling" with a glass hook (formed from a Pasteur pipet in a Bunsen burner flame), and resuspended in 250–500 mL of TE (*see* **Note 5**).

3.2.2. Determination of Copy Number by Quantitative Dot Blot

1. Denature 10 µL of chromosomal DNA by adding 2.5 µL of 1 *M* NaOH, and incubating at room temperature for 15 min.
2. Place samples on ice, neutralize with 2.5 µL 1 *M* HCl, and then dilute with 175 µL 10X SSC.
3. Dot each sample in triplicate onto a nitrocelluose filter using a dot-blot manifold, adding 30-µL aliquots of sample/well and applying a vacuum to draw the liquid through. Prepare duplicate filters, one for hybridization to a native gene control (e.g., 0.75-kb *Bgl*II fragment from *ARG4*) and one for hybridization to the foreign gene (e.g., 1.4 kb *Bgl*II–*Nhe*I fragment from fragment C gene).
4. Rinse the filters briefly in 2X SSC, allow to air-dry, and then place in a vacuum oven at 80°C for 1 h.
5. Hybridize the filters using standard techniques. Hybridization levels can be determined by phosphorimaging or alternatively by cutting out the dots, solubilizing in methanol, and scintillation counting. Triplicate counts are averaged, and the ratio of foreign gene/*ARG4* determined for each transformant. These values are then normalized to a known single-copy control (identified by Southern blot) to determine absolute copy number.

The yield of fragment C was also significantly improved by optimizing the induction conditions. Shake-flask induction is generally inefficient and yields can be dramatically improved in the fermenter. This is clearly illustrated with fragment C—the expression level in single-copy integrants was improved at least 10-fold in the fermenter, and even in the highest expressing strain (containing 14 gene copies), a 2.5-fold increase in yield was obtained (**Table 1, Fig.**

2). This effect is probably owing to the more efficient aeration in the fermenter. Shake-flask cultures are oxygen-limited, which can lead to variation in yield, even from the same transformant, especially for Mut^+ integrants, which have a higher oxygen demand.

A time-course for a fermenter induction of a high copy number Mut^s fragment C expressing clone is shown in **Fig. 3**. After an initial phase of batch growth, a period of glycerol-limited growth was carried out to increase biomass of the culture, and then the induction phase was initiated by replacing the glycerol feed with methanol (*see* **Subheading 3.2.3.**). The final level of fragment C was about 27% of total protein. It has been suggested that Mut^s strains require a longer induction period than Mut^+, but with fragment C maximal levels could be obtained within about 48 h of induction with either type of transformant. However, with Mut^s strains care must be taken to avoid the accumulation of toxic levels of methanol, which can result in decreased yields. This was observed with fragment C using induction periods longer than 48 hrs, but the problem could be avoided by reducing the methanol feed rate at this point.

A major advantage of *P. pastoris* compared to most other expression systems is the ease in obtaining very high cell density cultures. This is readily achieved during the glycerol-limited phase of fermenter inductions or, alternatively, with Mut^+ strains, which can be induced at low cell density, during the induction phase. In the fragment C induction shown in **Fig. 3**, the culture density reached about 100 g dry wt of cells/L. Thus, it can be calculated that the yield of fragment C in this induction was >12 g/L. Since the initial concentration was high (27% of total protein), purification of the product was relatively simple (**Fig. 4**). Although the purification was not optimized, several grams of purified fragment C were obtained, which would be expected to provide enough material for 10^5 vaccine doses. The fragment C process was not taken beyond 10 L scale, but experience with other foreign proteins suggests that *P. pastoris* inductions can be scaled up by several orders of magnitude without loss of yield (up to 10,000 l for insulin-like growth factor, IGF—*see* Chapter 11).

Fig. 3. *(opposite page)* Fermentation of a 14-copy Mut^s transformant expressing tetanus toxin fragment C. **(A)** Cell density (A_{600}), dissolved oxygen (dO_2), and fragment C (% of total protein) are plotted against time (0 h = start of induction). Arrow A indicates addition of inoculum, arrow B indicates glycerol-limited feed, and arrow C indicates methanol feed. Fermentation of a 14-copy Mut^s transformant expressing tetanus toxin fragment C. **(B)** Coomassie blue-stained SDS-polyacrylamide gel showing protein extracts from cells taken at different times during induction. The arrow indicates the position of fragment C. For comparison, extract from the same transformant induced in a shake-flask is shown in lane 3, and extract from an untransformed control strain is shown in lane 2.

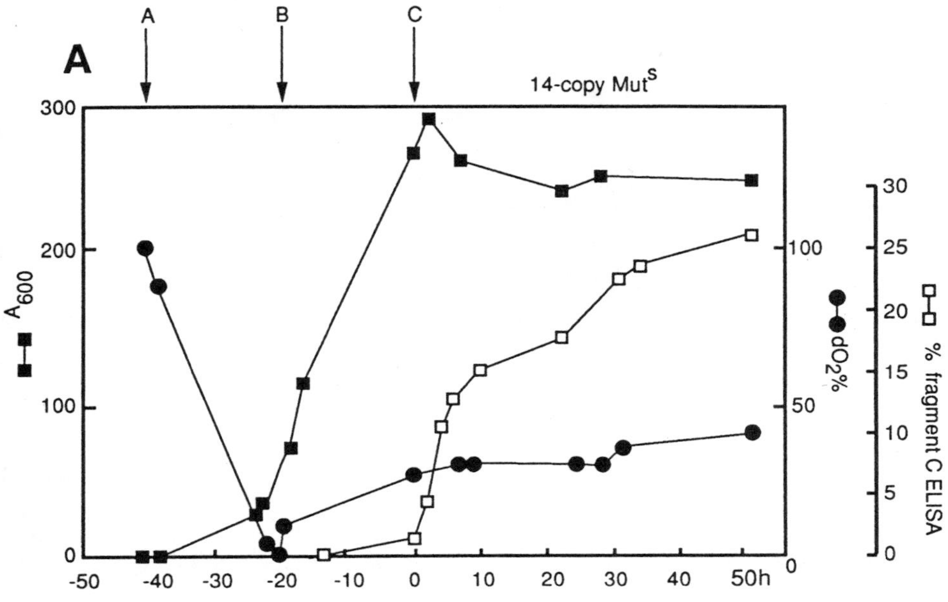
A
B
C
A
14-copy Mut^S
300
200
100
0
-50
-40
-30
-20
-10
0
10
20
30
40
50h
A600
dO2%
100
50
0
% fragment C ELISA
30
25
20
15
10
5
0

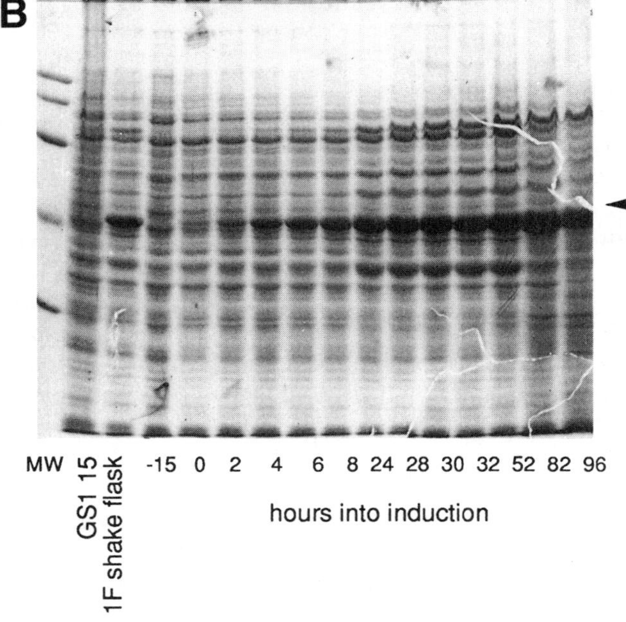
B
MW
GS1 15
1F shake flask
-15 0 2 4 6 8 24 28 30 32 52 82 96
hours into induction

Fig. 3.

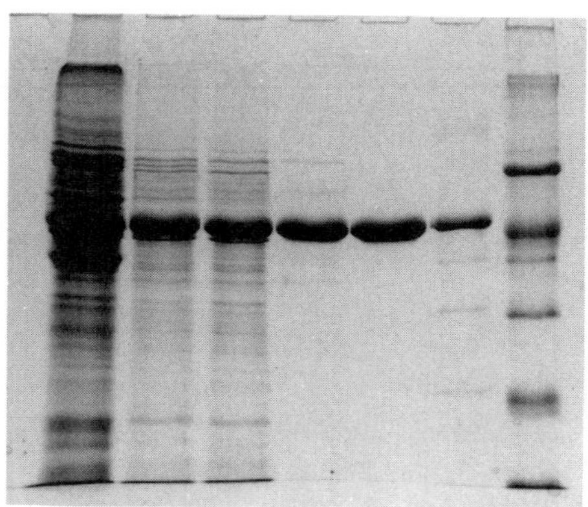

Fig. 4. Purification of fragment C from *P. pastoris*. Coomassie blue-stained gel showing aliquots from different stages of the purification. Lane 1, crude extract: lane 2, acid-clarified extract; lane 3, 0.2 μ filtrate; lane 4, eluate from zinc chelating Sepharose chromotography; lane 5, Q-Sepharose eluate; lane 6, fragment C from *C. tetani*, lane 7 mol-wt markers.

3.2.3. Large-Scale Fragment C Inductions

1. Inoculate 50 mL YNB containing 2% glycerol with cells from a single colony of the transformant, and incubate overnight with shaking in a conical flask at 30°C.
2. Add this culture to the fermenter containing 1 L of 5X basal salts plus 4 mL of PTM1 salts and 5% (v/v) glycerol.
3. Allow to grow at 30°C until glycerol is exhausted (24–30 h; *see* **Note 6**) maintaining dissolved O_2 above 20% by adjusting aeration and agitation, and maintaining the pH at 5.0 by the addition of 50% (v/v) ammonium hydroxide.
4. Initiate a limited glycerol/PTM1 feed (50% v/v glycerol containing 12 mL/L PTM1) at 12 mL/h, and continue for 17–24 h maintaining pH, temperature, and dissolved O_2 as before. The culture should grow to a density of about 100 g/L (dry wt of cells) (*see* **Note 7**). Cell density can be monitored by measuring A_{600}. (One A_{600} unit is approximately equivalent to 0.33 g/L, although different spectrophotometers will vary and each should be calibrated individually.)
5. Induce the culture by replacing the glycerol feed with a methanol/PTM1 feed (100% methanol with 12 mL/L PTM1) at 1 mL/h. Gradually increase the methanol feed rate over a period of 6 h to 6 mL/h, and continue the fermentation for a further 46–92 h. With Mut^s strains, the methanol feed rate should be reduced to 1 mL/h 48 h after the beginning of induction (*see* **Note 8**).

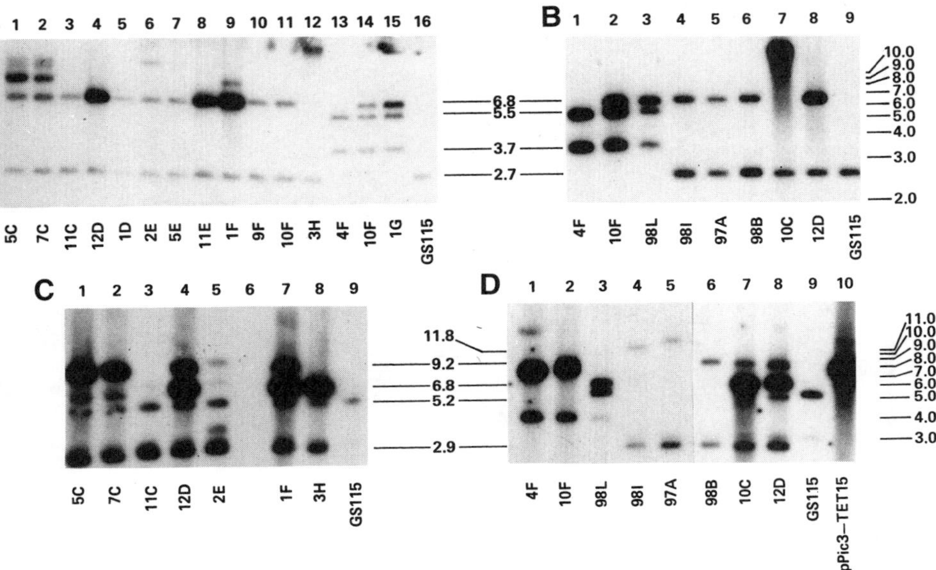

Fig. 5. Southern blot analysis of chromosomal DNA from different integrants from a fragment C transplacement transformation. **(A)** and **(B)** show *Bgl*II-cut DNA hybridized with a *HIS4* probe, **(C)** and **(D)** show *Eco*RI-cut DNA hybridized to an *AOX1* probe. DNA from Mut^s integrants are shown in (A) and (C), and DNA from Mut$^+$ integrants is shown in (B) and (D).

3.3. Chromosomal DNA Analysis of Fragment C Transformants—A Mechanism for Multicopy Transplacement

In order to understand how multicopy transplacement had occurred, a detailed analysis of chromosomal DNA from different fragment C-expressing transformants was performed (**Fig. 5**). To confirm the presence of the integrated expression cassettes, chromosomal DNA from different transformants was cut with the same restriction enzyme as that used to generate the linear DNA fragment prior to transformation, i.e., *Bgl*II (**Fig. 6**). After electrophoresis and Southern blotting, these DNA samples were hybridized to a *HIS4*-specific probe. To analyze the structure of the *AOX1* locus, chromosomal DNA was digested with *Eco*RI, which does not cut within *AOX1*, but has a single site within the fragment C gene (**Fig. 6**). These samples were subsequently hybridized to an *AOX1*-specific probe. The results of this analysis are shown in **Fig. 5**, and the sizes of the predicted bands are given in **Fig. 6**.

As expected, single-copy transplacements gave two *Bgl*II bands of similar intensity with the *HIS4* probe—a 2.7-kb band containing the chromosomal *his4*

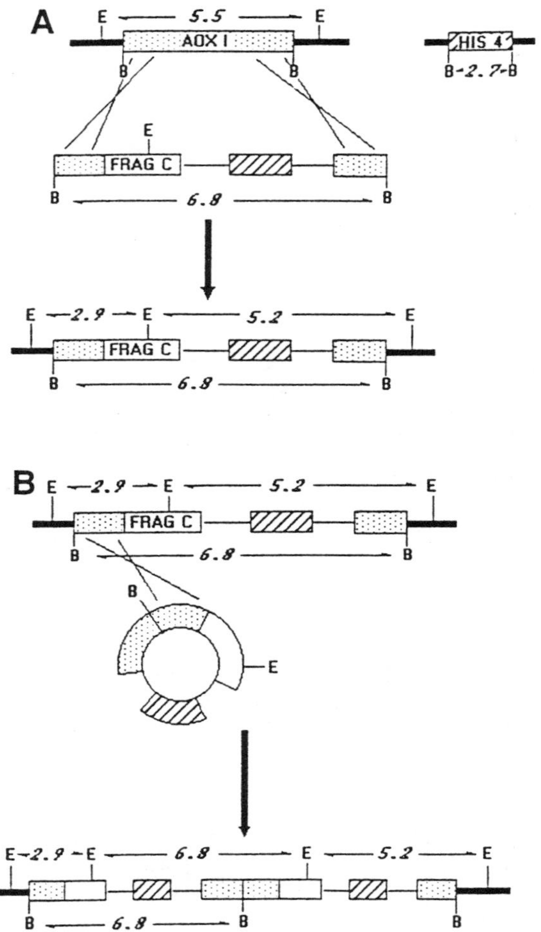

Fig. 6. Predicted structures and molecular events for **(A)** single-copy transplacement, **(B)** multicopy transplacement, **(C)** single crossover integration of circularized transplacement cassette into *AOX1* and **(D)** single-crossover integration of circularized transplacement cassette into *HIS4*.

allele and a 6.8-kb band containing the expression cassette (e.g., **Fig. 5A**, lane 3). Several of the clones gave an intense 6.8-kb band, clearly showing that they contained multiple copies of the expression cassette (e.g., **Fig. 5A**, lanes 4 and 8). Some clones gave additional larger bands (e.g., **Fig. 5A**, lanes 1, 2, 6, 9, 12), which were consistent in size with the loss of one or more *Bgl*II sites—presumably by exonucleolytic trimming in vivo prior to integration.

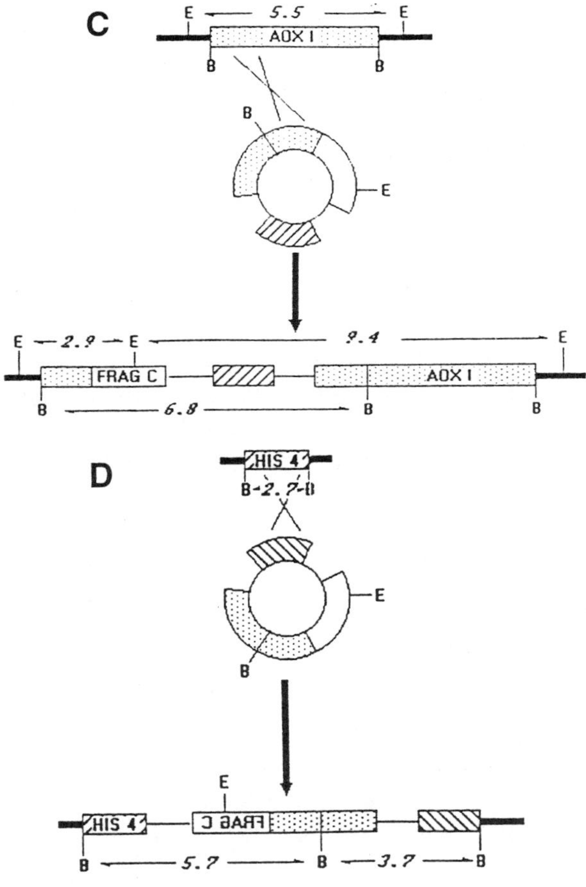

Fig. 6. *(cont.)*

In an untransformed strain, the *AOX1* probe gave a single 5.5-kb *Eco*RI band (**Fig. 5C**, lane 9). In single-copy transplacements, this band was disrupted giving two bands of 2.9 and 5.2 kb (e.g., **Fig. 5C**, lane 3). Multicopy transplacement transformants gave the same two bands, confirming disruption of *AOX1*, and an additional band of about the same size as the expression cassette (e.g., **Fig. 5C**, lanes 4, 7, and 8). This indicates that all copies of the expression cassette are present at the *AOX1* locus and that they are orientated in head-to-tail configuration. Only in one case was the pattern of bands consistent with a head-to-head (predicted band 4.1 kb) and tail-to-tail (predicted band 9.4 kb) arrangement (**Fig. 5C**, lane 5). This suggests that multicopy transplacement does not usually occur by multimerization prior to insertion, since expression cassettes would be randomly orientated. The presence of head-to-tail tandem arrays is consistent with circularization of expression cassettes prior to insertion by single crossover. Several transplacement transformants also contained multiple copies of the intact vector (9.2 kb, **Fig. 5C**, lanes 1, 2, 4, and 7).

Transplacement transformations with *P. pastoris* unexpectedly yield a high proportion of Mut^+ transformants. To understand further the events involved in these transformations, we analyzed chromosomal DNA from His^+, Mut^+ fragment C transfomants. The majority were found not to contain foreign DNA and may have arisen by *HIS4* gene conversion. However, Mut^+ transformants were identified that contained multiple copies of the transplacement cassette in tandem head-to-tail arrays at *AOX1* (9.4- and 2.9-kb *Eco*RI fragments with *AOX1* probe, e.g., **Fig. 5D**, lane 7). Others contained multiple copies in head-to-tail arrays inserted at *HIS4* (5.5- and 6.8-kb *Eco*RI bands with *AOX1* probe, 5.7- and 3.7-kb *Bgl*II bands with *HIS4* probe, e.g., **Fig. 5D**, lane 3 and **Fig. 5B**, lane 3). Both types of transformant are consistent with the mechanism proposed for multicopy transplacement, i.e., circularization of transplacement cassettes prior to single-crossover insertion. However, in the Mut^+ transformants these events occur independently of double-crossover transplacement, such that an intact copy of *AOX1* remains.

This analysis suggests that transplacing DNA fragments can circularize prior to insertion, presumably by self-ligation in vivo (**Fig. 6**). Insertion of multiple copies can then occur, either in addition to transplacement by a linear copy to give multicopy transplacement, or in the absence of transplacement to give multicopy Mut^+ transformants. This mechanism suggests that clonal variation should be a general phenomenon in transplacement transformations. Nevertheless, we have found that it occurs at variable frequency and that high-copy transformants can be quite rare. Accordingly, we have developed methods for screening for high DNA copy number (*see* Chapters 5 and 14), which have been used routinely to obtain high-level expression strains. These methods are useful for any gene product, but particularly where it may be difficult to screen transformants directly for expression.

3.3.1. Southern Analysis of Chromosomal DNA from Fragment C Transformants

1. Digest 10 μL of chromosomal DNA from each transformant and an untransformed control with *Bgl*II. Digest a second 10-μL aliqout of each with *Eco*RI.
2. Electrophorese the samples on 0.7% agarose gels together with suitable mol-wt markers. (Run *Bgl*II and *Eco*RI digests on separate gels.) Transfer to nylon membranes using a standard Southern blotting technique.
3. Hybridize the filter with *Bgl*II-cut DNA to an *HIS4*-specific probe (e.g., 0.6-kb *Kpn*I fragment from *HIS4*) and the filter with *Eco*RI-cut DNA to an *AOX1*-specific probe (e.g., 1.4-kb *Cla*I fragment from pPIC3).

4. The *HIS4*-probed filter can be used to estimate copy number by comparing the intensity of the native *HIS4 Bgl*II band (2.7 kb) with that of the expression cassette. Single integrants are confirmed by the presence of two *Eco*RI bands with the *AOX1* probe—the undisrupted *AOX1* gene gives a single band (5.5 kb). Multiple integrants contain a third band of the same size as the integrating DNA.

3.4. Summary

1. The yield of fragment C was only modestly affected by Mut phenotype, and the site and type of integration event.
2. Fragment C accumulation was closely correlated with gene dosage and maximal expression levels required high gene copy number.
3. Yields were greatly increased in controlled fermenters, compared to shake-flasks, owing to the high cell density achieved and to an increased efficiency of induction (2.5- to 10-fold).
4. In fermenter inductions of a 14-copy strain, fragment C accumulated to 27% of total protein, giving an estimated yield of 12 g/L.
5. Considerable clonal variation in the level of expression occurred with transplacement transformants, and this was owing to a diversity of different integration events and to differences in gene copy number.
6. These multicopy transplacement events occur by in vivo circularization of transforming DNA fragments followed by repeated single-crossover integration. This is presumably a general phenomenon, such that it should be possible to obtain multicopy integrants from all *P. pastoris* transplacement transformations.

4. Notes

1. Aeration of *P. pastoris* cultures is a key consideration, since the efficiency of growth and induction is reduced when oxygen is limited. To avoid this in small-scale inductions, incubate with vigorous shaking, keep the culture volume to a minimum (e.g., 5–10 mL), and use large flasks (e.g., >100 mL, preferably baffled).
2. When inducing shake-flask cultures for >1 d, add an additional 0.5% methanol on the second day.
3. The optimal induction period may vary for different proteins, and it is recommended that a time-course be carried out to maximize yields.
4. The efficiency of spheroplast formation can be monitored using phase-contrast microscopy by placing a drop of cells on a microscope slide together with a drop of 1% (w/v) SDS—unlysed cells appear phase-bright, whereas lysed "ghosts" are phase-dark.
5. The "spooling" method is preferable for harvesting chromosomal DNA, since better-quality DNA for restriction analysis is obtained. If the DNA precipitate is too loose to be transferred in this way, it can be spun down, resuspended, and precipitated a second time when it should be possible to remove it.
6. In fermenter inductions, the end of the initial phase (i.e., batch growth) can easily be monitored as the dissolved O_2 level begins to rise sharply when the glycerol becomes exhausted.

7. In order to acheive optimal yield in the fermenter, the cell density at induction can be varied by changing the duration of the glycerol-limited phase. For example, the yield of secreted proteins may be improved by early induction, allowing secretion to occur during a period of active growth. The methanol-utilization phenotype (Mut) should also be considered. Mut^+ strains can be induced at low cell density, since they continue to grow during the induction phase. Mut^s strains should be allowed to achieve high density before induction.
8. For efficient fermenter induction of Mut^+ strains, after an initial period to adapt the culture to methanol utilization, the aim is to achieve maximal methanol feed rates while maintaining sufficient aeration. If dissolved O_2 levels cannot be maintained above 20%, the methanol feed rate should be reduced. This will avoid the accumulation of toxic levels of of methanol. To avoid this problem with Mut^s strains, the methanol feed rate should be reduced as indicated.

References

1. Sreekrishna, K., Nelles, L., Potenz, R., Cruze, J., Mazzaferro, P., Fish, W., Motohiro, F., Holden, K., Phelps, D., Wood, P., and Parker, K. (1989) High-level expression, purification, and characterisation of recombinant human tumour necrosis factor synthesised in the methylotrophic yeast *Pichia pastoris*. *Biochemistry* **28,** 4117–4125.
2. Barr, K. A., Hopkins, S. A., and Sreekrishna, K. (1992) Protocol for efficient secretion of HSA developed from *Pichia pastoris*. *Pharm. Eng.* **12,** 48–51.
3. Clare, J. J., Rayment, F. B., Ballantine, S. P., Sreekrishna, K., and Romanos, M. A. (1991) High level expression of tetanus toxin fragment C in *Pichia pastoris* strains containing multiple tandem integrations of the gene. *Bio/Technology* **9,** 455–460.
4. Romanos, M. A., Scorer, C. A., and Clare, J. J. (1992) Foreign gene expression in yeast: a review. *Yeast* **8,** 423–488.
5. Romanos, M. A., Makoff, A. J., Fairweather, N. F., Beesley, K. M., Slater, D. E., Rayment, F. B., Payne, M. M., and Clare, J. J. (1991) Expression of tetanus toxin fragment C in yeast: gene synthesis is required to eliminate fortuitous polyadenylation sites in AT-rich DNA. *Nucleic Acids Res.* **19,** 1461–1467.
6. Romanos, M. A., Clare, J. J., Beesley, K. M., Rayment, F. B., Ballantine, S. P., Makoff, A. J., Dougan, G., Fairweather, N. F., and Charles, I. G. (1991) Recombinant Bordetella pertussis pertactin (P69) from the yeast *Pichia pastoris*: high-level production and immunological properties. *Vaccine* **9,** 901–906.
7. Cregg, J. M., Tschopp, J. F., Stillman, C., Siegel, R., Akong, M., Craig, W. S., Buckholz, R. G., Madden, K. R., Kellaris, P. A., Davies, G. R., Smiley, B. L., Cruze, J., Torregrossa, R., Velicelebi, G., and Thill, G. P. (1987) High-level expression and efficient assembly of hepatitis B surface antigen in the methylotrophic yeast *Pichia pastoris*. *Bio/Technology* **5,** 479–485.

14

Expression of EGF and HIV Envelope Glycoprotein

Jeff Clare, Carol Scorer, Rich Buckholz, and Mike Romanos

1. Introduction

This chapter reviews the expression of two proteins that are naturally secreted: murine epidermal growth factor (mEGF) and the 120-kDa envelope protein from human immunodeficiency virus (HIV-1 ENV). Although the initial aims of these projects were to examine the secretion in *Pichia pastoris* of a simple polypeptide and a complex glycoprotein, ultimately the work yielded insights into a wide range of issues affecting gene expression in *P. pastoris*, from foreign gene transcription to multicopy selection and glycosylation.

mEGF was used to determine the utility of the α-factor leader in *P. pastoris*, to examine the effect of gene dosage, and to compare secretion efficiency with *Saccharomyces cerevisiae* ***(1)***. To do this, the α-factor vector pPIC9 was constructed, which is now widely used for protein secretion in *P. pastoris*. Correct and efficient processing of the α-factor leader was demonstrated, and although the product was proteolytically unstable, it was possible to overcome this and achieve yields of 0.5 g/L of product. mEGF-expressing transformants showed clonal variation in product yield, and a general dot-blot screening method for the isolation of multicopy, high-expressing clones was developed.

Initially, the aim of the HIV-1 ENV project was to secrete an immunologically active form of the glycoprotein for vaccine development, in the expectation that the *P. pastoris*-derived protein would not be hyperglycosylated. In fact, several unforeseen obstacles were encountered in this work. The native HIV-1 ENV gene was not transcribed in *P. pastoris* owing to the presence of multiple fortuitous terminator sequences; gene synthesis to increase GC content was required to remove these and achieve efficient expression. The secreted product was then found to be hyperglycosylated and not recognized by antibodies to the native protein ***(2)***. ENV was therefore expressed intracellularly,

From: *Methods in Molecular Biology, Vol. 103:* Pichia *Protocols*
Edited by: D. R. Higgins and J. M. Cregg © Humana Press Inc., Totowa, NJ

developing a set of vectors (pPIC3K and pPIC9K) in order to use G418 selection of multicopy transformants *(3)*. These vectors enabled the isolation of a series of ENV-expressing transformants with progressively increasing copy number that were used to investigate the relationship between foreign gene copy number and mRNA level.

2. Materials

2.1. *P. pastoris* Strains and Expression Vectors

The strains used were GS115 (*his4*) and KM71 (*his4 aox1*::*ARG4*). The vectors used were pPIC3, pPIC3K, pPIC9, and pPIC9K (or closely related derivatives), which are described in Chapter 5.

2.2. Materials for Small-Scale Inductions

Yeast nitrogen base without amino acids (YNB, Difco, Detroit, MI) medium contained 4 ng/mL biotin and 2% (w/v) glycerol or 1% (v/v) methanol. YNB/methanol induction medium also contained 1% casamino acids and 0.1 *M* sodium phosphate, pH 6.0.

2.3. Apparatus and Reagents for Semiquantitative DNA Dot-Blot Screen

1. Sterile 96-well microtiter plates with lids (Falcon, Becton Dickenson, NJ).
2. YPD medium.
3. Multichannel pipet.
4. 96-Well dot-blot manifold (e.g., "minifold," Schleicher and Schuell, Keene, NH).
5. Vacuum pump.
6. Nitrocellulose filters.
7. 2.5% Mercaptoethanol, 50 m*M* EDTA, pH 9.0
8. Zymolyase-100T (ICN, Costa Mesa, CA), 3 mg/mL: Make fresh.
9. 0.1 *M* NaOH, 1.5 *M* NaCl.
10. 20X SSC stock solution: 3 *M* sodium chloride, 0.3 *M* sodium citrate.

2.4. Apparatus and Reagents for RNA Preparation

1. 15-mL polypropylene tubes with caps, and 1.5-mL mirocentrifuge tubes soaked overnight in 0.1% diethylpyrocarbonate (DEPC) and then autoclaved.
2. Platform vortex mixer (e.g., IKA-Vibrax-VXR, Sartorius, Epsom, Surrey, UK) with test tube rack.
3. Glass beads (0.45 mm, acid-washed, Sigma, St. Louis, MO): Soak in concentrated HCl, rinse thoroughly, and bake for 16 h at 150°C.
4. DEPC-treated water: Add 0.1% DEPC to distilled water, mix, loosen lid, leave overnight, and then autoclave.
5. Guanidinium thiocyanate solution: Dissolve 50 g guanidinium thiocyanate, 0.5 g *N*-laurylsarcosinate in 80 mL water. Add 2.5 mL 1 *M* sodium citrate, pH 7.0, 0.7 mL 2-mercaptoethanol. Make up to 100 mL with water and filter. Store at 4°C.

6. Phenol/chloroform/isoamyl alcohol: A 25:24:1 mixture of phenol equilibrated with 0.1 *M* Tris-HCl, pH 8.0, chloroform, and isoamyl alcohol.
7. Chloroform/isoamyl alcohol: A 24:1 mixture.
8. 1 *M* acetic acid (made in DEPC-treated water).
9. 3 *M* sodium acetate pH 4.5 (made in DEPC-treated water).

2.5. Reagents for Deglycosylation with Endoglycosidase H

1. Endoglycosidase H (Boehringer Mannheim, Mannheim, Germany).
2. Endo H buffer: 200 m*M* monobasic sodium phosphate, pH 5.5, 10 m*M* 2-mercapto-ethanol, 1% SDS.
3. 5X protease inhibitor cocktail: 20 m*M* EDTA, 20 m*M* EGTA, 20 m*M* PMSF, 10 mg/mL pepstatin, 10 mg/mL leupeptin, 10 mg/mL chymostatin, 10 mg/mL antipain. Store at –20°C.

2.6. Apparatus and Reagents for Isolating a Series of Transformants with Increasing Vector Copy Number

1. YPD medium.
2. Sterile 96-well microtiter plates with lids (Falcon, Becton Dickenson, NJ).
3. Multichannel pipet.
4. G418 (Geneticin, Gibco, Gaithusburg, MD) stock solution, 50 mg/mL. Filter-sterilize, and store at –20°C.

3. Methods

3.1. Murine Epidermal Growth Factor

mEGF was used in an evaluation of the secretory capabilities of the *P. pastoris* expression system ***(1)***, since it had been found that the human EGF was very efficiently secreted by baker's yeast, *S. cerevisiae* ***(4)***. For this work, the mEGF coding region was fused to sequences encoding the prepro leader peptide from *S. cerevisiae* α-factor, and yields were compared in single-copy *P. pastoris* integrants and in *S. cerevisiae* using a 2μ-based GAL7-promoter vector (**Table 1**). With both hosts >90% of the mEGF expressed was exported to the culture medium. Subsequent N-terminal sequence analysis showed that the leader peptide was efficiently and accurately cleaved in both cases. This was one of the first demonstrations that the *S. cerevisiae* α-factor leader is functional and authentically processed in *P. pastoris*.

Similar levels of mEGF were secreted by both yeast hosts when grown in shake flasks in rich medium (6–7 mg/L). However, in minimal medium, yields were significantly lower. For *S. cerevisiae*, this was simply owing to a reduced culture density, but in *P. pastoris*, poor yield was caused by nonspecific proteolytic degradation. Many fungal hosts produce extracellular proteases, but the extent of the problem is dependent not only on the host itself, but also on the susceptibility of the secreted protein. For example, secreted fungal proteins

Table 1
Secretion of mEGF by Yeast Transformants in Different Media

	YNB μg/mL	μg/10^8 cells	YPD μg/mL	μg/10^8 cells
S. cerevisiae (multicopy plasmid)	0.6	3.7	7.4	3.2
P. pastoris (single-copy integrant)	0.07	0.02	6.0	1.8

tend to be very resistant to degradation in heterologous hosts, probably because their native hosts produce extracellular proteases. The secretion of mEGF and numerous other polypeptides (IGF; *see* Chapter 11; HIV-1 ENV, **ref. *2***) suggests that extracellular proteases may be a greater problem for heterologous secretion in *P. pastoris* than in *S. cerevisiae*. Nevertheless, simple measures can often be taken to minimize proteolysis. Altering the culture conditions can have dramatic effects—for example, as described above, mEGF is largely protected from degradation in rich medium (**Table 1**). However, rich medium is undefined, which may lead to inconsistent yields, and its complexity presents difficulties for protein purification. A more convenient solution is to supplement minimal medium with casamino acids or peptone. Buffering the growth medium to a pH value where degradation is reduced can also be very effective. For mEGF, the addition of 1% casamino acids and buffering to to pH 6.0 increased yields dramatically and stabilized the protein against proteolytic cleavage (**Fig. 1**). The use of protease-deficient pep4 host strains has also been used with success to increase secreted protein levels (e.g., IGF, *see* Chapter 11).

3.1.1. Small-Scale Induction of mEGF

1. Inoculate 10 mL of YNB containing 2% glycerol with cells from a single colony of transformant and incubate overnight with shaking in a universal container at 30°C. (A_{600} should be about 5.)
2. Dilute the starter culture into 100 mL of YNB containing 2% glycerol to an A_{600} of 0.25 (*see* **Note 1**). Incubate at 30°C with vigorous shaking for 6–8 h to obtain an exponentially growing culture.
3. Harvest the cells by centrifugation (4°C, 2000*g*, 5 min). Resuspend the cell pellet in 20 mL of sterile water. Harvest the cells by centrifugation, and resuspend in 5 mL of YNB containing 1% casamino acids, 0.1 *M* sodium phosphate (pH 6.0), and 1% methanol. Incubate at 30°C with vigorous shaking for 1–5 d (*see* **Notes 2–4**). (A_{600} should be 5–10).
4. Centrifuge the culture (4°C, 2000*g*, 5 min), and retain both the culture supernatant and the cell pellet for analysis (*see* **Note 5**).

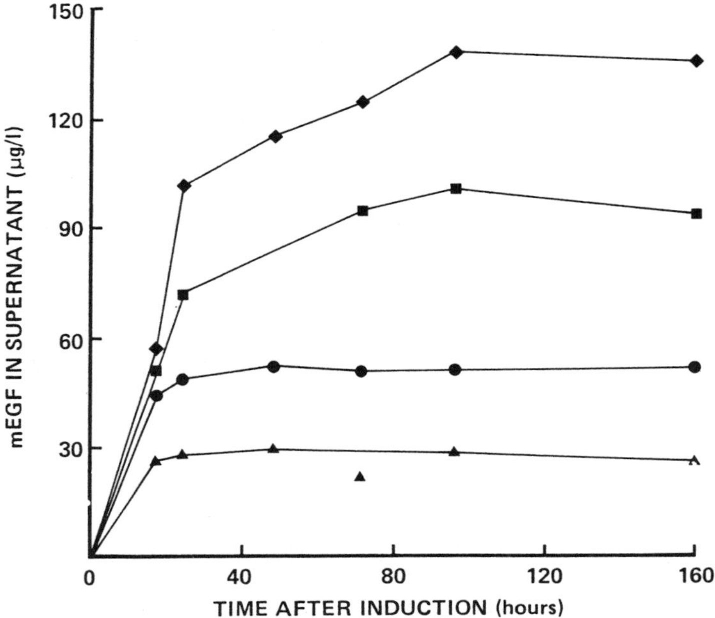

Fig. 1. Time-course for mEGF secretion in shake-flask cultures of multicopy strains. Cultures were induced in YNB/methanol (pH 6.0) plus casamino acids, and at induction, the cells were concentrated 20-fold by centrifugation (*see* **Note 4**). Strains containing 4 copies (▲), 9 copies (●), 13 copies (■), 19 copies (◆) of the mEGF gene are shown.

3.1.2. Isolation of Multicopy mEGF Transformants

mEGF was another early example of the clonal variation described in Chapter 5 for fragment C. Since the mechanism deduced from the fragment C analysis implied this variation should be a general occurrence, we developed a rapid DNA dot-blot screening technique (*see* **Subheading 3.1.2.1.**) and applied it to mEGF transplacement transformants. Strains containing up to 19 copies were obtained, and as with fragment C, both Mut$^+$ and Mut^s multicopy events were detected. A close correlation between gene dosage and the level of secreted mEGF was found (**Fig. 1**, **Table 2**). In shake flasks, the yield varied from 1.5 mg/L for single-copy integrants up to 48.7 mg/L for the 19-copy strain. There was no intracellular accumulation, even at the highest gene dosage, showing that mEGF is secreted particularly efficiently by this yeast. In contrast, there are now several examples where increased copy number does not enhance secretion levels of a foreign protein, and in some cases, e.g., bovine lysosyme, this can actually lead to reduced yields *(5)*. This is probably because, at high rates of expression, inefficently secreted proteins accumulate within the

Table 2
Secretion of mEGF by Multicopy *P. pastoris* Strains

	Copy number					
	1	2	4	9	13	19
Shake flask (μg/mL)	1.5	2.5	8.5	23.2	35.6	48.7
Fermenter (μg/mL)	33.6	ND	ND	355	402	447

secretory apparatus leading to blockage of the secretory pathway. In such cases, yields can be maximized by determining the optimal gene dosage.

3.1.2.1. Semiquantitative DNA Dot-Blot Screen

1. Inoculate independently selected His$^+$ transformants into 150 μL of YPD in the wells of a 96-well microtiter plate, and incubate static for 2 d at 30°C.
2. Resuspend the settled cells using a multichannel pipet, and subculture into a second microtiter plate by transferring 10 μL from each well into 150 μL of YPD and incubating for 1 d at 30°C (*see* **Note 6**).
3. After resuspending the settled cells, filter 50-μL aliquots onto nitrocellulose membranes using a dot-blot manifold and a multichannel pipet.
4. Lyse the cells, and pretreat for hybridization by placing the membranes sequentially on top of 3MM papers soaked in the following solutions:
   a. 2.5% Mercaptoethanol, 50 m*M* EDTA, pH 9.0, for 15 min.
   b. 3 mg/mL zymolyase for 2–3 h at 37°C in a covered dish (*see* **Note 7**).
   c. 0.1 *M* NaOH, 1.5 *M* NaCl for 5 min.
   d. 2X SSC for 5 min (twice).
5. Air-dry the membranes briefly, and then bake in a vacuum oven at 80°C for 1 h.
6. Hybridize membranes to a foreign gene probe (e.g., 0.39-kb *Bgl*II–*Nhe*I fragment from mEGF gene) using standard techniques. Multicopy transformants should give a strong hybridization signal.

3.1.3. Large-Scale Induction and Analysis of mEGF

Like many intracellularly expressed proteins (e.g., fragment C; *see* Chapter 13), the level of secreted mEGF produced in fermenter inductions was significantly higher than that in shake flasks (**Table 2**). The volumetric yield was about 10-fold higher with the 19-copy strain and 20-fold higher with a single-copy integrant. The highest yield was almost 0.5 g/L, and even at this level of expression, >90% was secreted to the medium. The *P. pastoris*-secreted mEGF was purified and characterized by N-terminal sequencing, amino acid analysis, and HPLC/FPLC analysis. In addition to full-length EGF, two other forms were identified that resulted from the specific cleavage of one or two C-terminal amino acids. This is likely to be due to processing by a KEX1-like protease which removes the C-terminal Arg residue and an uncharacterised carboxypeptidase which cleaves before the penultimate residue, Leu. These processed forms of EGF are also found in mammalian cells and are known to be biologically active.

3.2. HIV ENV

Posttranslational modification of proteins by the addition of carbohydrate side chains occurs in all eukaryotic cells. However, the pattern of glycosylation in *S. cerevisiae* differs in several respects to that of mammalian systems. It was therefore of considerable interest to determine the glycosylation pattern of foreign glycoproteins secreted by *P. pastoris*. The principal differences between mammalian and *S. cerevisiae* glycoprotein are that yeast outer chains are significantly longer, and are composed entirely of mannose, whereas mammalian glycoproteins have side chains that are relatively short and can contain other sugars, such as fucose, galactose, and sialic acid. Thus, when mammalian glycoproteins are expressed in *S. cerevisiae*, they tend to be hyperglycosylated with respect to the native form ***(6)***. A further significant difference is the presence of antigenic terminal α-1,3-linked mannose on glycoproteins from *S. cerevisiae*. Thus, mammalian glycoproteins produced in *S. cerevisiae* are not usually suitable for therapeutic use.

In contrast to *S. cerevisiae*, bulk glycoprotein from *P. pastoris* was found to have a short average length of carbohydrate side chains ***(7)***. Initial studies also demonstrated that *S. cerevisiae* invertase was not hyperglycosylated when produced in *P. pastoris* ***(8)***, and lacked and the antigenic terminal α-1,3-linkages ***(9)***. Furthermore, secreted fragment C and Epstein-Barr virus gp350, which are hyperglycosylated in *S. cerevisiae*, were found to be glycosylated to a lesser extent in *P. pastoris* (Clare et al., unpublished data). However, both proteins were immunologically inactive. To examine further the nature of glycosylation in *P. pastoris*, we also expressed HIV-1 ENV, and several aspects of this work are of general relevance to the *P. pastoris* expression system.

3.2.1. Premature Termination of Transcription Within HIV-1 ENV

HIV-1 ENV had been secreted in *S. cerevisiae*, but was found to be hyperglycosylated and immunologically inactive ***(10)***. However, in preliminary experiments with *P. pastoris*, no ENV could be detected. Northern blot analysis indicated that no full-length transcript (1900 nt) was produced, but instead there was an abundant species of about 700 nt (**Fig. 2**). This transcript hybridized to a probe from the *AOX1* untranslated region (UTR) but not to a downstream probe, suggesting it was shortened at the 3'-end. The truncated transcript was still bound by oligo (dT) cellulose and was presumably the result of pre-

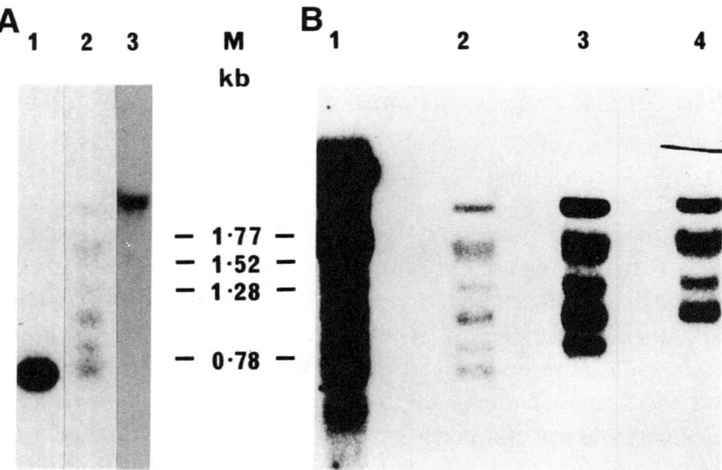

Fig. 2. Northern blot analysis of mRNA from native, mutated, and partly synthetic HIV-1 ENV genes. **(A)** Total RNA from transformants containing different versions of the ENV gene probed with ENV DNA: lane 1, native ENV gene; lane 2, mutated ENV gene; lane 3, partly synthetic ENV gene. **(B)** Total RNA from the mutated ENV gene probed with synthetic oligonucleotides designed to hybridize to different regions of the ENV mRNA: lane 1, nt 71–86; lane 2, 174–209; lane 3, 699–729; lane 4, 794–824.

mature termination of transcription within the coding region. Inspection of the DNA sequence close to the 3'-end of the transcript revealed a sequence, ATTATTTTATAAA, which resembled a consensus site, TTTTTATA, implicated in transcription termination in yeast *(11)*. This site was altered by site-directed mutagenesis to TTTCTTCTACAAG. When tested, this construct no longer gave rise to the abundant 700-nt transcript, but instead a series of longer, lower abundance transcripts were observed (**Fig. 2**). The shorter transcripts did not hybridize to 3'-probes, implying that a 5'-nested set of truncated transcripts with downstream end points at different sites within the gene were produced. Thus, inactivation of the major premature termination site merely revealed several weaker ones further downstream.

We had previously observed this phenomenon with fragment C expression in *S. cerevisiae (12)*. This was found to be a consequence of the unusually high AT content of this gene (71%) and the problem was solved by expressing a synthetic gene in which the GC content was increased. This is consistent with the finding that all the consensus sites for transcription termination in yeast identified to date are AT-rich. Thus, fortuitous termination sites are more likely to occur in AT-rich DNA, although the presence of such DNA is not sufficient to cause premature termination. The ENV gene is not particularly AT-rich, but nevertheless contains regions with high AT content. We therefore synthesized the relevant portion of the gene (i.e., downstream from the initial fortuitous termination site), eliminating any AT-rich runs and optimizing codons to that are found in highly expressed genes in yeast. The overall GC content of this region was increased from 37.6–43.6%. Expression of this semisynthetic gene resulted in the production of abundant full-length transcripts (**Fig. 2**).

This problem of premature termination of transcription has now been found with a number foreign genes expressed in yeast ***(13)***. The data given here for HIV-1 ENV demonstrate that it can be a species-specific phenomenon and that it occurs in genes that are not AT-rich. For several reasons, e.g., difficulties in unambiguously identifying termination sites, the occurrence of multiple sites, and so forth, the simplest solution is often to synthesize chemically sections of the gene, as described here.

3.2.1.1. Preparation of RNA from HIV-1 ENV Transformants

1. Inoculate 10 mL of YNB containing 2% glycerol with cells from a single colony of transformant, and incubate overnight with shaking in a universal container at 30°C. (A_{600} should be about 5.)
2. Dilute the starter culture into 50 mL of YNB containing 2% glycerol to an A_{600} of 0.25 (*see* **Note 8**). Incubate at 30°C with vigorous shaking for 6–8 h to obtain an exponentially growing culture.
3. Harvest the cells by centrifugation (4°C, 2000*g*, 5 min). Resuspend the cell pellet in 50 mL sterile water. Harvest the cells by centrifugation, and resuspend in the same volume of YNB containing 1% methanol. Incubate at 30°C with vigorous shaking for 8 h.
4. Harvest the cells by centrifugation (4°C, 2000*g*, 5 min). Resuspend the cell pellet in 10 mL ice-cold water. Centrifuge once again, resuspend the pellet in 3 mL of guanidinium thiocyanate, and transfer to a 15-mL tube.
5. Add glass beads to two-thirds the height of the meniscus, place the tubes in a platform vortex mixer in a cold room, and vortex at full speed for 15 min.
6. Add an equal volume of phenol/chloroform/isoamyl alcohol, vortex again for 5 min, and centrifuge (10,000*g*, 15 min). Aspirate the aqueous (upper) phase, and transfer it to a new tube.
7. Re-extract the aqueous phase repeatedly with phenol/chloroform/isoamyl alcohol, as in step 6, until there is little or no material at the interface.
8. Extract once with an equal volume of chloroform/isoamyl alcohol.
9. Transfer the aqueous layer to a clean tube and precipitate the RNA by adding 0.025 vol of 1 *M* acetic acid and 0.75 vol of ethanol. Leave at –20°C for at least 30 min.
10. Collect the precipitate by centrifugation (10,000*g*, 15 min), and drain off excess liquid. Vortex the pellet in 70% ethanol (in DEPC-treated water), centrifuge, and remove excess liquid.

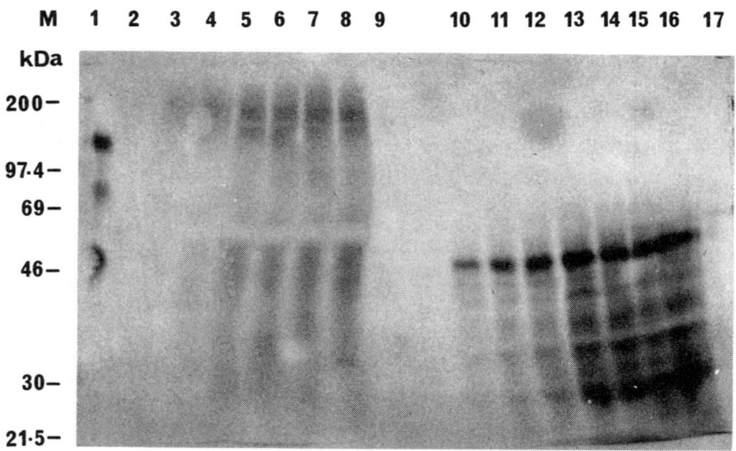

Fig. 3. Secretion of HIV-1 ENV. Western blot analysis of extracts from a single-copy α-factor/ENV transformant induced in the fermenter. Samples for analysis were taken at different time points following induction. Lane 1, CHO cell-derived gp120; lane 2, 0 h; lane 3, 3 h; lane 4, 6 h; lane 5, 18 h; lane 6, 24 h; lane 7, 28 h; lane 8, 48 h; lane 9, untranformed control; lanes 10–17, as lanes 2–9, except treated with endoglycosidase H.

11. Dissolve the pellet in 0.4 mL DEPC-treated water. Transfer the solution to a microfuge tube, and re-extract once with phenol/chloroform/isoamyl alcohol and once with chloroform/isoamyl alcohol. Precipitate by adding 0.1 vol of 3 *M* sodium acetate and 2.5 vol of ethanol.
12. Collect the precipitate by centrifugation in a microcentrifuge (12,000*g*, 15 min).
13. Remove excess liquid, and resuspend the RNA in 100 µL DEPC-treated water. Store at –70°C.

3.2.2. Expression Analysis of HIV-1 ENV

Western blot analysis of culture supernatants from single-copy and multicopy transformants expressing ENV fused to the α-factor prepro leader indicated that low levels of protein were secreted. Instead of the expected 120-kDa protein, the product was found to be very heterogeneous in size (~30–200 kDa, **Fig. 3**). Deglycosylation by digestion with endoglycosidase H (*see* **Subheading 3.2.2.1.**) indicated that this heterogeneity was owing to both hyperglycosylation and proteolytic degradation. Proteolysis could be reduced by buffering the culture to pH 3.0 in shake flasks or pH 5.0 in the fermenter, although some degradation still occurred. There is some evidence to suggest that the secretion signal can affect the kinetics of secretion, which may influence the extent of glycosylation ***(5)***. However, we obtained similar results

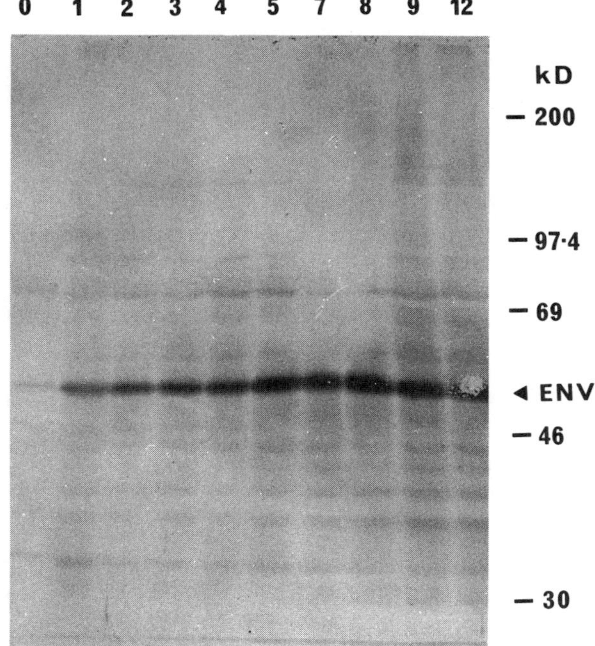

Fig. 4. Intracellular expression of HIV-1 ENV. Western blot analysis of extracts from transformants with increasing gene copy number.

using the *S. cerevisiae* SUC2 secretion signal, which lacks a pro region. The *P. pastoris*-secreted ENV failed to bind to a number of ENV-specific monoclonal antibodies (MAb), including one directed against the major neutralizing domain of HIV-1. Thus, it is clear that some mammalian glycoproteins are hyperglycosylated when expressed in *P. pastoris*, and this can significantly affect their immunological properties.

3.2.2.1. Deglycosylation of HIV-1 ENV Using Endoglycosidase H

1. To 25 µL of sample (containing up to 100 µg of protein), add 5 µL of Endo H buffer, and incubate at 100°C for 5 min.
2. Cool sample on ice, and then add 10 µL of 5X protease inhibitors and 10 µL (10 mU) of Endo H. Incubate at 37°C overnight prior to SDS-PAGE.

We also examined ENV production when expressed without a secretion signal (**Fig. 4**). A series of transformants each containing a different vector copy number were isolated (*see* **Subheading 3.2.3.**), and expression levels were compared. Single-copy integrants gave low levels, but in contrast to the results with the secreted form, levels were greatly increased in multicopy integrants.

In high-density fermenter, cultures of a 12-copy transformant ENV accumulated to about 2.5% of total protein yielding about 1.25 g/L. This level of expression contrasts sharply with the accumulation of alcohol oxidase, which can exceed 30% of total protein from a single gene copy. Since ENV transcripts are significantly more abundant than *AOX1* in the 12-copy strain (*see* **Subheading 3.2.3.** and **Fig. 5B**) it is clear that other factors in addition to promoter strength, such as protein stability, are important in determining high levels of accumulation. Alcohol oxidase is particularly stable in vivo, since it can form extensive lattice-like complexes.

3.2.3. Isolation and mRNA Analysis of a Series of HIV-1 ENV Transformants with Increasing Copy Number

In order to develop improved methods for obtaining multicopy transformants, we constructed vectors containing the Tn903 kan^R gene, which confers resistance to G418 in *P. pastoris* (pPIC3K and pPIC9K, *see* Chapter 5). To establish the relationship between copy number and the level of resistance, we transformed GS115 with the vector pPIC3K-ENV and tested randomly selected His^+ transformants for growth on different concentrations of G418. The data shown in **Table 3** demonstrates that there is a good correlation between gene dosage and the minimum level of G418 required to inhibit growth. This finding enabled the development of a method for the selection of high-copy transformants (*see* Chapter 5). In addition, these vectors can be used to obtain a series of transformants that contain progressively increasing foreign gene copy number by testing for inhibition of growth on medium containing increasing concentrations of G418, as described here for ENV. This may be very useful for optimizing expression in cases where high gene dosage may be detrimental, for example, with proteins that are inefficiently secreted.

3.2.3.1. G418 Growth-Inhibition Screen

1. Transform cells by electroporation or spheroplasting.
2. Inoculate independently selected His^+ transformants into 150 μL of YPD in the wells of a 96-well microtiter plate, and incubate static for 2 d at 30°C.
3. Resuspend the settled cells using a multichannel pipet, and subculture into a second microtiter plate by transferring 10 μL from each well into 150 μL of YPD and incubating for 1 d at 30°C. This is necessary to obtain cultures of uniform density.
4. Resuspend the settled cells, and spot 3-μL aliquots of each transformant, using a multichannel pipet, onto 15-cm Petri dishes containing YPD with increasing concentrations of G418 (e.g., 0.25, 0.5, 0.75, 1.00, 1.50, 2.00, 4.00 mg/mL).
5. Prepare chromosomal DNA, and analyze copy number (*see* Chapter 13) in transformants having a range of different minimum inhibitory G418 concentrations.

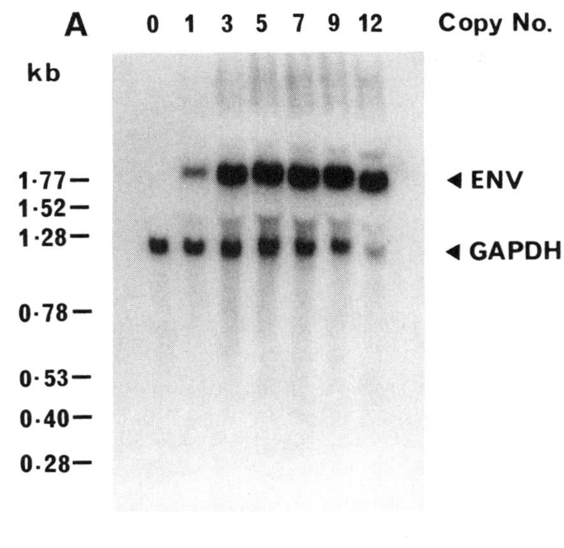

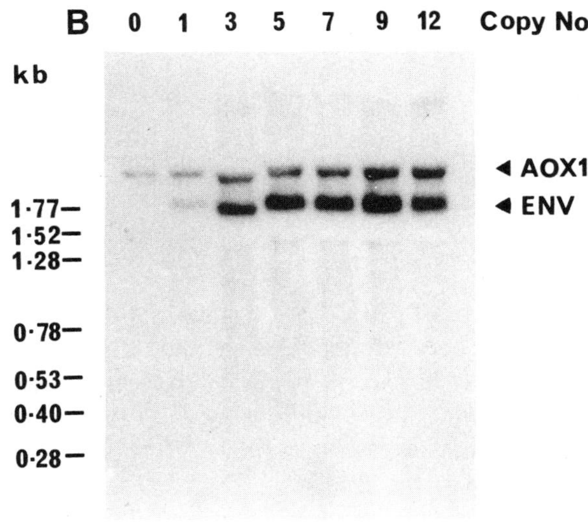

Fig. 5. Northern blot analysis of mRNA from a series of transformants containing increasing copy number of the HIV-1 ENV gene. **(A)** Total RNA hybridized to both ENV and *P. pastoris* GAPDH DNA probes to show ENV mRNA levels relative to an internal standard. **(B)** Total RNA hybridized to probes corresponding to the 5'- and 3'-untranslated regions of *AOX1* to compare ENV and *AOX1* mRNA levels.

Table 3
The Use of G418 Resistance to Isolate a Series of HIV-1 ENV Transformants Containing Progressively Increasing Gene Copy Number

Transformant	G418 concentration,[a] mg/mL	Copy number
G2	<0.25	1
G12	<0.25	0
A6	0.50	1
C6	0.50	1
A1	0.75	2
A2	0.75	2
D1	1.00	2
D5	1.00	3
E8	1.50	4
G4	1.50	3
E5	1.75	5
F8	1.75	5
A4	2.00	7
E12	2.00	5
A8	>4.00	9
F11	>4.00	12
G9	>4.00	9
G10	>4.00	8
A12	>4.00	7

[a]This was the minimum G418 concentration required to inhibit growth completely.

The copy number series of pPIC3K-ENV transformants allowed an investigation into the relationship between gene dosage and mRNA levels of a heterologous gene. The extremely high level of alcohol oxidase produced from the *AOX1* gene had suggested that only a single copy of *AOX1* expression vectors might be necessary for high-level expression of foreign genes. In certain examples, this was found to be the case, e.g., HbsAg ***(14)***, invertase ***(8)***, but in others, e.g., TNF ***(15)***, fragment C, multiple copies were required. A possible explanation for this is that many foreign genes are not transcribed as efficiently as *AOX1*. In *S. cerevisiae*, it is well documented that homologous genes give much higher levels of transcript compared to heterologous ones even when transcription is driven by the same upstream promoter ***(16,17)***. To see if this was also true in *P. pastoris*, we compared HIV-1 ENV transcript levels in single and multicopy transformants to that of *AOX1*.

Figure 5A shows that when measured against a glyceraldehyde 3-phosphate dehydrogenase gene (*GAPDH*) internal control, ENV mRNA levels increase progressively with copy number. The abundance of ENV mRNA in different copy number strains is compared to endogenous *AOX1* mRNA levels in **Fig. 5B**. In single-copy, ENV mRNA is only slightly less abundant than *AOX1*, whereas at three copies, it exceeds it by severalfold. Similar results were obtained with fragment C (data not shown), suggesting that in *P. pastoris*, foreign transcripts are produced almost as efficiently as homologous ones, and that unlike certain inducible promoters in *S. cerevisiae* (e.g., GAL promoters), transcription from multiple copies of the *AOX1* promoter does not appear to be significantly limited, e.g., by the level of transactivating factors.

3.3. Summary

1. The *S. cerevisiae* α-factor prepro leader is functional and is correctly processed in *P. pastoris*.
2. *P. pastoris* has a high secretory capacity, but yields can be severely reduced by extracellular proteases. This problem can be reduced by altering the medium composition, e.g., adjusting the pH or by adding casamino acids.
3. A rapid DNA dot-blot technique can be used for mass screening of transformants to obtain high-copy-number, high-expressing strains.
4. For mEGF, which is an efficiently secreted protein, there was a good correlation between gene dosage and yield, and maximum levels were obtained at high copy number.
5. Vectors conferring resistance to G418 have been developed for the selection of high-copy-number transformants. These vectors can also be used to isolate a series of transformants with increasing copy number for optimizing the expression of genes where high copy number may be detrimental.
6. The HIV-1 ENV gene was not expressed in *P. pastoris* owing to fortuitous termination of transcription within AT-rich regions. This is a species-specific phenomenon, since full-length HIV-1 ENV transcripts are produced in *S. cerevisiae*. The problem was overcome by synthesizing the relevent portion of the gene with increased GC content.
7. ENV was hyperglycosylated and immunologically inactive when secreted by *P. pastoris*. The yield was reduced by extracellular proteases, but like mEGF, this could be significantly improved by altering the pH of the culture medium and by adding casamino acids.
8. In single-copy integrants, transcripts from the semisynthetic HIV-1 ENV gene were almost as abundant as endogenous *AOX1*. Transcript levels increased progressively with increasing copy number, showing that the *AOX1* promoter is not greatly limited by the level of trans-activating factors.

4. Notes

1. The concentration of a secreted product in the medium is generally proportional to the density of the culture. Thus, to increase the concentration of a secreted

protein, a large culture (e.g., 100 mL) is grown, and the cells are concentrated at the beginning of the induction by centrifuging and resuspending in a smaller volume of induction medium (e.g., 5 mL).

2. Aeration of *P. pastoris* cultures is a key consideration, since the efficiency of growth and induction is reduced when oxygen is limited. To avoid this in small-scale inductions, incubate with vigorous shaking, keep the culture volume to a minimum (e.g., 5–10 mL), and use large flasks (e.g., >100 mL, preferably baffled).
3. When inducing shake-flask cultures for >1 d, add an additional 0.5% methanol on the second day.
4. The optimal induction period may vary for different proteins, and it is recommended that a time-course be carried out to maximize yields.
5. To determine the efficiency of protein secretion, the cell pellet is retained for analysis as well as the culture supernatant. If necessary, proteins in the culture supernatant can be concentrated by ultrafiltration using Centricon (Amicon, Lexington, MA) tubes according to the manufacturer's instructions.
6. Subculturing into a second microtiter plate is necessary to obtain cultures of uniform density and that are suitable for spheroplasting.
7. The efficiency of spheroplast formation can be monitored using phase-contrast microscopy by placing a drop of cells on a microscope slide together with a drop of 1% (w/v) SDS—unlysed cells appear phase-bright whereas lysed "ghosts" are phase-dark.
8. A 50-mL culture should yield 150–300 μg RNA.

References

1. Clare, J. J., Romanos, M. A., Rayment, F. B., Rowedder, J. E., Smith, M. A., Payne, M. M., Sreekrishna, K., and Henwood, C. A. (1991) Production of mouse epidermal growth factor in yeast: high-level secretion using *Pichia pastoris* strains containing multiple gene copies. *Gene* **105,** 205–212.
2. Scorer, C. A., Buckholz, R. G., Clare, J. J., and Romanos, M. A. (1993) The intracellular production and secretion of HIV-1 envelope protein in the methlyotrophic yeast *Pichia pastoris*. *Gene* **136,** 111–119.
3. Scorer, C. A., Clare, J. J., McCombie, W. R., Romanos, M. A., and Sreekrishna, K. (1994) Rapid selection using G418 of high copy number transformants of *Pichia pastoris* for high-level foreign gene expression. *Bio/Technology* **12,** 181–184.
4. Brake, A. J., Merryweather, J. P., Coit, D. G., Heberlein, U. A., Masiarz, F. R., Mullenbach, G. T., Urdea, M. S., Valenzuela, P., and Barr, P. J. (1988) Characterization of recombinant human epidermal growth factor produced in yeast. *Biochemistry* **27,** 797–802.
5. Thill, G. P., Davis, G. R., Stillman, C., Holtz, G., Brierley, R., Engel, M., Buckholtz, R., Kinney, J., Provow, S., Vedvick, T., and Seigel, R. S. (1990) Positive and negative effects of multi-copy integrated expression vectors on protein expression in *Pichia pastoris*, in *Proceedings of the 6th International Symposium on Genetics of Microorganisms, vol. II* (Heslot, H., Davies, J., Florent, J., Bobichon, L., Durand, G., and Penasse, L., eds.), Societe Francaise de Microbiologie, Paris, France, pp. 477–490.

6. Schultz, L. D., Tanner, J., Hofmann, K. J., Emini, E. A., Condra, J. H., Jones, R. E., Kieff, E., and Ellis, R. W. (1987) Expression and secretion in yeast of a 400-kDa envelope glycoprotein derived from Epstein-Barr virus. *Gene* **54,** 113–123.
7. Grinna, L. S. and Tschopp, J. F. (1989). Size distribution and general structural features of N-linked oligosaccharides from the methylotrophic yeast, *Pichia pastoris*. *Yeast* **5,** 107–115.
8. Tschopp, J. F., Sverlow, G., Kosson, R., Craig, W., and Grinna, L. (1987) High level secretion of glycosylated invertase in the methylotrophic yeast, *Pichia pastoris*. *Bio/Technology* **5,** 1305–1308.
9. Trimble, R. B., Atkinson, P. H., Tschopp, J. F., Townsend, R. R., and Maley, F. (1991) Structure of oligosaccharides on *Saccharomyces* SUC2 invertase secreted by the methylotrophic yeast *Pichia pastoris*. *J. Biol. Chem.* **266,** 22,807–22,817.
10. Hitzeman, R. A., Chen, C. Y., Dowbenko, D. J., Renz, M. E., Lui, C., Pai, R., Simpson, N. J., Kohr, W. J., Singh, A., Chisolm, V., Hamilton, R., and Chang, C. N. (1990) Use of heterologous and homologous signal sequences for secretion of heterologous proteins from yeast. *Methods Enzymol.* **185,** 421–440.
11. Henikoff, S. and Cohen, E. H. (1984) Sequences responsible for transcription termination on a gene segment in *Saccharomyces cerevisiae*. *Mol. Cell. Biol.* **4,** 1515–1520.
12. Romanos, M. A., Makoff, A. J., Fairweather, N. F., Beesley, K. M., Slater, D. E., Rayment, F. B., Payne, M. M., and Clare, J. J. (1991) Expression of tetanus toxin fragment C in yeast: gene synthesis is required to eliminate fortuitous polyadenylation sites in AT-rich DNA. *Nucleic Acids Res.* **19,** 1461–1467.
13. Romanos, M. A., Scorer, C. A., and Clare, J. J. (1992) Foreign gene expression in yeast: a review. *Yeast* **8,** 423–488.
14. Cregg, J. M., Tschopp, J. F., Stillman, C., Siegel, R., Akong, M., Craig, W. S., Buckholz, R. G., Madden, K. R., Kellaris, P. A., Davies, G. R., Smiley, B. L., Cruze, J., Torregrossa, R., Velicelebi, G., and Thill, G. P. (1987) High-level expression and efficient assembly of hepatitis B surface antigen in the methylotrophic yeast *Pichia pastoris*. *Bio/Technology* **5,** 479–485.
15. Sreekrishna, K., Nelles, L., Potenz, R., Cruze, J., Mazzaferro, P., Fish, W., Motohiro, F., Holden, K., Phelps, D., Wood, P., and Parker, K. (1989) High-level expression, purification, and characterisation of recombinant human tumour necrosis factor synthesised in the methylotrophic yeast *Pichia pastoris*. *Biochemistry* **28,** 4117–4125.
16. Chen, C. Y., Oppermann, H., and Hitzeman, R. A. (1984). Homologous versus heterologous gene expression in the yeast, *Saccharomyces cerevisiae*. *Nucleic Acids Res.* **12,** 8951–8970.
17. Mellor, J., Dobson, M. J., Roberts, N. A., Kingsman, A. J., and Kingsman, S. M. (1985). Factors affecting heterologous gene expression in *Saccharomyces cerevisiae*. *Gene* **33,** 215–226.

15

Expression of an Integral Membrane Protein, the $5HT_{5A}$ Receptor

H. Markus Weiß, Winfried Haase, and Helmut Reiländer

1. Introduction

Research in the field of membrane proteins has undergone explosive growth during the last decade, primarily owing to the influence of the powerful techniques of modern molecular biology. Membrane proteins fulfill essential functions, such as communication, selective transport of metabolites and ions, and energy transformation. It is estimated that one-third of the genes of an organism encode integral membrane proteins *(1)*. We are just now beginning to understand the molecular structures of this group of proteins and how they function within the confines of the cellular membranes. Among the different families of membrane proteins, the so-called G protein-coupled receptors (GPCRs) comprise the largest family. From the viewpoint of pharmacology, this family is of great importance, since about 60% of all pharmaceuticals known today mediate their effects via interaction with GPCRs. Therefore, much progress has been made in the characterization of the pharmacological and biochemical properties, as well as the signal transduction mechanisms of the GPCRs. Nevertheless, in order to understand the function and molecular dynamics of these receptors, detailed structural information will be needed. Despite the steady progress in understanding of GPCRs, solid three-dimensional (3D) structural data are still missing. To date, the crystallization and 3D determination have been successfully performed on only a handful of membrane proteins. All these structural determinations were performed on membrane proteins that are naturally highly expressed and can be purified in large quantities from their natural sources. Conventionally, the GPCRs are investigated in or purified from tissues. The procedure for isolation and purification from natural sources tends to be laborious and time-consuming, if possible at all. On the other hand, milligram amounts

From: *Methods in Molecular Biology, Vol. 103:* Pichia *Protocols*
Edited by: D. R. Higgins and J. M. Cregg © Humana Press Inc., Totowa, NJ

of purified homogenous receptors are an absolute prerequisite for detailed structural characterization. A further problem for the study of receptors is the existence of multiple receptor subtypes. Thus, most sources provide mixtures of different receptor subtypes, which makes the accurate biochemical, biophysical, pharmacological analysis as well as the identification of subtype-specific agents extremely difficult if not impossible. The requirement for large amounts of protein for crystallization and subsequent structural determination can be overcome by the overexpression of these membrane proteins in an artificial system. The search for good expression systems for membrane proteins has proven difficult owing to poor yields of active protein.

For heterologous expression of a eukaryotic membrane protein, a sensitive assay for the protein is irreplaceable. Normally, the protein cannot be detected as a Coomassie band after SDS-PAGE of crude membrane preparations, and even if this were possible, the functionality of the heterologously expressed protein would still need to be demonstrated. One way to follow heterologous expression is the Western blot analysis technique, which requires antibodies directed against the recombinant protein. If no specific mono- or polyclonal antibodies for the protein of choice are available, a peptide or polypeptide "tag" can be genetically engineered to the protein's coding region. In some cases, this artificial tag also allows for purification of the recombinant protein. Listed in **Table 1** are three tags for which monoclonal antibodies (MAb) are commercial available and that have been successfully used for the detection/purification of recombinant proteins.

Immunoblot analysis using the tag-specific antibodies allows the determination of the apparent molecular weight of a protein and also allows information about aggregation and proteolysis of the heterologously expressed protein to be obtained. In combination with tunicamycin or glycosidases, the extent of glycosylation of the heterologously expressed protein can be examined. Additionally, immunohistochemical methods allow localization of the recombinant protein within the cell (*see* Chapter 16). Nevertheless, an assay demonstrating the functionality of the recombinant protein is essential. In the case of membrane receptors like the GPCRs, ligand-gated ion channels or tyrosine kinase receptors, a ligand binding assay with a radioactively labeled compound is the method of choice. Binding assays are easy to perform, very sensitive, and provide direct information about proper folding and, therefore, functionality of a heterologously expressed protein. For further pharmacological characterization of receptors, ligand-displacement measurements can be performed.

In this chapter, we describe methods used for overproduction and subsequent initial biochemical and pharmacological analysis of the mouse $5HT_{5A}$ receptor, a member of the GPCR family ***(2,3)***, in the unicellular methylotrophic yeast *Pichia pastoris*. As previously reported, three different plasmids for the

able 1
eptide Tags for the Detection and/or Purification of Recombinant Protein

ag	Immunological detectability	Use for purification	Comments
yc	MAb 9E10, very sensitive	No	The c-*myc* tag now is included in the pPICZ vector series
lis_6	MAb, not very sensitive	Yes Immobilized metal affinity chromatography (IMAC)	Different metal chelate resins are commercially available; the His_6 tag is included in the pPICZ vector series
lag	MAbM1 and MAbM2; M1 antibody is sensitive, binding is dependent on Ca^{2+} and N-terminal localization of the epitope; M2 antibody recognizes epitope independently of its localization, but is not as sensitive as M1	Yes M1 and M2 antibody columns	In combination, the two antibodies allow for analysis of N-terminal processing, if the target protein has been fused to a signal sequence

heterologous expression of this serotonin receptor have been constructed ***(4)***. One of the constructs synthesizes an unmodified form of the protein, whereas the other two constructs add a c-*myc* tag to the $5HT_{5A}$ receptor along with either the *P. pastoris* acid phosphatase (PHO1) or *Saccharomyces cerevisiae* α-factor prepro (α-mating factor) signal sequences to direct the proteins into the endoplasmic reticulum (ER). The PHO1 and α-factor signal sequences have distinctive properties. The PHO1 sequence is cleaved in the ER, whereas the α-factor prepro sequence depends on both the signal peptidase activity of the ER and proper cleavage by the KEX2 protease in the Golgi. The plasmid construct utilizing the α-factor prepro sequence of *S. cerevisiae* resulted in the highest level of receptor expression. As already described for heterologous expression of another GPCR in *S. cerevisiae* ***(5)***, the use of a protease-deficient strain is advantageous in most cases. For heterologous expression in *P. pastoris*, we recommend the use of the protease-deficient strain SMD1168.

2. Materials

2.1. Small-Scale Expression

1. 50-mL conical tubes.
2. 100-mL baffled Erlenmeyer flasks.

3. 250-mL Erlenmeyer flasks.
4. BMGY medium: 1% yeast extract, 2% peptone, 100 m*M* potassium phosphate, pH 6.0, 1.34% yeast nitrogen base with ammonium sulfate and without amino acid, 0.00004% biotin, and 1% glycerol.
5. BMMY medium: same as BMGY, except with 0.5% methanol in place of glycerol.
6. Bench-top centrifuge suitable for 50-mL Falcon tubes.

2.2. Membrane Preparation

1. Ultracentrifuge, rotors, and corresponding ultracentrifuge tubes for volumes ≥2 mL.
2. Acid-washed glass beads with a diameter of ~500 μ (Sigma, St. Louis, MO).
3. Breaking buffer: 50 m*M* sodium phosphate, pH 7.4, 1 m*M* EDTA, 5% glycerol.
4. Membrane suspension buffer: 50 m*M* Tris-HCl, pH 7.4, 150 m*M* NaCl, 1 m*M* EDTA.
5. Protease inhibitor stock solutions: protease inhibitors are diluted to 1X in the respective buffers immediately before use. Be careful handling protease inhibitors, since they are usually very toxic. This is especially true for PMSF.
   a. 100X PMSF: 100 m*M* PMSF in DMSO (store at –20°C, stable for at least 6 mo).
   b. 100X chymostatin: 6 mg/mL chymostatin in DMSO (store at –20°C, stable for about 1 mo).
   c. 500X leupeptin: 0.25 mg/mL leupeptin in H_2O (store at –20°C, stable for about 6 mo).
   d. 500X pepstatin: 0.35 mg/mL pepstatin in methanol (store at –20°C, stable for about 1 mo).

2.3. Ligand Binding Assay

1. Device for rapid vacuum filtration (e.g., Cell Harvester Model M30, Brandel, Inc., Gaithersburg, MD).
2. GF/F glass filters (Whatman, Maidstone, UK).
3. 0.3% Polyethylenimine in water.
4. Assay buffer: 50 m*M* Tris-HCl, pH 7.4, 150 m*M* NaCl.
5. [*N*-methyl-^{3}H]LSD (DuPont/NEN, Dreieich, Germany).
6. Inhibitor stock solution: Dissolve 10.6 mg serotonin/HCl in 1 mL H_2O. Prepare this solution fresh every time. Serotonin/HCl can be purchased from Research Biochemicals International (Natick, MA).
7. Liquid scintillation cocktail (Rotiszint® *Eco* plus, Carl Roth, Karlsruhe, Germany, or similar product).
8. Liquid scintillation counter (e.g., Tri-Carb 1500 Liquid Scintillation Analyzer, Canberra Packard, Dreieich, Germany, or equivalent).

2.4. Polyacrylamide Gel Electrophoresis (PAGE) and Immunoblot Analysis

1. Equipment for PAGE and blotting device can be purchased from Pharmacia (Freiburg, Germany), Bio-Rad (Hercules, CA), Hoeffer Scientific (San Francisco, CA), or from another supplier.
2. 10% Polyacrylamide gels (self-made or purchased).

3. Molecular-weight markers from Serva (Heidelberg, Germany).
4. Peptidase: *N*-glycosidase F (PNGaseF) from Boehringer Mannheim (Mannheim, Germany) or New England Biolabs (Beverly, MA).
5. 10X PNGaseF reaction buffer: 500 m*M* sodium phosphate, pH 7.5. Usually, the buffer is supplied with the enzyme.
6. 2X sample buffer: 250 m*M* Tris-HCl, pH 8.1, 25% glycerol, 12.5% β-mercaptoethanol, 7.5% SDS, and 0.01% bromphenol blue.
7. PonceauS solution: 0.2% PonceauS in 3% TCA (aqueous solution).
8. 0.45-μm Nitrocellulose filters (Schleicher and Schuell, Dassel, Germany) or poly(vinylidene difluoride) (PVDF) membranes (Immobilon™ P, d = 0.45 μm, Millipore, Eschborn, Germany).
9. 5X blotting buffer: 190 m*M* glycine, 50 m*M* Tris-HCl.
10. 1X blotting buffer: 100 mL 5' blotting buffer + 100 mL methanol + 300 mL water.
11. 10X phosphate-buffered saline (PBS) with sodium azide: dissolve 80 g NaCl, 2 g KCl, 2 g KH_2PO_4, 11.6 g Na_2HPO_4, 1 g NaN_3 in 1 L H_2O (pH will be ≈7.45).
12. Antibody buffer: 1X PBS with 0.025% Tween 20.
13. Washing buffer: 1X PBS with 0.1% Triton X-100.
14. MAb 9E10 was either directly purchased (Invitrogen, San Diego, CA, or Cambridge Research, Cambridge, UK) or was purified from the culture supernatant of the corresponding monoclonal cell line derived from the American Tissue Culture Collection (Rockville, MD; no. ATCC CRL-1729=myc1-9E10.2). Store purified and concentrated antibody at –20°C. Supernatants from antibody producing hybridoma cells can be collected and stored at 4°C until purification over an appropriate affinity column. Add Na-azide to ≈0.01% to the supernatants to prevent microbial growth.
15. Goat-antimouse IgG alkaline phosphatase conjugate from Sigma or Boehringer Mannheim (Mannheim, Germany).
16. 5-Bromo-4-chloro-3-indolylphosphate-*p*-toluidinium salt (BCIP) stock solution: 50 mg BCIP/mL dimethylformamide (store at –20°C).
17. Nitroblue tetrazolium chloride (NBT): 50 mg NBT/mL 70% dimethylformamide (store at –20°C).
18. AP-buffer: 100 m*M* Tris-HCl, pH 9.5, 100 m*M* NaCl, 5 m*M* $MgCl_2$.

2.5. Preparation of Genomic DNA

1. Detergent buffer: 2% (v/v) Triton X-100, 1% (w/v) SDS, 100 m*M* NaCl, 1 m*M* EDTA, 10 m*M* Tris-HCl, pH 8.0. The buffer can be stored at room temperature for several months.
2. 100% Ethanol.
3. 1 mg/mL RNase A (DNase-free) from Boehringer Mannheim.
4. 4 *M* ammonium acetate, pH 7.5.
5. Phenol/chloroform/isoamyl alcohol 25:24:1 (store at 4°C).
6. 50-mL conical tubes.
7. 100X TE: 1 *M* Tris-HCl, pH 8.0, 0.1 *M* EDTA. Sterilize the buffer by autoclaving, and store at room temperature.

8. Nylon membrane (e.g., Biodyne A Transfer Membrane, 0.2 μm, Pall Corp., East Hills, NY).
9. Kit for the production of a radioactive labeled probe. This can be obtained from Boehringer Mannheim, Stratagene (San Diego, CA), or New England Biolabs. Probe can be prepared by in vitro transcription, nick-translation, or random-labeling methods.

3. Methods

3.1. Screening of Recombinant P. pastoris *Clones for [^{3}H]LSD Binding Sites*

3.1.1. Growth and Induction of Clones

Large numbers of selected yeast clones (His$^+$ transformants for which the Mut phenotype has been tested) can be easily screened for expression of $5HT_{5A}$, since LSD specifically binds to whole cells expressing the receptor. The clone that appeared to have the highest number of specific binding sites was selected for subsequent propagation, membrane preparation, receptor characterization, and scaling up.

1. Pick single yeast-transformed colonies previously characterized for Mut phenotype into BMGY medium (10 mL BMGY medium in a conical tube for Mut$^+$ clones or 50 mL in a 250-mL flask for Mut^s clones), and let the cells grow overnight to an OD_{600} of 2–6 in an incubator shaker at 30°C and 250–300 rpm.
2. Harvest cells by centrifugation at 2000*g* for 5 min. Resuspend Mut^s cells in 10 mL BMMY medium. Dilute Mut$^+$ cells to an OD_{600} of 1 in BMMY medium.
3. Transfer 10–15 mL of the resuspended cells into a 100-mL baffled flask, and allow them to grow in an incubator shaker as described above. For maintenance of induction, supplement methanol to the cultures every 24 h. During induction, O_2 levels in the cultures should be kept as high as possible. Therefore, do not seal the culture flasks tightly. Maximum receptor density typically was observed at 24–48 h after induction.
4. Remove 20-μL samples from the cultures at 24–48 h after induction, and determine the OD_{600} at either this time-point or, to analyze the time-course of expression, at multiple time-points.
5. Harvest cells by centrifugation in a microcentrifuge tube at 2000*g* for a few minutes. Freeze the pellets in ethanol/dry ice or liquid nitrogen, and store at –80°C.
6. Calculate culture densities (1 OD_{600} corresponds to ~5×10^7 cells/mL), and estimate the number of receptors as described in **Subheading 3.1.3.**

3.1.2. Small-Scale Membrane Preparation

Depending on the number of positions available in the ultracentrifuge rotor, membranes from six or eight different clones can be prepared in parallel. Membranes from nonexpressing control strains should also be prepared.

1. Harvest the yeast cells from 10–15 mL cultures 24–48 h after induction by centrifugation (2000*g*, 4°C, 5 min) in 12- or 15-mL centrifuge tubes.
2. Resuspend the cell pellet in 2.5 mL of ice-cold breakage buffer freshly supplemented with the protease inhibitor mix (without PMSF), and add 2 mL of cold acid-washed glass beads. Store the tubes on ice (*see* **Note 1**).
3. Vigorously vortex the tubes for 30 s, and then cool for a minimum of 30 s on ice. Repeat the vortexing and cooling steps eight times. Add PMSF after every other vortexing step (PMSF has a short half-life in aqueous solution).
4. Remove unbroken cells and cell debris by centrifugation at 1500–2000*g* for 5 min at 4°C.
5. Carefully transfer supernatants (1.5–2 mL) to ultracentrifuge tubes, and centrifuge the cell-free extracts at 100,000*g* for 30 min at 4°C.
6. Resuspend each membrane-enriched pellet in 500 µL of cold membrane suspension buffer supplemented with 1X protease inhibitor mix.
7. Determine the protein concentration of the preparation using a common protein assay (Bradford, Bio-Rad; BCA protein test, Pierce, Rockford, IL). With a 15-mL culture grown to a final OD_{600} of 10–12, the protein concentration should be in the range of 9–13 mg protein/mL suspension.
8. For storage, divide the membrane suspension into three or four aliquots, and freeze in either dry ice/alcohol or liquid N_2. Store the raw membranes at –80°C.

3.1.3. Ligand Binding Assay

In a radioligand binding assay, total binding of the radiolabeled ligand to the sample as well as nonspecific binding in the presence of high concentrations of an unlabeled competitor is determined. Total binding sites in a probe are measured in the presence of $10 \times K_d$ of the radiolabeled ligand, whereas nonspecific binding sites are measured with the unlabeled ligand added to a final concentration of $\approx 1000 \times K_i$ (*see* **Note 2**). The bound radioligand is then separated from the free radioligand by rapid vacuum filtration ***(6)***. To determine specific binding, the disintegrations per min owing to nonspecific binding are subtracted from total disintegrations per minute. The number of binding sites can then be calculated directly using the specific activity of the labeled compound. For further information behind the theory of binding assays, data processing, and analysis, *see* Hulme and Birdsall ***(7,8)***.

1. Keep raw membranes (from **Subheading 3.1.2.**) or cell pellets (from **Subheading 3.1.1.**) on ice. Ensure that the membranes are completely thawed before use. It is important to include proper controls (i.e., cells or membranes from a nonrecombinant yeast strain and/or cells or membranes from a strain transformed with the vector only).
2. Prepare a fresh solution of serotonin/HCl, and hold on ice. Prepare sufficient serotonin/HCl inhibitor stock solution to provide 20 µL for each sample in the test.

3. Soak the appropriate number of GF/F filter sheets (for Brandel cell harvester) or GF/F filters (for small vacuum devices) in 0.3% polyethylenimine solution for a minimum of 1 h.
4. Suspend cells from 20 µL of culture in 300 µL of ice-cold binding assay buffer, or dilute 6 µL of raw membrane suspension in 300 µL of this buffer and hold on ice.
5. For each measurement, prepare five assay tubes in a rack on ice. Pipet 10 µL of the serotonin/HCl solution and 390 µL of binding assay buffer into two of the five assay tubes and 400 µL of the buffer into the remaining three tubes. Add 50 µL of cells or membranes to each assay tube.
6. Dilute the [^{3}H]LSD in binding assay buffer to a concentration of 190 n*M* LSD, and add 50 µL of this solution to each tube.
7. Mix by vortexing (1–2 s), and incubate the tubes at 30°C for ≥30 min.
8. Separate bound from unbound [^{3}H]LSD by rapid filtration through GF/F filters, and wash each filter two times with ~4 mL of cold water.
9. Transfer the filters into scintillation tubes, and add 4 mL of liquid scintillation cocktail.
10. Samples can be counted immediately, but for the most accurate results, the samples should be held for ~24 h at room temperature before counting for 5 min each.

3.2. Analysis of N-Linked Glycosylation on the Receptor with Peptide:N-Glycosidase F

Proteolytic enzymes from *P. pastoris* can cause serious protein degradation and loss of enzyme activity (*see* **Note 3**). Addition of the SDS to membrane proteins prior to SDS-PAGE denatures proteins, allowing proteases to degrade them rapidly in many instances (*see* **Note 4**). For soluble proteins, this problem can be overcome by boiling of the samples. However, membrane proteins tend to aggregate when boiled, making difficult their subsequent detection via Western analysis. Instead of boiling, protease inhibitors can be added to the samples. However, the diversity of proteases complicates the situation so that the correct protease inhibitor mix must usually be determined empirically for each foreign protein. Protein degradation initiated by addition of SDS to *P. pastoris* membranes containing the protein could be efficiently inhibited by PMSF and chymostatin (**Fig. 1**).

The following procedures utilize proteinase inhibitors as part of an experiment to analyze N-linked glycosylation on myc-tagged $5HT_{5A}$ receptor.

1. Incubate membranes (45–60 µg of membrane protein) with 500–1000 U of PNGaseF in 1X PNGaseF reaction buffer for 30 min at 37°C in a 1.5-mL microfuge tube. The final volume of the assay should be 30 µL. A second aliquot of the same membrane preparation and a sample of membranes prepared from nontransformed or vector only-transformed yeast cells should be treated the same way, except without enzyme.

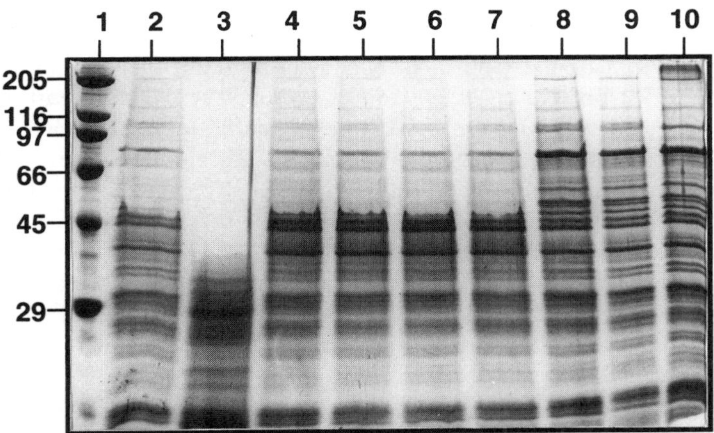

Fig. 1. Protein degradation in crude membrane preparations prepared from *P. pastoris* and analyzed by SDS-PAGE. Addition of SDS containing sample buffer to crude membranes from *P. pastoris* resulted in severe protein degradation, especially in proteins in the range >50 kDa. This degradation was not prevented by incubation of samples on ice (lane 2). Protein degradation was strongly inhibited by either boiling samples (lane 10), or by addition of PMSF (lane 9) or chymostatin (lane 8). The protease inhibitors Pefabloc (lane 4), *N*-tosyl-L-phenylalanine chloromethyl ketone (TPCK) (lane 5), *N*α-*p*-tosyl-L-lysine chloromethyl ketone (TLCK) (lane 6), and leupeptin (lane 7) did not inhibit degradation. When samples were incubated at room temperature or at 37°C, degradation was even more severe. Degradation was also exacerbated by conversion of ATP to ADP by hexokinase and deoxyglucose (lane 3). Thus, protein degradation does not appear to be ATP-dependent, indicating the ubiquitin system does not play a role.

2. Place the membrane protein samples on ice, add 1 µL of PMSF stock solution, immediately transfer 31 µL of sample buffer, and mix.
3. Load 15–20 µg of protein (~20 µL)/lane along with a commercially available mol-wt marker (boiled for 3 min) onto a 10% SDS-gel. Electrophorese samples into two identical gels, one for Coomassie or silver staining and one for electroblotting and subsequent immunoblot analysis.
4. For immunoblotting, transfer the proteins from the gel to a nitrocellulose membrane or a PVDF membrane, using either a semidry blotting apparatus (blot for 60–80 min with a current of ≈1 mA/cm^2 of membrane) or equivalent equipment. If a nitrocellulose membrane is utilized, it should be soaked for at least 1 h in the blotting buffer before use.
5. After electrotransfer, stain the membrane with Ponceau S solution for 30–60 s, and then destain in water until the mol-wt marker is visible. Mark the marker bands with a ballpoint pen, and then completely destain in water.

6. Block the membrane in antibody buffer with 4% nonfat dry milk powder for 1 h at room temperature.
7. Apply monoclonal 9E10 (anti-*myc*) antibody in 10 mL of antibody buffer with 1% BSA either for 1 h at room temperature or overnight at 4°C with gentle agitation.
8. Wash the membrane three times with washing buffer for 15 min each.
9. Apply antimouse secondary alkaline phosphatase-coupled antibodies at a dilution of 1:2000 in 10 mL of antibody buffer with 1% BSA for 1 h at room temperature.
10. Wash three times as described above.
11. For color development, add 10 mL of AP buffer, freshly supplemented with 66 μL of NBT and 33 μL of BCIP.
12. After appearance of stained bands, wash the membrane extensively with water. Dry the membrane for documentation.

3.3. Rapid Preparation of Genomic DNA

A protocol for rapid preparation of *S. cerevisiae* genomic DNA ***(9)***, was adapted to *P. pastoris*. The resulting *P. pastoris* genomic DNA can be used for the detection of multicopy events by dot-blot hybridization as well as for analysis of insertion of the expression cassette via PCR. Quantitative PCR is also a possibility to quantify the number of integrants (*see* **Note 5**).

1. Grow recombinant clones in 10 mL of YPD in 50-mL conical tubes overnight (30°C, 250 rpm) to an $OD_{600} > 3$.
2. Collect cells by centrifugation at 1500*g* for 5 min at room temperature.
3. Discard the supernatant, resuspend the cells in 500 μL of sterile water, and transfer the suspension to a 1.5-mL microfuge tube.
4. Centrifuge for 5 s at maximum speed, discard the supernatant, and resuspend the cells in 200 μL of detergent buffer.
5. Add 0.3 g of glass beads and 200 μL of phenol/chloroform/isoamyl alcohol, and vortex at maximum speed for 3 min. Three to four tubes can be handled at one time. Wear gloves when working with phenol/chloroform/isoamyl alcohol. Fix labels on the Eppendorf tubes with tape, and work under a fume hood, since tubes sometimes do not seal tightly owing to glass beads sticking in the lid (*see* **Note 6**).
6. Add 200 μL of 1X TE, and vortex briefly.
7. For separation of the phases, centrifuge for 5 min at maximum speed, and carefully transfer the upper aqueous layer to a new 1.5-mL microcentrifuge tube.
8. Precipitate nucleic acids by addition of 1 mL of 100% ethanol, invert the tubes several times, and again centrifuge for 3 min at maximum speed.
9. Remove the supernatant carefully, air-dry (do not use a vacuum for drying), and dissolve the pellet in 400 μL of TE.
10. Add 30 μL of 1 mg/mL RNase A, mix gently, and incubate for 5 min at 37°C.
11. Precipitate the genomic DNA by addition of 10 μL of 4 *M* ammonium acetate and 1 mL of 100% ethanol, mix by inversion, and centrifuge for 3 min at maximum speed.

12. Wash the pellet with 500 µL of 70% ethanol, dry, and dissolve the DNA in 60 µL of 1X TE.
13. Determine the DNA concentration at 260 and 280 nm using a UV spectrophotometer. The concentration of the DNA should be 1–2 µg/µL. The quality of the genomic DNA can quickly be checked on an agarose minigel.
14. Determination of the copy number in each clone is done by DNA hybridization using a ^{33}P-labeled DNA probe. As a final step, draw a 1.5 × 1.5 cm grid onto a nylon membrane using a ballpoint pen.
15. Carefully drop 15 µg of genomic DNA from each clone onto a field on the grid. Also add wild-type genomic DNA and genomic DNA prepared from a clone that contains only a single copy of the expression cassette as internal controls. As a positive control, add 15 ng of expression plasmid DNA to one field.
16. Fix the DNA to the nylon membrane by UV illumination (90 s with low-energy UV radiation).
17. Label the DNA probe, and perform hybridization as described with the kit for in vitro radioactive labeling of the probe. Nonradioactive labeling methods are not suitable for quantitative determinations of copy number.
18. After autoradiography, cut the membrane in 1.5 × 1.5 cm pieces, transfer the pieces into a scintillation tube, and add 4 mL of liquid scintillation cocktail. After 24 h, count in a liquid scintillation counter (maximum β-energy of ^{33}P is 0.249 MeV).

4. Notes

1. For large-scale membrane preparations, cell breakage is the most critical step. We use the cell-disintegrator-C (BIOmatik, Rodgau, Germany), which is a continuous bead-beating device that is cooled effectively during breakage of the cells. To break cells from a 1–2 L culture using this device, fill with maximum amount of glass beads (~80 mL) and run at maximum speed (2500–3000 rpm). Recycle the cell suspension (~25% wet cells/volume of breaking buffer) once at a pump speed of 0.7 mL/min. With $5HT_{5A}$, we noted ≥30% loss in specific activity during this process. Another cell disruption system is the cell homogenizer from B. Braun Biotech Int. (Melsungen, Germany), which works well with *S. cerevisiae* and, therefore, should also work with *P. pastoris*. The Microfluidizer from Constant Systems Ltd. (Warwick, UK) is reported to be very efficient and may work better than the bead-base disruption devices. This latter system is most expensive, however.
2. For the initial screening of large numbers of clones using the radioligand binding assay, one can work with a concentration of the labeled ligand lower than $10 \times K_d$ ($3–5 \times K_d$) which significantly lowers the cost. If the K_d of the ligand to a certain receptor is unknown, the K_d must be calculated from a saturation curve by nonlinear regression. If the K_d is published for the target receptor, determine a suitable ligand concentration by adding ligand to 0.1–10 × the K_d. For quantitative measurements, maximal ligand concentrations should be at least $5 \times K_d$, whereas the receptor concentration in the test should be $<0.5\ K_d$.

3. Epitope tags used for detection and/or purification of target proteins and for localization studies may be sensitive to proteolytic degradation. To overcome this problem, express the tagged protein in a protease-deficient strain, change the location of the tag on the target protein, or fuse protein to a different tag.
4. Test whether the specific immunoblot signal for the protein is still present after boiling samples (usually for about 3 min) prior to SDS-PAGE. If signal for the protein is still present, then routinely boil the sample before SDS-PAGE. For PNGase F digestion, boil the target protein prior to treatment with PNGaseF. For this, heat the membrane protein samples in 0.5% SDS and 1% β-mercaptoethanol for 10 min. Since PNGaseF is inhibited by SDS, add the nonionic detergent NP-40 to 1% together with PNGaseF into the deglycosylation reaction. The detergent solution is usually supplied with the enzyme. Enzymatic deglycosylation by PNGaseF can be inhibited by incomplete denaturation of the target protein. An alternative means of studying glycosylation of a protein is with tunicamycin (Sigma), which inhibits in vivo the first step in the synthesis of the core oligosaccharide on dolichol-phosphate and subsequent N-linked glycosylation. Prepare a stock solution of 1 mg/mL tunicamycin in 0.1 *N* NaOH, which can be stored at –20°C for several months without loss of activity. Add tunicamycin to the cell culture at a final concentration of 15 µg of tunicamycin/mL of culture shortly after induction of the cells with methanol.
5. Heterologous expression of the $5HT_{5A}$ receptor was higher in yeast clones bearing multiple copies of the expression vector. As described for certain secreted proteins, the optimal copy number is sometimes not the highest for best expression levels ***(10)***. For expression of the $5HT_{5A}$ receptor, the optimal copy number appears to be about 3 ***(11)***. For selection of clones bearing higher copy numbers, the G418 resistance scheme may be used ***(12)***. However, if an efficient colony assay is available to identify high-level expression clones, it may be more efficient to screen a large number of clones directly for protein expression. Spheroplasting in this case is the transformation method of choice, since it produces a higher rate of multicopy clones than other methods.
6. During preparation of genomic DNA, a 3-min vortexing step may be insufficient to break the yeast cells, especially when using a multitube vortexer. It is possible to increase the time to 4–5 min, but this most probably will result in further shearing of the genomic DNA. Monitor cell breakage in a light microscope and limit vortexing time to that needed to break at least 50% of the cells.

References

1. Schatz, G. and Dobberstein, B. (1996) Common principles of protein translocation across membranes. *Science* **271,** 1519–1526.
2. Plassat, J. L., Boschert, U., Amlaiky, N., and Hen, R. (1992) The mouse 5-HT5 receptor reveals a remarkable heterogeneity within the $5HT_{1D}$ receptor family. *EMBO J.* **11,** 4779–4786.
3. Matthes, H., Boschert, U., Amlaiky, N., Grailhe, R., Plassat, J. L., Muscatelli, F., Mattei, M. G., and Hen, R. (1992) Mouse 5-hydroxytryptamine$_{5A}$ and 5-

hydroxytryptamine$_{5B}$ receptors define a new family of serotonin receptors: cloning, functional expression, and chromosomal localization. *Mol. Pharmacol.* **43,** 313–319.

4. Weiß, H. M., Haase, W., Michel, H., and Reiländer, H. (1995) Expression of functional mouse 5-HT$_{5A}$ serotonin receptor in the methylotrophic yeast *Pichia pastoris*: pharmacological characterization and localization. *FEBS Lett.* **377,** 451–456.
5. Sander, P., Grünewald, S., Bach, M., Haase, W., Reiländer, H., and Michel, H. (1994) Heterologous expression of the human D$_{2S}$ dopamine receptor in protease-deficient *Saccharomyces cerevisiae* strains. *Eur. J. Biochem.* **226,** 697–705.
6. Wang, J.-X., Yamamura, H. I., Wang, W., and Roeske, W. R. (1992) The use of the filtration technique in in vitro radioligand binding assays for membrane-bound and solubilized receptors, in *Receptor–Ligand Interactions: A Practical Approach* (Hulme, E. C., ed.), IRL, Oxford, pp. 213–234.
7. Hulme, E. C. and Birdsall, N. J. M. (1992) Strategy and tactics in receptor-binding studies, in *Receptor–Ligand Interactions: A Practical Approach* (Hulme, E. C., ed.), IRL, Oxford, pp. 63–176.
8. Hulme, E. C. and Birdsall, N. J. M. (1992) Receptor preparations for binding studies, in *Receptor–Ligand Interactions: A Practical Approach* (Hulme, E. C., ed.), IRL, Oxford, pp. 177–212.
9. Ausubel, F. M., Brent, R., Kingston, R. E., Moore, D. D., Smith, J. A., Seidman, J. G., and Struhl, K. (1989) *Current Protocols in Molecular Biology*, Wiley, New York.
10. Romanos, M. (1995) Advances in the use of *Pichia pastoris* for high-level gene expression. *Curr. Opin. Biotechnol.* **6,** 527–533.
11. Weiß, M., Haase, W., Michel, H., and Reilånder, H. (1998) Comparative biochemical and pharmacological characterization of the mouse 5HT$_{5A}$ serotonin receptor and the human β_2 adrenergic receptor produced in the methylotrophic yeast *Pichia pastoris*. *Biochem. J.*, in press.
12. Scorer, C. A., Clare, J. J., McCombie, W. R., Romanos, M. A., and Sreekrishna, K. (1994) Rapid selection using G418 of high copy number transformants of *Pichia pastoris* for high-level foreign gene expression. *Biotechnology* **12,** 181–184.

16

Localization of the *myc*-Tagged $5HT_{5A}$ Receptor by Immunogold Staining of Ultrathin Sections

Winfried Haase, H. Markus Weiß, and Helmut Reiländer

1. Introduction

Immunohistochemistry is a powerful technique to localize proteins in tissues and cultured cells as well as in fractions of subcellular compartments, like mitochondria, endoplasmic reticulum, vesicles, and membranes. In most cases, the detection of a specific protein occurs by a sandwich approach, consisting of a specific primary antibody directed against the target protein and a secondary antibody directed against the primary antibody, which is coupled to a marker for subsequent visualization. Therefore, a key prerequisite for the successful localization of a target protein is the availability of a high-quality specific primary antibody preparation. The production of specific polyclonal antibodies or monoclonal antibodies (MAb) can be time-consuming and cumbersome, and must be performed for each new target protein. Furthermore, membrane proteins are often poorly antigenic. Additional difficulties may arise when an antibody preparation raised against the target protein has a high background of nonspecific antibodies or when the antibodies, despite working well for Western blot analysis, do not work well for immunohistochemical studies.

For immunolocalization of recombinant proteins, most of these problems can be overcome by expressing the protein as a fusion with another protein or a peptide tag (*see* Chapter 15) for which highly specific antibodies are available. During the construction of a cloning vector for heterologous expression, the coding region of the target protein is fused to sequences encoding one of these tag proteins or peptides. We found that the c-*myc* tag, which is recognized by the mouse MAb 9E10, is ideally suited for immunolocalizations in yeasts as well as other organisms ***(1–4)***. This tag is extremely stable, tolerating the constraints of harsh sample preparation needed when samples are prepared

From: *Methods in Molecular Biology, Vol. 103:* Pichia *Protocols*
Edited by: D. R. Higgins and J. M. Cregg © Humana Press Inc., Totowa, NJ

for examination in an electron microscope. Epitope tagging is an efficient method for the localization of membrane proteins, and also of secreted and intracellularly located proteins. With such a method, questions, such as where the recombinant protein is localized, whether the protein is correctly targeted, and how abundant the recombinant protein is, can be answered.

Depending on the desired resolution in the localization study, either light or electron microscopic immunostaining, involving either pre- or postembedding labeling procedures, can be performed. However, in the case of yeast cells like *Pichia pastoris*, problems occur owing to the rigid cell wall. Therefore, light and electron microscopic pre-embedding immunostaining cannot be performed directly on untreated cells. The yeast cells must be spheroplasted prior to the application of pre-embedding techniques. For an overview of low-resolution immunofluorescence techniques as well as the preparation of yeast spheroplasts, *see* Pringle et al. ***(5)***. However, spheroplasting may result in the degradation of proteins located on the surface of the plasma membrane. Thus, for localization of a recombinant protein in a yeast cell, the postembedding labeling method is usually preferred ***(6)***. With this method, insertion of a *myc*-tag at either an extra- or intracellular domain of a recombinant protein is satisfactory for reaction with the 9E10 antibody.

Here, we present an electron microscopic postembedding immunostaining method that has been successfully used for the detection and localization of the *myc*-tagged $5HT_{5A}$ serotonin receptor heterologously expressed in *P. pastoris* (*4*; **Fig. 1**). The method should be generally applicable to any *myc*-tagged recombinant protein. For additional information on methods of microscopic immunogold staining, *see* Griffiths ***(7)*** and Hayat ***(8)***.

2. Materials

2.1. Yeast Strain

The methylotrophic yeast *P. pastoris* (Invitrogen, San Diego, CA) bearing recombinant plasmid pPIC9-$5HT_{5A}$ myc was used for the expression of the mouse $5HT_{5A}$ receptor. The serotonin receptor was tagged with the c-*myc* at the N-terminus as described previously ***(4)***.

2.2. Fixation and Embedding

Chemicals should be analytical grade, and can be obtained from either Merck (Darmstadt, Germany) or Sigma (St. Louis, MO).

1. Glutaraldehyde.
2. *Para*-formaldehyde.
3. Uranyl acetate.
4. Lead citrate.
5. Sodium *meta*-periodate.
6. Sodium cacodylate.

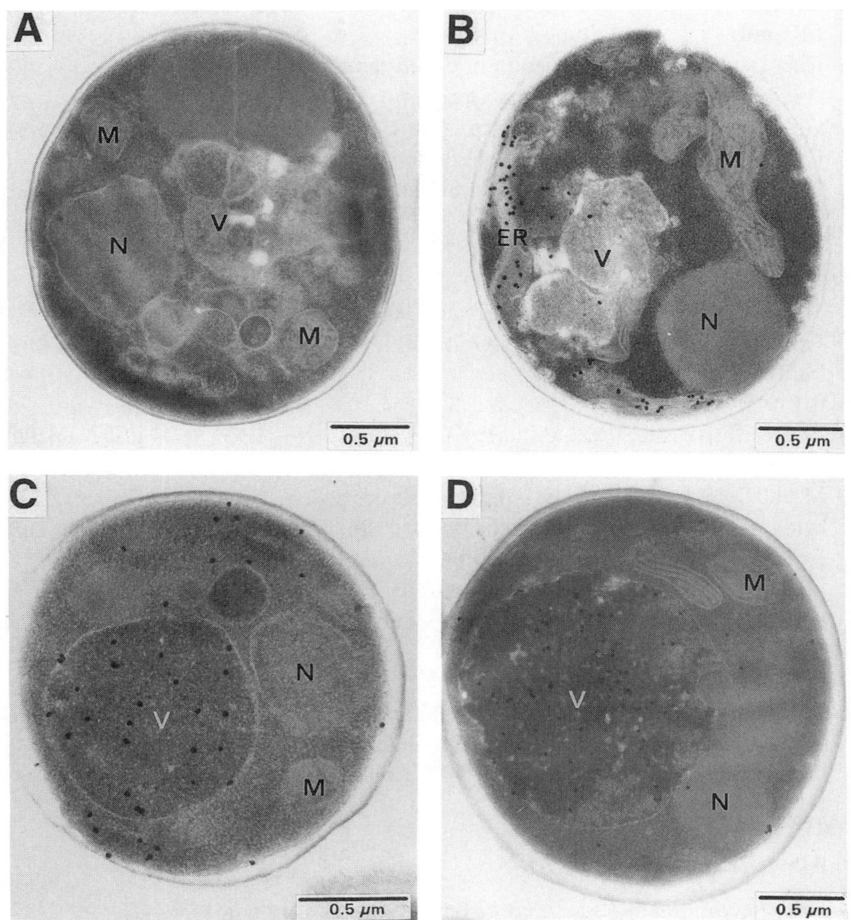

Fig. 1. Postembedding immunogold staining of *P. pastoris* SMD 1163 cells heterologously expressing the *myc*-tagged $5HT_{5A}$ receptor. The *myc*-tagged receptor was visualized by reaction with the MAb 9E10 and secondary goat antimouse polyclonal antibodies coupled to gold particles. Gold particles were enlarged by silver enhancement. No labeling is observed in yeast cells expressing the nontagged receptor **(A)**. In cells expressing the *myc*-tagged receptor, specific labeling occurs either over the endoplasmic reticulum **(B)**, most probably during the early phase of expression, or over the vacuole **(C,D)**. ER = endoplasmic reticulum; M = mitochondrium; N = nucleus; V = vacuole. Bars = 0.5 μm.

7. Tween-20.
8. Triton X-100.
9. BSA.

10. Glycine.
11. Ethanol.
12. Cacodylate buffer: 0.1 *M* sodium cacodylate-HCl, pH 7.2.
13. Agar-agar (Serva, Heidelberg, Germany).
14. Gelatin capsules (Pohl-Boskamp GmbH & Co., Hohenlockstedt, Germany).
15. Embedding medium LR White (London Resin Co., Basingstoke, UK).
16. Fixation solution: 4% *para*-formaldehyde/2.5% glutaraldehyde in cacodylate buffer (*see* **Note 1**).

2.3. Immunogold Staining

1. Nickel grids (Plano GmbH, Wetzlar, Germany).
2. Formvar® (Serva, Heidelberg, Germany).
3. Parafilm.
4. 0.1 *M* HCl.
5. PBS buffer: 8 g/L NaCl, 0.2 g/L KCl, 1.44 g/L Na_2HPO_4, 0.24 g/L NaH_2PO_4.
6. Filters (sterilized and stored at room temperature).
7. MAb 9E10 (either purchased from Invitrogen or Cambridge Research, Cambridge, UK, or purified from the culture supernatant of the monoclonal cell line from the American Tissue Culture Collection, No. ATCC CRL-1729=myc1-9E10.2). Store purified and concentrated antibody at –20°C. Supernatants from antibody-producing hybridoma cells can be collected and stored at 4°C until purified. Add Na-azide to ~0.01% to the supernatants to prevent microbial contamination.
8. Secondary goat antimouse polyclonal antibody coupled to 10-nm gold particles (Amersham Buchler, Braunschweig, Germany).
9. Silver enhancement kit (IntenSE™M, Amersham Buchler).

3. Methods

3.1. Fixation and Embedding

1. Grow recombinant *P. pastoris* as described in Chapter 15.
2. Harvest cells ~25 h after induction of expression, and immediately fix.
3. Resuspend yeast cell pellets (e.g., a pellet from a 10 mL culture, $OD_{600} \approx 20$, which corresponds to ~1×10^9 cells) in fixation solution, and incubate for 5 h at room temperature.
4. Centrifuge briefly, discard the supernatant (fixative), resuspend in cacodylate buffer supplemented with 2% glycine, and incubate the pellet at 4°C overnight.
5. To avoid repeated centrifugation during the different dehydration steps and medium changes during embedding in plastic resins, encapsulate the cells in agar–agar. For this, boil 2% agar–agar in water until a clear solution is obtained. Cool the agar solution in a water bath to about 40°C, and then pour the agar as a thin layer onto a prewarmed glass plate (e.g., a microscope slide).
6. Collect the fixed cells by a brief centrifugation, and add a compact piece of cell pellet with a minimum of residual fluid into the still molten 2% agar on the slide. Stir with a fine needle. The volume of the cells should be kept as small as possible.

7. Cool the slide to 4°C, and carefully cut small (2 × 2 mm) cubes from the solidified agar with a scalpel.
8. Dehydrate the cell sample-containing agar cubes in an Eppendorf tube by incubation in increasing concentrations of ethanol (50, 70, 90%). Incubate the cubes for 15 min at each step, changing the ethanol three times during each incubation at each step.
9. Infiltrate the cubes in a microcentrifuge tube with 1 vol of LR White resin and 2 vol of 90% ethanol. Then treat with equal volumes of resin and ethanol followed by 2 vol resin and 1 vol ethanol, each twice for 2 h. Subsequently, incubate overnight in 100% LR White resin. On the next day exchange the resin three times with 3 h incubations at room temperature at each point.
10. Embed the cubes in dried gelatin capsules (BEEM capsules can also be used for embedding). Fill the capsules up to the top with resin and close carefully. Avoid the trapping of air; otherwise polymerization will be incomplete.
11. Polymerize the resin in the gelatine capsules by incubation at 55°C for about 24 h (*see* **Note 2**).

3.2. Immunogold Staining

1. Perform postembedding immunogold staining on ultrathin sections (<0.1 μm) mounted on Formvar®-coated nickel grids. The Formvar® coating has the following advantages:
   a. It allows the use of grids that have a large viewing area;
   b. It allows small sections to adhere to grids efficiently;
   c. It allows grids to float by surface tension on surfaces of incubation solutions; and
   d. It inhibits interactions between the colloidal gold particles and the grid material.

   With respect to the last point, nickel or gold grids are preferable to copper grids.
2. For subsequent immunolabeling of the sections, float the grids with the sections side facing downward on droplets of ~20–30 μL of solutions. Normally, the droplets are placed on sheets of parafilm. The grids are carefully transferred with tweezers from one drop of solution to the next drop. Carefully, adsorb fluid from one solution with filter paper before moving grid to the next solution. However, avoid drying of the grid surface.
3. Treat the grids at room temperature with the following solutions in the listed order:
   a. Water-saturated sodium *meta*-periodate for 1 h.
   b. Water three times for 3 min each.
   c. 0.1 *N* HCl for 10 min.
   d. Water three times for 3 min each.
   e. PBS + 1% BSA + 0.5% Tween 20 + 0.5% Triton X-100 for 15 min.
   f. PBS + 0.1% BSA + 0.05% Tween 20 twice for 10 min each.
   g. Primary MAb 9E10 (0.5 μg/mL in PBS with 0.1% BSA) for 2 h (*see* **Note 3**).
   h. PBS + 0.1% BSA + 0.05% Tween 20 four times for 5 min each.
4. Dilute the secondary goat antimouse polyclonal antibody coupled to colloidal gold 1:50 in PBS, and incubate with the grids for about 1 h at room temperature.

The secondary antibody must be directed against the species from which the primary antibody was obtained, i.e., when a mouse MAb is used as a primary antibody, as is the case here, an antimouse gold-coupled secondary antibody from either goat, rabbit, or rat must be applied.

5. After incubation with the secondary antibody, treat the grids at room temperature with the following solutions in the listed order:
   a. PBS + 0.1% BSA + 0.05% Tween 20 twice for 5 min each.
   b. PBS three times for 5 min each.
   c. 1% Glutaraldehyde in PBS for 10 min.
   d. Water three times for 5 min each.
6. Before the dry sections can be visualized in an electron microscope, they must be double-contrasted. For this, incubate the grids with 2% uranyl acetate for 2 min and then 1% lead citrate for 1 min with a washing step for 1 min in water following each treatment ***(9)*** (*see* **Note 4**).
7. Examine the sections in an electron microscope.

4. Notes

1. We recommend the inclusion of *para*-formaldehyde in the fixative, because its small size allows it to penetrate the cells faster than glutardialdehyde. For each antigen to be analyzed, the optimal concentration of glutardialdehyde (0.1–2.5%) in fixative and duration of fixation time must be empirically determined to balance retention of antigenicity and good structural preservation. Sodium cacodylate buffer is routinely used in the fixative, but other buffer systems may be used as well. Fixation by osmiumtetroxide should be avoided, since this compound interferes with the antigenicity of membrane-bound proteins.
2. There are several embedding media commercially available. For postembedding immunostaining, hydrophilic media like LR White (London Resin Company), Lowicryl (Chemische Werke Lowi GmbH & Co, Waldkraiburg, Germany), and Unicryl (British BioCell International, Cardiff, UK) are recommended. Epoxy resins are less well suited, since they tend to interact with the protein, resulting in a loss of antigenicity. They also require total dehydration of the samples. Sometimes, good results can also be obtained with the low-viscosity epoxy resin formulated by Spurr ***(10)***, but sections prepared as described above will have low contrast. With the LR White resin, polymerization can be achieved either by application of heat (55°C for ~24 h) or by illumination with UV light (Sylvania/Blacklite-Blue). During the UV treatment (15 h), the temperature can be kept at 22°C or lower.
3. To avoid nonspecific binding of polyclonal antibodies, a protein A-purified IgG fraction should be used; affinity-purified antibodies are highly recommended. In addition, nonspecific binding sites must be blocked prior to incubation. Blocking with 1% BSA in most cases is sufficient, but supplements of Tween 20 and/or Triton X-100 are preferable. However, some antibodies are inactivated by detergents. Fish gelatine (Science Services, München, Germany) and nonfat dry milk powder, which is available in many drugstores, are alternatives to BSA and give

acceptable results. Nonspecific binding also depends on the concentrations of the antibodies applied. Therefore, a series of antibody dilutions should be tested. To find the optimal antibody dilution, it may help to compare background labeling over intracellular compartments like mitochondria with labeling over structures where the antigen is expected to reside.

4. Gold particles of 5 nm, coupled to the secondary antibody, are preferred because higher labeling densities can be achieved. For better visualization of these small gold particles, we recommend enlargement of the gold particles by silver enhancement. For this, we use the kit from Amersham and follow the protocol for the enhancement provided with the kit. However, to slow the silver reaction process, cool the two components of the kit to 4°C before mixing (1:1), and incubate the grids on droplets placed on an ice-cooled metal plate covered with a sheet of parafilm for approx 4 min.

References

1. Bach, M., Sander, P., Haase, W., and Reiländer, H. (1996) Pharmacological and biochemical characterization of the mouse $5HT_{5A}$ serotonin receptor heterologously produced in the yeast *Saccharomyces cerevisiae*. *Receptors and Channels* **4,** 129–139.
2. Lenhard, T., Maul, G., Haase, W., and Reiländer, H. (1996) A new set of versatile transfer vectors for the heterologous expression of foreign genes using the baculovirus system. *Gene* **169,** 187–190.
3. Grünewald, S., Haase, W., Reiländer, H., and Michel, H. (1996) Glycosylation, palmitoylation and localization of the human D2S receptor in baculovirus-infected insect cells. *Biochemistry* **35,** 15,149–15,161.
4. Weiß, H. M., Haase, W., Michel, H., and Reiländer, H. (1995) Expression of functional mouse 5-HT_{5A} serotonin receptor in the methylotrophic yeast *Pichia pastoris*: pharmacological characterization and localization. *FEBS Lett.* **377,** 451–456.
5. Pringle, J. R., Adams, A. E., Drubin, D. G., and Haarer, B. K. (1991) Immunofluorescence methods for yeast. *Methods Enzymol.* **194,** 565–602.
6. Sander, P., Grünewald, S., Bach, M., Haase, W., Reiländer, H., and Michel, H. (1994) Heterologous expression of the human D_{2S} dopamine receptor in protease-deficient *Saccharomyces cerevisiae* strains. *Eur. J. Biochem.* **226,** 697–705.
7. Griffiths, G. (1993) *Fine Structure Immunocytochemistry*. Springer-Verlag, Berlin.
8. Hayat, M. A., ed. (1995) *Immunogold-Silver Staining: Principles, Methods, and Applications*. CRC, Boca Raton, FL.
9. Reynolds, E. S. (1963) The use of lead citrate at high pH as an electron-dense stain in electron microscopy. *J. Cell Biol.* **17,** 208–212.
10. Spurr, A. R. (1969) A low-viscosity epoxy resin embedding medium for electron microscopy. *J. Ultrastruct. Res.* **26,** 31–43.

17

Appendix

Foreign Proteins Expressed in *P. pastoris*

Protein (organism)	Mode[b]	Ref.
Bacteria		
β-Galactosidase (*Escherichia*)	I	*1*
β-Lactamase (*Escherichia*)	I	*2*
Pertussis pertactin protein (*Bordetella*)	I	*3*
Streptokinase (*Streptomyces*)	I	*4*
Tetanus toxin C fragment (*Clostridium*)	I	*5*
Accessory cholera enterotoxin (*Vibrio*)	S	*6*
D-Alanine carboxypeptidase (*Bacillus*)	S	*7*
Subtilisin inhibitor (*Streptomyces*)	S	*8*
Fungi		
Catalase T (*Saccharomyces*)	I	*9*
Laccase (*Tramates*)	I	*10*
Alt a 1 allergen (*Alternaria*)	S	*11*
Catalase (*Aspergillus*)	S	*12*
β-Cryptogein (*Phytophthora*)	S	*13*
Dipeptidyl-peptidases IV and V (*Aspergillus*)	S	*14,15*
Glucoamylase (*Aspergillus*)	S	*16*
β-Glucosidase (*Candida*)	S	*17*
Invertase (*Saccharomyces*)	S	*18*
Lipase (*Geotrichum*)	S	*19*
α1,2-Mannosyltransferase (*Saccharomyces*)	S	*20*
Pectate lyase (*Fusarium*)	S	*21*
Vitamin B2-aldehyde-forming enzyme (*Schizophyllum*)	S	*22*
Plants		
Glycolate oxidase (Spinach)	I	*9,23*
Hexose oxidase (Red Alga)	I	*24*
Hydroxynitrile lyase (*Hevea*)	I	*25*
Nitrate reductase (*Arabidopsis*)	I	*26*
Phosphoribulokinase (Spinach)	I	*27*

From: *Methods in Molecular Biology, Vol. 103:* Pichia *Protocols*
Edited by: D. R. Higgins and J. M. Cregg

Protein (organism)	Mode[b]	Ref.
Phytochromes A and B (Potato)	I	***28***
α-Amylase isozymes (Barley)	S	***29***
Cyn d 1 allergen (Bermuda grass)	S	***30***
α-Galactosidase (Coffee bean)	S	***31***
Group I allergen (Timothy grass)	S	***32***
Invertebrates		
Green fluorescent protein (Jellyfish)	I	***33***
Bm86 antigen (Tick)	I	***34***
Dragline silk protein (Spider)	I	***35***
Luciferase (Fire fly)	I	***36***
Acetylcholinesterase (Electric eel)	S	***37***
ADP-ribosyl cyclase (*Aplysia*)	S	***38***
Angiotensin-converting enzyme (Fruit fly)	S	***39***
Anticoagulant pepetide (Tick)	S	***40***
Ghilanten (Leech)	S	***41***
Hirudin (Leech)	S	***42***
Vertebrates (other than humans)		
Carnitine palmitoyltransferases (Rat)	I	***43***
17 α-Hydroxylase/C17,20-lyase (Shark)	I	***44***
Leukocyte 12-lipoxygenase (Pig)	I	***45***
Multifunctional enzyme (Rat)	I	***46***
NO synthase reductase domain (Rat)	I	***47***
Polyomavirus large T antigen (Mouse)	I	***48***
Reovirus λ1 core protein (Mouse)	I	***49***
Acetylcholinesterase H and T subunits (Rat)	S	***50***
α-*N*-acetylgalactosaminidase (Chicken)	S	***51***
Angiotensin converting enzyme (Rabbit)	S*	***52***
Antifreeze protein type II (Sea raven)	S	***53***
Carbonic anhydrase inhibitor (Pig)	S	***54***
β-Casein (Cow)	S	***55***
Cholesteryl ester transfer protein (Rabbit)	S	***56***
Complement regulator (Rat)	S	***57***
Enterokinase catalytic domain (Cow)	S	***58***
Epidermal growth factor (Mouse)	S	***59***
Follicle-stimulating hormone β-subunit (Cow)	S	***60***
Gelatinase B (Mouse)	S	***61***
Herpesvirus type 1 glycoprotein D (Cow)	S	***62***
High-mobility group 1 (HMG1) protein (Rat)	S	***63***
Interferon-τ (Goat)	S	***64***
Intestinal peptide transporter (PepT1) (Rabbit)	S*	***65***
Lysozyme (Cow)	S	***66***
Macrophage inflammation protein 2 (Mouse)	S	***67***

Protein (organism)	Mode[b]	Ref.
Major urinary protein complex (Mouse)	S	*68*
α-Mannosidase (Mouse)	S	*69*
Neural cell adhesion molecule, first immunoglobulin-like domain (Mouse)	S	*70*
Opsin (Cow)	S*	*71*
5-HT5A Serotonin receptor (Mouse)	S*	*72*
Sialoglycoprotein type I (Rat)	S	*73*
Single-chain Fv fragments (Mouse)	S	*74*
Humans		
Caspase 3	I	*75*
Carnitine palmitoyltransferase I	I	*76*
CD40 ligand	I	*77*
Cytomegalovirus ppUL44 antigen	I	*78*
dsRNA-specific editase 1	I	*79*
Hepatitis B surface antigen	I	*80*
Hepatitis B surface antigen-HIV gp41 epitope chimera	I	*81*
Hepatitis E virus ORF3 protein	I	*82*
α-Mannosidase	I	*83*
NonO nucleic acid binding protein	I^+	*84*
Prolyl 4-hydroxylase	I	*85*
Protein kinase C	I	*86*
Proteinase inhibitor 6	I	*87*
Proteinase inhibitor 8	I	*88*
Tumor necrosis factor	I	*89*
Amyloid precursor protein	S	*90*
Amyloid β-protein precursor-like protein-2 (Kunitz-type proteinase inhibitor domain)	S	*91*
Angiostatin (kringles 1-4 of plasminogen)	S	*92*
Aprotinin analog	S	*93*
B7 Costimulatory molecules and CTLA-4 counter receptor	S	*94*
B7-2 Extracellular domain-scFv chimeric protein	S	*95*
Cathepsin E	S	*96*
Cathepsin K	S	*97*
Cathepsin L propepetide	S	*98*
Collagen type III	S	*85*
Decay-accelerating factor (echovirus receptor)	S	*99*
Dengue virus structural protein	S	*100*
Epidermal growth factor	S	*101*
Fas ligand (soluble form)	S	*102*
Fibrinogen γ chain C-terminus	S	*103*

Protein (organism)	Mode[b]	Ref.
Fibroblast collagenase	S	***104***
HIV-1 envelope protein	S	***105***
Influenza virus neuraminidase head domain	S	***106***
Insulin-like growth factor 1	S	Chapter 11
Interleukin-6 (cytokine receptor domain)	S	***107***
Leukemia inhibitory factor (LIF) receptor component gp130	S	***108***
Lymphocyte surface antigen CD38	S	***109***
MHC class II heterodimers (soluble form)	S	***110***
Monocyte chemotactic protein	S	***111***
μ-Opioid recepter	S*	***112***
Procathepsin B	S	***113***
Proteinase 3	S	***114***
Serum albumin	S	***115***
Tissue-type plasminogen activator (kringle 2 domain)	S	***116***
Thrombomodulin	S	***117***
Transferrin (N-terminal domain)	S	***118***
Urokinase-type plasminogen activator-anexin V chimeras	S	***119***
Vaccinia virus complement control protein	S	***120***

[a]I=Intracellular; I^+ = Endoplasmic reticulum resident protien; S = Secreted plasma membrane protein.

[b]Published reports of foreign protiens expressed in Pichia pastoris and listed in medline (http://www.ncbionlm.nih.gov/htbn-post/Entrez/query?db=m_d) through December 1997.

References

1. Tschopp, J. F., Brust, P. F., Cregg, J. M., Stillman, C. A., and Gingeras, T. R. (1987) Expression of the *lacZ* gene from two methanol-regulated promoters in *Pichia pastoris*. *Nucleic Acids Res.* **15**, 3859–3876.
2. Waterham, H. R., Digan, M. E., Koutz, P. J., Lair, S. L., and Cregg, J. M. (1997) Isolation of the *Pichia pastoris* glyceraldehyde-3-phosphate deydrogenase gene and regulation and use of its promoter. *Gene* **186**, 37–44.
3. Romanos, M. A., Clare, J. J., Beesley, K. M., Rayment, F. B., Ballantine, S. P., Makoff, A. J., Dougan, G., Fairweather, N. F., and Charles, I. G. (1991) Recombinant *Bordetella pertussis* pertactin (P69) from the yeast *Pichia pastoris*: high-level production and immunological properties. *Vaccine* **9**, 901–906.
4. Hagenson, M. J., Holden, K. A., Parker, K. A., Wood, P. J., Cruze, J. A., Fuke, M., Hopkins, T. R., and Stroman, D. W. (1989) Expression of streptokinase in *Pichia pastoris* yeast. *Enzyme Microb. Technol.* **11,** 650–656.
5. Clare, J. J., Rayment, F. B., Ballantine, S. P., Sreekrishna, K., and Romanos, M. A. (1991) High-level expression of tetanus toxin fragment C in *Pichia pastoris* strains containing multiple tandem integrations of the gene. *Bio/Technology* **9**, 455–460.

6. Trucksis, M., Conn, T. L., Fasano, A., and Kaper, J. B. (1997) Production of *Vibrio cholerae* accessory cholera enterotoxin (Ace) in the yeast *Pichia pastoris. Infect. Immun.* **65**, 4984–4988.
7. Despreaux, C. W. and Manning, R. F. (1993) The dacA gene of *Bacillus stearothermophilus* coding for D-alanine carboxypeptidase: cloning, structure and expression in *Escherichia coli* and *Pichia pastoris. Gene* **131,** 35–41.
8. Markaryan, A., Beall, C. J. and Kolattukudy, P. E. (1996) Inhibition of *Aspergillus serine* proteinase by *Streptomyces subtilisin* inhibitor and high-level expression of this inhibitor in *Pichia pastoris. Biochem. Biophys. Res. Commun.* **220,** 372–376.
9. Payne, M. S., Petrillo, K. L., Gavagan, J. E., Wagner, L. W., DiCosimo, R., and Anton, D. L. (1995) High-level production of spinach glycolate oxidase in the methylotrophic yeast *Pichia pastoris*: engineering a biocatalyst. *Gene* **167,** 215–219.
10. Jönsson, L. J., Saloheimo, M., and Penttilä, M. (1997) Laccase from the white-rot fungus *Tramates versicolor*: cDNA cloning of *lcc1* and expression in *Pichia pastoris. Curr. Genet.* **32,** 425–430.
11. de Vouge, M. W., Thaker, A. J., Curran, I. H., Zhang, L., Muradia, G., Rode, H., and Vijay, H. M. (1996) Isolation and expression of a cDNA clone encoding an *Alternaria alternata* Alt a 1 subunit. *Int. Arch. Allergy Immunol.* **111**, 385–395.
12. Calera, J. A., Paris, S., Monod, M., Hamilton, A. J., Debeaupuis, J. P., Diaquin, M., Lopez-Modrano, R., Leal, F., and Latge, J. P. (1997) Cloning and disruption of the antigenic catalase gene of *Aspergillus fumigatus. Infect. Immun.* **65,** 4718–4724.
13. O'Donohue, M. J., Boissy, G., Huet, J. C., Nespoulous, C., Brunie, S., and Pernollet, J. C. (1996) Overexpression in *Pichia pastoris* and crystallization of an elicitor protein secreted by the phytopathogenic fungus, *Phytophthora cryptogea. Protein Expr. Purif.* **8,** 254–261.
14. Beauvais, A., Monod, M., Wyniger, J., Debeaupuis, J. P., Grouzmann, E., Brakch, N., Svab, J., Hovanessian, A. G., and Latgé, J. P. (1997) Dipeptidyl-peptidase IV secreted by *Aspergillus fumigatus*, a fungus pathogenic to humans. *Infect. Immun.* **65**, 3042–3047.
15. Beauvais, A., Monod, M., Debeaupuis, J. P., Diaquin, M., Kobayashi, H. and Latgé, J. P. (1997) Biochemical and antigenic characterization of a new didpeptidyl-peptidase isolated from *Aspergillus fumigatus. J. Biol. Chem.* **272,** 6238–3244.
16. Fierobe, H.-P., Mirgorodskaya, E., Frandsen, R. P., Roepstorff, P. and Svensson, B. (1997) Overexpression and characterization of *Aspergillus awamori* wild-type and mutant glucoamylase secreted by the methylotrophic yeast *Pichia pastoris*: comparison with wild-type recombinant glucoamylase produced using *Saccharomyces cerevisiae* and *Aspergillus niger* as hosts. *Protein Expr. Purif.* **9,** 159–170.
17. Skory, C. D., Freer, S. N., and Bothast, R. J. (1996) Expression and secretion of the *Candida wickerhamii* extracellular β-glucosidase gene, *bglB*, in *Saccharomyces cerevisiae. Curr. Genet.* **30**, 417–422.
18. Tschopp, J. F., Sverlow, G., Kosson, R., Craig, W., and Grinna, L. (1987) High-level secretion of glycosylated invertase in the methylotrophic yeast, *Pichia pastoris. Bio/Technology* **5**, 1305–1308.

19. Holmquist, M., Tessier, D. C., and Cygler, M. (1997) High-level production of recombinant *Geotrichum candidum* lipases in yeast *Pichia pastoris*. *Protein Expr. Purif.* **11,** 35–40.
20. Romero, P. A., Lussier, M., Sdicu, A. M., Bussey, H., and Herscovics, A. (1997) Ktr1p is an α-1,2-mannosyltransferase of *Saccharomyces cerevisiae*. Comparison of the enzymic properties of soluble recombinant Ktr1p and Kre2p/Mnt1p produced in *Pichia pastoris*. *Biochem. J.* **321**, 289–295.
21. Guo, W., González-Candelas, L., and Kolattukudy, P. E. (1996) Identification of a novel pelD gene expressed uniquely in planta by *Fusarium solani* f. sp. pisi (*Nectria haematococca*, mating type VI) and characterization of its protein product as an endo-pectate lyase. *Arch. Biochem. Biophys.* **332,** 305–312.
22. Chen, H. and McCormick, D. B. (1997) Riboflavin 5'-hydroxymethyl oxidation. Molecular cloning, expression and glycoprotein nature of the 5'-aldehyde-forming enzyme from *Schizophyllum commune*. *J. Biol. Chem.* **272,** 20,077–20,081.
23. Payne, M. S., Petrillo, K. L., Gavagan, J. E., DiCosimo, R., Wagner, L. W., and Anton, D. L. (1997) Engineering *Pichia pastoris* for biocatalysis: co-production of two active enzymes. *Gene* **194,** 179–182.
24. Hansen, O. C. and Stougaard, P. (1997) Hexose oxidase from the red alga *Chondrus crispus*. Purification, molecular cloning, and expression in *Pichia pastoris*. *J. Biol. Chem.* **272,** 11,581–11,587.
25. Hasslacher, M., Schall, M., Hayn, M., Bona, R., Rumbold, K., Luckl, J., Griengl, H., Kohlwein, S. D., and Schwab, H. (1997) High-level intracellular expression of hydroxynitrile lyase from the tropical rubber tree *Hevea brasiliensis* in microbial hosts. *Protein Expr. Purif.* **11,** 61–71.
26. Su, W., Huber, S. C., and Crawford, N. M. (1996) Identification *in vitro* of a post-translational regulatory site in the hinge 1 region of *Arabidopsis* nitrate reductase. *Plant Cell* **8**, 519–527.
27. Brandes, H. K., Hartman, F. C., Lu, T. Y., and Larimer, F. W. (1996) Efficient expression of the gene for spinach phosphoribulokinase in *Pichia pastoris* and utilization of the recombinant enzyme to explore the role of regulatory cysteinyl residues by site-directed mutagenesis. *J. Biol. Chem.* **271,** 6490–6796.
28. Ruddat, A., Schmidt, P., Gatz, C., Braslavsky, S. E., Gartner, W., and Schaffner, K. (1997) Recombinant type A and B phytochromes from potato. Transient absorption spectroscopy. *Biochemistry* **36**, 103–111.
29. Juge, N., Andersen, J. S., Tull, D., Roepstorff, P., and Svensson, B. (1996) Overexpression, purification, and characterization of recombinant barley α-amylases 1 and 2 secreted by the methylotrophic yeast *Pichia pastoris*. *Protein Expr. Purif.* **8,** 204–214.
30. Smith, P. M., Suphioglu, C., Griffith, I. J., Theriault, K., Knox, R. B., and Singh, M. B. (1996) Cloning and expression in yeast *Pichia pastoris* of a biologically active form of Cyn d 1, the major allergen of Bermuda grass pollen. *J. Allergy Clin. Immunol.* **98**, 331–343.
31. Zhu, A., Monohan, C., Zhang, Z., Hurst, R., Leng, L., and Goldstein, J. (1995) High-level expression and purification of coffee bean α-galactosidase produced in the yeast *Pichia pastoris*. *Arch. Biochem. Biophys.* **324**, 65–70.

32. Petersen, A., Grobe, K., Lindner, B., Schlaak, M., and Becker, W. M. (1997) Comparison of natural and recombinant isoforms of grass pollen allergens. *Electrophoresis* **18**, 819–825.
33. Mononsov, E. Z., Wenzel, T. J., Luers, G. H., Heyman, J. A. and Subramani, S. (1996) Labeling of peroxisomes with green fluorescent protein in living *P. pastoris* cells. *J. Histochem. Cytochem.* **44,** 581–589.
34. Rodriguez, M., Rubiera, R., Penichet, M., Montesinos, R., Cremata, J., and de la Fuente, J. (1994) High level expression of the *B. microplus* Bm86 antigen in the yeast *Pichia pastoris* forming highly immunogenic particles for cattle. *J. Biotechnol.* **33**, 135–146.
35. Fahnestock, S. R. and Bedzyk, L. A. (1997) Production of synthetic spider dragline silk protein in *Pichia pastoris. Appl. Microbiol. Biotechnol.* **47,** 33–39.
36. McCollum, D., Monosov, E., and Subramani, S. (1993) The *pas8* mutant of *Pichia pastoris* exhibits the peroxisomal protein import deficiencies of Zellweger syndrome—the PAS8 protein binds to the COOH-terminal tripeptide peroxisomal targeting signal, and is a member of the TPR protein family. *J. Cell Biol.* **121**, 761–774.
37. Simon, S. and Massouli, J. (1997) Cloning and expression of acetylcholinesterase from electrophorus. Splicing pattern of the 3' exons *in vivo* and in transfected mammalian cells. *J. Biol. Chem.* **272**, 33,045–33,055.
38. Munshi, C. and Lee, H. C. (1997) High-level expression of recombinant *Aplysia* ADP-riboxyl cyclase in *Pichia pastoris* by fermentation. *Protein Expr. Purif.* **11,** 104–110.
39. Williams, T. A., Michaud, A., Houard, X., Chauvet, M. T., Soubrier, F., and Corvol, P. (1996) *Drospholia melanogaster* angiotensin I-converting enzyme expressed in *Pichia pastoris* resembles the C domain of the mammalian homologue and does not require glycosylation for secretion and enzymic activity. *Biochem. J.* **318**, 125–131.
40. Laroche, Y., Storme, V., De Meutter, J., Messens, J. and Lauwereys, M. (1994) High-level secretion and very efficient isotopic labeling of tick anticoagulant peptide (TAP) expressed in the methylotrophic yeast, *Pichia pastoris. Biotechnol NY* **12,** 1119–1124.
41. Brandkamp, R. G., Sreekrishna, K., Smith, P. L., Blankenship, D. T., and Cardin, A. D. (1995) Expression of a synthetic gene encoding the anticoagulant-antimetastatic protein ghilanten by the methylotropic yeast *Pichia pastoris. Protein Expr. Purif.* **6,** 813–820.
42. Rosenfeld, S. A., Nadeau, D., Tirado, J., Hollis, G. F., Knabb, R. M., and Jia, S. (1996) Production and purification of recombinant hirudin expressed in the methylotrophic yeast *Pichia pastoris. Protein Expr. Purif.* **8,** 476–482.
43. de Vries, Y., Arvidson, D., Waterham, H. R., Cregg, J. M. and Woldegiorgis, G. (1997) Functional characterization of mitochondrial carnitine palmitoyltransferases I and II expressed in the yeast *Pichia pastoris. Biochemistry* **36,** 5285–5292.
44. Trant, J. M. (1996) Functional expression of recombinant spiny dogfish shark (*Squalus acanthias*) cytochrome P450c17 (17 α-hydroxylase/C17,20-lyase) in yeast (*Pichia pastoris*). *Arch. Biochem. Biophys.* **326**, 8–14.

45. Reddy, R. G., Yoshimoto, T., Yamamoto, S., and Marnett, L. J. (1994) Expression, purification, and characterization of porcine leukocyte 12-lipoxygenase produced in the methylotrophic yeast, *Pichia pastoris. Biochem. Biophys. Res. Commun.* **205**, 381–388.
46. Qin, Y. M., Poutanen, M. H., Helander, H. M., Kvist, A. P., Siivari, K. M., Schmitz, W., Conzelmann, E., Hellman, U., and Hiltunen, J. K. (1997) Peroxisomal multifunctional enzyme of β-oxidation metabolizing D-3-hydroxyacyl-CoA esters in rat liver: molecular cloning, expression and characterization. *Biochem. J.* **321**, 21–28.
47. Gachhui, R., Presta, A., Bentley, D. F., AbuSoud, M. N., McArthur, R., Brudvig, G., Ghosh, D. K. and Strehr, D. J. (1996) Characterization of the reductase domain of rat neuronal nitric oxide synthase generated in the methyotrophic yeast *Pichia pastoris.* Calmodulin response is complete within the reductase domain itself. *J. Biol. Chem.* **271,** 20,594–20,602.
48. Peng, Y. C. and Acheson, N. H. (1997) Production of active polyomavirus large T antigen in yeast *Pichia pastoris. Virus Res.* **49,** 41–47.
49. Bisaillon, M., Bergeron, J., and Lemay, G. (1997) Characterization of the nucleoside triphosphate phosphohydrolase and helicase activities of the reovirus lambda 1 protein. *J. Biol. Chem.* **272,** 18,298–18,303.
50. Morel, N. and Massoulié, J. (1997) Expression and processing of vertebrate acetylcholinesterase in the yeast *Pichia pastoris. Biochem. J.* **328,** 121–129.
51. Zhu, A., Monohan, C., Wang, Z. K., and Goldstein, J. (1996) Expression, purification and characterization of recombinant α-N-acetylgalactosaminidase produced in the yeast *Pichia pastoris. Protein Expr. Purif.* **8,** 456–462.
52. Sadhukhan, R., Sen, G. C., and Sen, I. (1996) Synthesis and cleavage—secretion of enzymatically active rabbit angiotensin-converting enzyme in *Pichia pastoris. J. Biol. Chem.* **271**, 18,310–18,313.
53. Loewen, M. C., Liu, X., Davies, P. L., and Daugulis, A. J. (1997) Biosynthetic production of type II fish antifreeze protein: fermentation by *Pichia pastoris. Appl. Microbiol. Biotechnol.* **48,** 480–486.
54. Wuebbens, M. W., Roush, E. D., Decastro, C. M., and Fierke, C. A. (1997) Cloning, sequencing, and recombinant expression of the porcine inhibitor of carbonic anhydrase: a novel member of the transferrin family. *Biochemistry* **36,** 4327–4336.
55. Choi, B.-K. and Jiménez-Flores, R. (1996) Study of putative glycosylation sites in bovine β-casein introduced by PCR-based site-directed mutagenesis. *J. Agric. Food Chem.* **44,** 358–364.
56. Kotake, H., Li, Q., Ohnishi, T., Ko, K. W., Agellon, L. B. and Yokoyama, S. (1996) Expression and secretion of rabbit plasma cholesteryl ester transfer protein by *Pichia pastoris. J. Lipid Res.* **37,** 599–605.
57. He, C., Alexander, J. J., Lim, A., and Quigg, R. J. (1997) Production of the rat complement regulator, Crry, as an active soluble protein in *Pichia pastoris. Arch. Biochem. Biophys.* **341,** 347–352.
58. Vozza, L. A., Wittwer, L., Higgins, D. R., Purcell, T. J., Bergseid, M., Collins-Racie, L. A., LaVallie, E. R., and Hoeffler, J. P. (1996) Production of a recombinant bovine enterokinase catalytic subunit in the methylotrophic yeast *Pichia pastoris. Bio/Technology* **14,** 77–81.

59. Clare, J. J., Romanos, M. A., Rayment, F. B., Rowedder, J. E., Smith, M. A., Payne, M. M., Sreekrishna, K., and Henwood, C. A. (1991) Production of mouse epidermal growth factor in yeast: high-level secretion using *Pichia pastoris* strains containing multiple gene copies. *Gene* **105**, 205–212.
60. Samaddar, M., Catterall, J. F., and Dighde, R. R. (1997) Expression of biologically active beta subunit of bovine follicle-stimulating hormone in the methylotrophic yeast *Pichia pastoris. Protein Expr. Purif.* **10,** 345–355.
61. Masure, S., Paemen, L., Van Aelst, I., Fiten, P., Proost, P., Billiau, A., Van Damme, J. and Opdenakker, G. (1997) Production and characterization of recombinant active mouse gelatinase B from eukaryotic cells and *in vivo* effects after intravenous administration. *Eur. J. Biochem.* **244,** 21–30.
62. Zhu, X., Wu, S., and Letchworth, G. J., III (1997) Yeast-secreted bovine herpesvirus type 1 glycoprotein D has authentic conformational structure and immunogenicity. *Vaccine* **15,** 679–688.
63. Mistry, A. R., Falciola, L., Monaco, L., Tagliabue, R., Acerbis, G., Knight, A., Harbottle, R. P., Soria, M., Bianchi, M. E., Coutelle, C., and Hart, S. L. (1997) Recombinant HMG1 protein produced in *Pichia pastoris*: a nonviral gene delivery agent. *Biotechniques* **22,** 718–729.
64. van Heeke, G., Ott, T. L., Strauss, A., Ammaturo, D., and Bazer, F. W. (1996) High yield expression and secretion of ovine pregnancy recognition hormone interferon-tau by *Pichia pastoris. J. Interferon Cytokine Res.* **16**, 119–126.
65. Doring, F., Theis, S., and Daniel, H. (1997) Expression and functional characterization of the mammalian intestinal peptide transporter PepT1 in the methylotropic yeast *Pichia pastoris. Biochem. Biophys. Res. Commun.* **232,** 656–662.
66. Digan, M. E., Lair, S. V., Brierley, R. A., Siegel, R. S., Williams, M. E., Ellis, S. B., Kellaris, P. A., Provow, S. A., Craig, W. S., Velicelebi, G., Harpold, M. M., and Thill, G. P. (1989) Continuous production of a novel lysozyme via secretion from the yeast, *Pichia pastoris. Bio/Technology* **7**, 160–164.
67. Jerva, L. F., Sullivan, G., and Lolis, E. (1997) Functional and receptor binding characterization of recombinant murine macrophage inflammatory protein 2: sequence analysis and mutagenesis identify receptor binding isotopes. *Protein Sci.* **6,** 1643–1652.
68. Ferrari, E., Lodi, T., Sorbi, R. T., Tirindelli, R., Cavaggioni, A., and Spisni, A. (1997) Expression of a lipocalin in *Pichia pastoris*: secretion, purification and binding activity of a recombinant mouse major urinary protein. *FEBS Lett.* **401,** 73–77.
69. Merkel, R. K., Zhang, Y., Ruest, P. J., Lal, A., Liao, Y. F., and Moremen, K. W. (1997) Cloning, expression, purification, and characterization of the murine lysosomal acid α-mannosidase. *Biochim. Biophys. Acta* **1336,** 132–146.
70. Kiselyov, V. V., Berezin, V., Maar, T. E., Soroka, V., Edvardsen, K., Schousboe, A., and Bock, E. (1997) The first immunoglobulin-like neural cell adhesion molecule (NCAM) domain is involved in double-reciprocal interaction with the second immunoglobulin-like NCAM domain and in heparin binding. *J. Biol. Chem.* **272,** 10,125–10,134.

71. Abdulaev, N. G., Popp, M. P., Smith, W. C., and Ridge, K. D. (1997) Functional expression of bovine opsin in the methylotrophic yeast *Pichia pastoris. Protein Expr. Purif.* **10,** 61–69.
72. Weiss, H. M., Haase, W., Michel, H., and Reilander, H. (1995) Expression of functional mouse 5-HT5A serotonin receptor in the methylotrophic yeast *Pichia pastoris*: pharmacological characterization and localization. *FEBS Lett.* **377**, 451–456.
73. Chen, Y. J., and Gonatas, N. K. (1997) The Golgi sialoglycoprotein MG160, expressed in *Pichia pastoris*, does not require complex carbohydrates and sialic acid for secretion and basic fibroblast growth factor binding. *Biochem. Biophys. Res. Commun.* **234,** 68–72.
74. Luo, D., Mah, N., Krantz, M., Wilde, K., Wishart, D., Zhang, Y., Jacobs, F. and Martin, L. (1995) V1-linker-Vh orientation-dependent expression of single chain Fv-containing an engineered disulfide-stabilized bond in the framework regions. *J. Biochem. Tokyo* **118,** 825–831.
75. Sun, J., Bottomley, S. P., Kumar, S., and Bird, P. I. (1997) Recombinant caspase-3 expressed in *Pichia pastoris* is fully activated and kinetically indistinguishable from the native enzyme. *Biochem. Biophys. Res. Commun.* **238,** 920–924.
76. Zhu, H., Shi, J., Cregg, J. M., and Woldegiorgis, G. (1997) Reconstitution of highly expressed human heart muscle carnitine palmitoyltransferase I. *Biochem. Biophys. Res. Commun.* **239,** 498–502.
77. McGrew, J. T., Leiske, D., Dell, B., Klinke, R., Krasts, D., Wee, S. F., Abbott, N., Armitage, R., and Harrington, K. (1997) Expression of trimeric CD40 ligand in *Pichia pastoris*: use of a rapid method to detect high-leel expressing transformants. *Gene* **187,** 193–200.
78. Battista, M. C., Bergamini, G., Campanini, F., Landini, M. P., and Ripaldi, A. (1996) Intracellular production of a major cytomegalovirus antigenic protein in the methylotrophic yeast *Pichia pastoris. Gene* **176,** 197–201.
79. Gerber, A., O'Connell, M. A., and Keller, W. (1997) Two forms of human double-stranded RNA-specific editase 1 (hRED1) generated by the insertion of an Alu cassette. *RNA* **3,** 453–463.
80. Cregg, J. M., Tschopp, J. F., Stillman, C., Siegel, R., Akong, M., Craig, W. S., Buckholz, R. G., Madden, K. R., Kellaris, P. A., Davis, G. R., Smiley, B. L., Cruze, J., Torregrossa, R., Velicelebi, G., and Thill, G. P. (1987) High-level expression and efficient assembly of hepatitis B surface antigen in the methylotrophic yeast, *Pichia pastoris. Bio/Technology* **5,** 479–485.
81. Eckhart, L., Raffelsberger, W., Ferko, B., Klima, A., Purtscher, M., Katinger, H. and Ruker, F. (1996) Immunogenic presentation of a conserved gp41 epitope of human immunodeficiency virus type 1 on recombinant surface antigen of hepatitis B virus. *J. Gen. Virol.* **77,** 2001–2008.
82. Lal, S. K., Tulasiram, P., and Jameel, S. (1997) Expression and characterization of the hepatitis E virus ORF3 protein in the methylotrophic yeast, *Pichia pastoris. Gene* **190,** 63–67.
83. Liao, Y. F., Lal, A., and Moremen, K. W. (1996) Cloning, expression, purification, and characterization of the human broad specificity lysosomal acid α-mannosidase. *J. Biol. Chem.* **271,** 28,348–28,458.

84. Yang, Y. S., Yang, M. C., Tucker, P. W., and Capra, J. D. (1997) NonO enchances the association of many DNA-binding proteins to their targets. *Nucleic Acids Res.* **25,** 2284–2292.
85. Vuorela, A., Myllyharju, J., Nissi, R., Pihlajaniemi, T., and Kivirikko, K. I. (1997) Assembly of human prolyl 4-hydroxylase and type III collagen in the yeast *Pichia pastoris*: formation of a stable enzyme tetramer requires coexpression with collagen and assembly of a stable collagen requires coexpression with prolyl 4-hydroxylase. *EMBO J.* **16,** 6702–6712.
86. Lima, C. D., Klein, M. G., Weinstein, I. B., and Hendrickson, W. A. (1996) Three-dimensional structure of human protein kinase C interacting protein 1, a member of the HIT family of proteins. *Proc. Natl. Acad. Sci. USA* **83,** 5357–5362.
87. Sun, J., Coughlin, P., Salem, H. H., and Bird, P. (1995) Production and characterization of recombinant human proteinase inhibitor 6 expressed in *Pichia pastoris. Biochim. Biophys. Acta* **1252,** 28–34.
88. Dahlen, J. R., Foster, D. C., and Kisiel, W. (1997) Expression, purification, and inhibitory properties of human proteinase inhibitor. *Biochemistry* **36,** 14,874–14,882.
89. Sreekrishna, K., Nelles, L., Potenz, R., Cruze, J., Mazzaferro, P., Fish, W., Fuke, M., Holden, K., Phelps, D., Wood, P., and Parker, K. (1989) High-level expression, purification, and characterization of recombinant human tumor necrosis factor synthesized in the methylotrophic yeast *Pichia pastoris. Biochemistry* **28,** 4117–4125.
90. Ohsawa, I., Hirose, Y., Ishiguro, M., Imai, Y., Ishiura, S., and Kohsaka, S. (1995) Expression, purification, and neurotrophic activity of amyloid precursor protein-secreted forms produced by yeast. *Biochem. Biophys. Res. Commun.* **213,** 52–58.
91. van Nostrand, W. E., Schmaier, A. H., Neiditch, B. R., Siegel, R. S., Raschke, W. C., Sisodia, S. S., and Wagner, S. L. (1994) Expression, purification, and characterization of the Kunitz-type proteinase inhibitor domain of the amyloid β-protein precursor-like protein-2. *Biochim. Biophys. Acta* **1209,** 165–170.
92. Sim, B. K., O'Reilly, M. S., Liang, H., Fortier, A. H., He, W., Madsen, J. W., Lapcevich, R., and Nacy, C. A. (1997) A recombinant human angiostatin protein inhibits experimental primary and metastic cancer. *Cancer Res.* **57,** 1329–1334.
93. Vedvick, T., Buckholz, R. G., Engel, M., Urcan, M., Kinney, J., Provow, S., Siegel, R. S., and Thill, G. P. (1991) High-level secretion of biologically active aprotinin from the yeast *Pichia pastoris. J. Ind. Microbiol.* **7,** 197–202.
94. Gerstmayer, B., Altenschmidt, U., Hoffmann, M., and Wels, W. (1997) Costimulation of T cell proliferation by a chimeric B7-2 antibody fusion protein specifically targeted to cells expressing the *erbB2* proto-oncogene. *J. Immunol.* **158,** 4584–4590.
95. Gerstmayer, B., Pessara, U., and Wels, W. (1997) Construction and expression in the yeast *Pichia pastoris* of functionally active soluble forms of the human costimulatory molecules B7-1 and B7-2 and the B7 counter-receptor CTLA-4. *FEBS Lett.* **407,** 63–68.

96. Yamada, M., Azuma, T., Matsuba, T., Iida, H., Suzuki, H., Yamamoto, K., Kohli, Y., and Hori, H. (1994) Secretion of human intracellular aspartic proteinase cathepsin E expressed in the methylotrophic yeast, *Pichia pastoris* and characterization of produced recombinant cathepsin E. *Biochim. Biophys. Acta* **1206**, 279–285.
97. Linnevers, C. J., McGrath, M. E., Armstrong, R., Mistry, F. R., Barnes, M. G., Klaus, J. L., Palmer, J. T., Katz, B. A., and Bromme, D. (1997) Expression of human cathepsin K in *Pichia pastoris* and preliminary crystallographic studies of an inhibitor complex. *Protein Sci.* **6,** 919–921.
98. Carmona, E., Dufour, E., Plouffe, C., Takebe, S., Mason, P., Mort, J. S., and Ménard, R. (1996) Potency and selectivity of the cathepsin Ll propeptide as an inhibitor of cysteine proteases. *Biochemistry* **35,** 8149–8157
99. Powell, R. M., Ward, T., Evans, D. J., and Almond, J. W. (1997) Interaction between echovirus 7 and its receptor, decay-accelerating factor (CD55): evidence for a secondary cellular factor in A-particle formation. *J. Virol.* **71,** 9306–9312.
100. Sugrue, R. J., Fu, J., Howe, J., and Chan, Y. C. (1997) Expression of the dengue virus structural proteins in *Pichia pastoris* leads to the generation of virus-like particles. *J. Gen. Virol.* **78,** 1861–1866.
101. Siegel, R. S., Buckholz, R. G., Thill, G. P., and Wondrack, L. M. (1990) Production of epidermal growth factor in methylogrophic yeast cells. European Patent Application W090/10697.
102. Tanaka, M., Suda, T., Yatomi, T., Nakamura, N., and Nagata, S. (1997) Lethal effect of recombinant human Fas ligand in mice pretreated with Propionibacterium acnes. *J. Immunol.* **158**, 2303–2309.
103. Côté, H. C., Pratt, K. P., Davie, E. W., and Chung, D. W. (1997) The polymerization pocked "a" within the carboxyl-terminal region of the gamma chain of human fibrinogen is adjacent to but independent from the calcium-binding site. *J. Biol. Chem.* **272,** 23,792–23,798.
104. Rosenfeld, S. A., Ross, O. H., Hillman, M. C., Corman, J. I., and Dowling, R. L. (1996) Production and purification of human fibroblast collagenase (MMP-1) expressed in the methylotrophic yeast *Pichia pastoris. Protein Expr. Purif.* **7**, 423–430.
105. Scorer, C. A., Buckholz, R. G., Clare, J. J., and Romanos, M. A. (1993) The intracellular production and secretion of HIV-1 envelope protein in the methylotrophic yeast *Pichia pastoris*. *Gene* **136**, 111–119.
106. Martinet, W., Saelens, X., Deroo, T., Neirynck, S., Contreras, R., Min Jou, W., and Fiers, W. (1997) Protection of mice against a lethal influenza challenge by immunization with yeast-derived recombinant influenza neuraminidase. *Eur. J. Biochem.* **247,** 332–338.
107. Vollmer, P., Peters, M., Ehlers, M., Yagame, H., Matsuba, T., Kondo, M., Yasukawa, K., Buschenfelde, K. H., and Rose-John, S. (1996) Yeast expression of the cytokine receptor domain of the soluble interleukin-6 receptor. *J. Immunol. Methods* **199**, 47–54.
108. Zhang, J. G., Owczarek, C. M., Ward, L. D., Howlett, G. J., Fabri, L. J., Roberts, B. A., and Nicola, N. A. (1997) Evidence for the formation of a heterotrimeric complex of leukaemia inhibitory factor with its receptor subunits in solution. *Biochem. J.* **325,** 693–700.

109. Fryxell, K. B., O'Donoghue, K., Graeff, R. M., Lee, H. C., and Branton, W. D. (1995) Functional expression of soluble forms of human CD38 in *Escherichia coli* and *Pichia pastoris*. *Protein Expr. Purif.* **6,** 329–336.
110. Kalandadze, A., Galleno, M., Foncerrada, L., Strominger, J. L. and Wucherpfennig, K. W. (1996) Expression of recombinant HLA-DR2 molecules. Replacement of the hydrophobic transmembrane region by a leucine zipper dimerization motif allows the assembly and secretion of soluble DR αβ heterodimers. *J. Biol. Chem.* **271,** 20,156–20,162.
111. Masure, S., Paemen, L., Proost, P., Van Damme, J., and Opdenakker, G. (1995) Expression of a human mutant monocyte chemotactic protein 3 in *Pichia pastoris* and characterization as an MCP-3 receptor antagonist. *J. Interferon Cytokine Res.* **15,** 955–963.
112. Talmont, F., Sidobre, S., Demange, P., Milon, A., and Emorine, L. J. (1996) Expression and pharmacological characterization of the human μ-opioid receptor in the methylotrophic yeast *Pichia pastoris*. *FEBS Lett.* **394**, 268–272.
113. Illy, C., Quraishi, O., Wang, J., Purisima, E., Vernet, T., and Mort, J. S. (1997) Role of the occluding loop in cathepsin B activity. *J. Biol. Chem.* **272,** 1197–1202.
114. Harmsen, M. C., Heeringa, P., van der Geld, Y. M., Huitema, M. G., Klimp, A., Tiran, A., and Kallenberg, C. G. (1997) Recombinant proteinase 3 (Wegener's antigen) expressed in *Pichia pastoris* is functionally active and is recognized by patient sera. *Clin. Exp. Immunol.* **110,** 257–264.
115. Ikegaya, K., Hirose, M., Ohmura, T., and Nokihara, K. (1997) Complete determination of disulfide forms of purified recombinant human serum albumin, secreted by the yeast *Pichia pastoris*. *Anal. Chem.* **69,** 1986–1991.
116. Nilsen, S. L., DeFord, M. E., Prorok, M., Chibber, B. A., Bretthauer, R. K., and Castellino, F. J. (1997) High-level secretion in *Pichia pastoris* and biochemical characterization of the recombinant kringle 2 domain of tissue-type plasminogen activator. *Biotechnol. Appl. Biochem.* **25,** 63–74.
117. White, C. E., Hunter, M. J., Meininger, D. P., White, L. R., and Komives, E. A. (1995) Large-scale expression, purification and characterization of small fragments of thrombomodulin: the roles of the sixth domain and of methionine 388. *Protein Eng.* **8**, 1177–1187.
118. Steinlein, L. M., Graf, T. N., and Ikeda, R. A. (1995) Production and purification of N-terminal half-transferrin in *Pichia pastoris*. *Protein Expr. Purif.* **6,** 619–624.
119. Okabayashi, K., Tsujikawa, M., Morita, M., Einaga, K., Tanaka, K., Tanabe, T., Yamanouchi, K., Hirama, M., Tait, J. F., and Fujikawa, K. (1996) Secretory production of recombinant urokinase-type plaminogen activator-annexin V chimerase in *Pichia pastoris*. *Gene* **177,** 69–76.
120. Wiles, A. P., Shaw, G., Bright, J., Perczel, A., Campbell, I. D., and Barlow, P. N. (1997) NMR studies of a viral protein that mimics the regulators of complement activation. *J. Mol. Biol.* **272,** 253–265.

Index